A Semi-Analytical Approach to Nonlinear Mechanical Engineering Issues

A Semi-Analytical Approach to Nonlinear Mechanical Engineering Issues

By

Bahram Jalili,

Payam Jalili,

Elyas Rostami

and Davood Domairry Ganji (MirGanj)

Cambridge Scholars
Publishing

A Semi-Analytical Approach to Nonlinear Mechanical Engineering Issues

By Bahram Jalili, Payam Jalili, Elyas Rostami and Davood Domairry Ganji (MirGanj)

This book first published 2024

Cambridge Scholars Publishing

Lady Stephenson Library, Newcastle upon Tyne, NE6 2PA, UK

British Library Cataloguing in Publication Data
A catalogue record for this book is available from the British Library

ISBN (10): 1-5275-5256-X
ISBN (13): 978-1-5275-5256-2

TABLE OF CONTENTS

CHAPTER 1

INTRODUCTION

1.1. Preface

Most users of differential equations in the mathematical and engineering sciences face the difficulty of being nonlinear. There has not been a particularly effective way to solve different kinds of nonlinear differential equations in diverse fields of study up until now.

It is important to note that although academics have recently developed a few semi-analytical techniques for solving a small number of groups of nonlinear differential equations, none can be applied to various nonlinear equations. To understand better, it is necessary to mention that some techniques now used for solving nonlinear equations include the homotopy analysis method (HAM)[1], homotopy perturbation method (HPM)[2], VIM[3], DTM[4], ADM[5], etc. However, each technique works well for a subset of nonlinear differential equations.

Additionally, the use of every approach has occasionally resulted in calculations that contain serious errors. As a result, only some of the nonlinear issues can be solved using the abovementioned methods, which do not apply to all nonlinear situations. To further elaborate, the second drawback of these procedures is the drawn-out and challenging process.

In this book, a practical method is presented for all the various types of nonlinear differential equations and various sets of nonlinear equations, allowing for straightforward analytical solutions to all the differential equations and the final solution of each differential equation as an algebraic function. By selecting an answer function for a differential equation with

---

[1] Homotopy Analysis Method
[2] Homotopy Perturbation Method
[3] Vibrational Iteration Method
[4] Differential Transformation Method
[5] Adomian Decomposition Method

constant coefficients, which may be obtained by applying initial or boundary conditions in a certain way, AGM seeks to solve all nonlinear differential equations algebraically.

Also, semi-analytical methods can be divided into two groups based on how they solve problems; for the sake of simplicity, we will refer to them as the Iterate-Base Method and the Trial Function-Base Method. The number of iterations is a crucial aspect that influences the solving procedures in iterate-based methods like (HPM), VIM, ADM, and others. Although we can use trial functions in this procedure that are based on our independent functions, to solve each step, we must first solve the stages that came before it. The reasons above make it clear that we will encounter issues that impede our problem-solving processes when iteration results in higher phases inaccessible to the relevant software.

Additionally, these approaches typically require more time to find solutions.

The primary element that affects the solving procedures in trial function-based methods like Collocation Method (CM)[6], least square method (LSM)[7], AGM, and others are trial function. This approach assumes a practical trial function with various constant coefficients based on the problem's boundary and initial conditions.

After that, we must solve the constant coefficients because of each approach's fundamental premise. Most of the time, a collection of polynomials can be solved to obtain the constant coefficients readily. Although the number of terms in our trial function in these approaches might be referred to as the number of needed iterations, it is essential to note that used constants will be obtained concurrently in solution procedures. Therefore, these techniques do not have iteration issues.

AGM has high efficiency and accuracy for solving nonlinear problems with high nonlinearity, according to recent achievements from this approach and the Trial Function-Base properties of this method. It is important to note that the following can be used as a summary of this method's superiority versus other approaches: In the process of solving differential equations, boundary conditions are required in the order of the differential equations; however, if the number of boundary conditions is less than the order of the differential equation, this method may result in the creation of additional

---

[6] Colocation Method

[7] Least Square Method

new boundary conditions for the internal differential equation and its derivatives.

1.2. The basic concept of AGM

Every engineering science branch uses linear or nonlinear differential equations as its governing equations to solve physics-based problems. It is necessary to apply sufficient boundary or beginning conditions to the physics of these problems and the mathematical formulation they have been given to solve the difficulties under consideration. We can understand the significance of these boundaries and initial conditions in determining the accuracy and efficiency of analytical methods achieving acceptable solutions due to the physics of problems since procedures of applying analytical methods for obtaining the solution of linear and nonlinear differential equations are not an exception from mentioned fact. The entire process has been defined in plain terms to make it easier to understand the supplied method in this book.

The general approach to a differential equation depends on the boundary conditions and is as follows:

$$p_k : f\left(u, u', u'', \ldots, u^{(m)}\right) = 0; \quad \mathrm{u} = u(x) \tag{1.1}$$

The nonlinear differential equation of p, which is a function of u, the parameter u, which is a function of x, and their derivatives are assumed as follows:

Boundary conditions:

$$\begin{cases} u(x) = u_0, u'(x) = u_1, \ldots, u^{(m-1)}(x) = u_{m-1} & \text{at} \quad \mathrm{x} = 0 \\ u(x) = u_{L_0}, u'(x) = u_{L_1}, \ldots, u^{(m-1)}(x) = u_{L_{m-1}} & \text{at} \quad \mathrm{x} = 0 \end{cases} \tag{1.2}$$

The sequence of letters in the nth order with constant coefficients that we assume as the solution of the first differential equation is taken into consideration to solve the first differential equation concerning the boundary conditions in x = L in Eq. (1.2) as follows:

$$u(x) = \sum_{i=0}^{n} a_i x^i = a_0 + a_1 x^1 + a_2 x^2 + \ldots + a_n x^n \tag{1.3}$$

The additional series statements in Eq. (1.3) lead to a more accurate solution for Eq (1.1). There are (n + 1) unknown coefficients that require (n + 1) equations to be stated to achieve the solution of the differential equation (1.1) about the series from degree (n). A group of (n+1) equations is solved using the boundary conditions of equation (1.2).

1.2.1. Applying the boundary conditions

(a) The boundary conditions are used in the following way to solve differential equation (1.3):

When x = 0:

$$\left.\begin{cases} u(0)=a_0=u_0 \\ u'(0)=a_1=u_1 \\ u''(0)=a_2=u_2 \\ \vdots \quad \vdots \quad \vdots \end{cases}\right\} \tag{1.4}$$

And when x= L:

$$\begin{aligned} &u(L)=a_0+a_1L+a_2L^2+\ldots+a_nL^n=u_{L_0} \\ &u'(L)=a_1+2a_2L+3a_3L^2+\ldots+na_nL^{n-1}=u_{L_1} \\ &u''(L)=2a_2+6a_3L+12a_4L^2+\ldots+n(n-1)a_nL^{n-2}=u_{L_{m-1}} \\ &\vdots \quad \vdots \quad \vdots \quad \vdots \quad \vdots \quad \vdots \end{aligned} \tag{1.5}$$

(b) The boundary conditions are applied to the differential Eq. (1.1) in the following manner after substituting Eq. (1.5) into Eq. (1.1):

$$\begin{aligned} &p_0 : f\,(u(0),u'(0),u''(0),\ldots,u^{(m)}(0)) \\ &p_1 : f\,(u(L),u'(L),u''(L),\ldots,u^{(m)}(L)) \\ &\vdots \quad \vdots \quad \vdots \quad \vdots \quad \vdots \end{aligned} \tag{1.6}$$

To create a system of equations with (n+1) equations and (n+1); unknowns, based on the selection of n; ($n \prec m$) phrases from Eq. (1.3), we must deal with several extra unknowns that are the identical coefficients of Eq (1.3). We must first apply the boundary conditions to the differential equations indicated above to solve this issue before deriving m times from Eq. (1.1) by the new unknowns.

$$
\begin{aligned}
&p'_k : f\,(u', u'', u''', \ldots, u^{(m+1)}) \\
&p''_k : f\,(u'', u''', u^{(IV)} \ldots, u^{(m+2)}) \\
&\vdots \qquad \vdots \qquad \vdots \qquad \vdots
\end{aligned}
\tag{1.7}
$$

(c) The boundary conditions in Eq. (1.7) are applied to the derivatives of the differential equation P_k as follows:

$$
p'_k : \begin{Bmatrix} f\,(u'(0), u''(0), u'''(0), \ldots, u^{(m+1)}(0)) \\ f\,(u'(L), u''(L), u'''(L), \ldots, u^{(m+1)}(L)) \end{Bmatrix}
\tag{1.8}
$$

$$
p''_k : \begin{Bmatrix} f\,(u''(0), u'''(0), \ldots, u^{(m+2)}(0)) \\ f\,(u''(L), u'''(L), \ldots, u^{(m+2)}(L)) \end{Bmatrix}
\tag{1.9}
$$

It is possible to construct (n+ 1) equations from Eq. (1.4) to Eq. (1.9) to determine the (n +1) unknown coefficients of Eq. (1.3), such as a_0, a_1, $a_{2,}$..., and a_n. Finding the coefficients of Eq. (1.1) will lead to solving the nonlinear differential equation (1.3).

CHAPTER 2

THE APPLICATION OF AGM IN SOLID MECHANICS

2.1. Swinging Oscillation of a Solid Circular Sector Object

In this problem, a new and innovative semi-analytical method called AGM has been applied to solve nonlinear equations of the semicircular oscillator.

Consider a homogeneous solid circular sector object with angle α and radius R, as shown in Fig. 2.1.1, that rolls in an oscillatory motion back and forth on a flat, stationary support with no sliding effect. When α becomes radian, no oscillatory swinging motion will occur. It may be easily verified that the governing equation of the oscillation is as follows:

$$(\frac{3}{2}R^2 - \frac{4\sin(\alpha)}{3\alpha}R\cos(\theta))\ddot{\theta} +$$

$$\mathrm{R}(\frac{2R\sin(\alpha)}{3\alpha}\sin(\theta))\dot{\theta}^2 + (\frac{2\sin(\alpha)}{3\alpha}g)\sin(\theta)=0 \qquad (2.1.1)$$

$$\theta(0)=\mathrm{A}\,,\ \dot{\theta}(0)=0$$

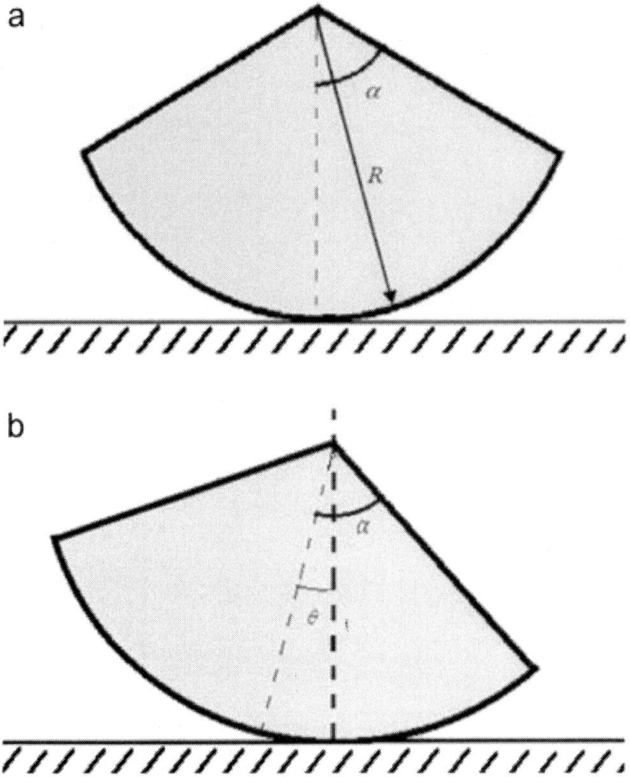

Fig. 2.1.1. Geometric parameters of the homogeneous solid circular sector body

By substitution of the relatively accurate approximations $\sin(\theta) \approx \theta - \frac{\theta^3}{3!}$ and $\cos(\theta) \approx 1 - \frac{\theta^2}{2!}$ Into Eq. (2.1.1), the governing equation would be in the following order:

$$\left(\frac{3}{2}R^2 - \frac{4\sin(\alpha)}{3\alpha}\left(1-\frac{\theta^2}{2!}\right)\right)\ddot{\theta} +$$

$$\mathrm{R}\left(\frac{2R\sin(\alpha)}{3\alpha}\left(\theta-\frac{\theta^3}{3!}\right)\dot{\theta}^2 + \left(\frac{2\sin(\alpha)}{3\alpha}g\right)\left(\theta-\frac{\theta^3}{3!}\right)=0 \qquad (2.1.2)$$

$$\theta(0)=\mathrm{A}\ ,\ \dot{\theta}(0)=0$$

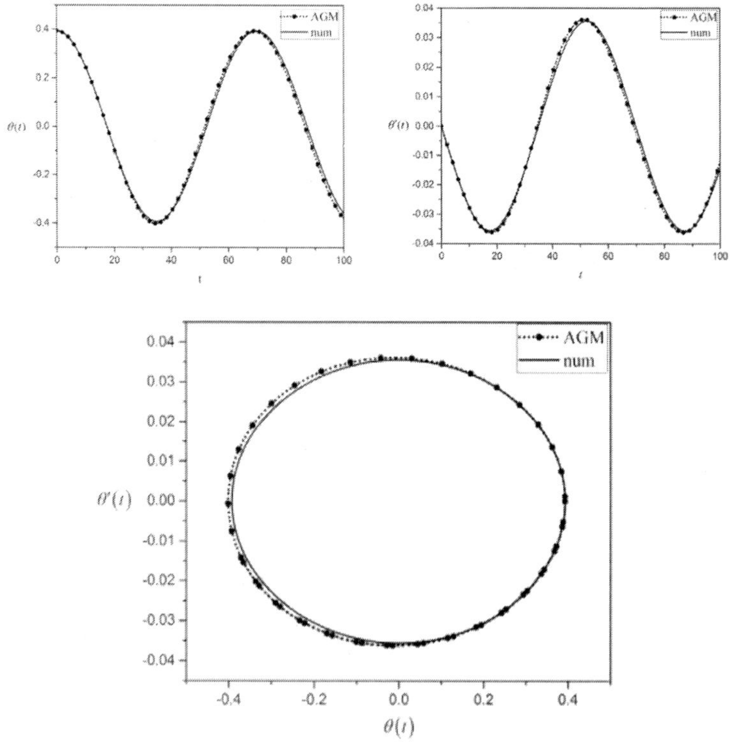

Fig. 2.1.2. (a, b, c) A comparison between the solutions obtained by AGM and the numerical method for R = 15, α = 2π/3.

According to the basic idea of the proposed method, first, we rewrite Eq. (2.1.1) in the following order:

$$\begin{aligned} &\mathrm{G(t)}:(\frac{3}{2}R^{2}-\frac{4\sin(\alpha)}{3\alpha}(1-\frac{\theta^{2}}{2!}))\ddot{\theta}+ \\ &\mathrm{R}(\frac{2R\sin(\alpha)}{3\alpha}(\theta-\frac{\theta^{3}}{3!})\dot{\theta}^{2}+(\frac{2\sin(\alpha)}{3\alpha}g)(\theta-\frac{\theta^{3}}{3!})=0 \end{aligned} \tag{2.1.3}$$

For solving the nonlinear differential equation by AGM, it is necessary to consider a function as the solution to the presented problem as follows:

$$\theta(t)=e^{-at}(b_0\cos(\omega t+\phi_0)+b_1\cos(2\omega t+\phi_1)) \tag{2.1.4}$$

The term (e^{-at}) in Eq. (2.1.4) indicates a damping component in the oscillating system. Since no damping components exist in the mentioned example, the term (a) in Eq. (2.1.4) will automatically be zero after applying the initial conditions in AGM. Moreover, the constant coefficient (b), the initial vibrational phase (φ), and the angular frequency (ω) can be computed by applying the initial conditions. According to the above theory, the solution to the problem is assumed as follows:

$$\theta(t)=b_0\cos(\omega t+\phi_0)+b_1\cos(2\omega t+\phi_1) \tag{2.1.5}$$

Regarding the proposed physical model, there are no boundary conditions, so the constant coefficients of Eq. (2.1.5) are acquired only concerning the given initial conditions presented in Eq. (2.1.3). Notably, initial or boundary conditions are applied in two manners in the following form:

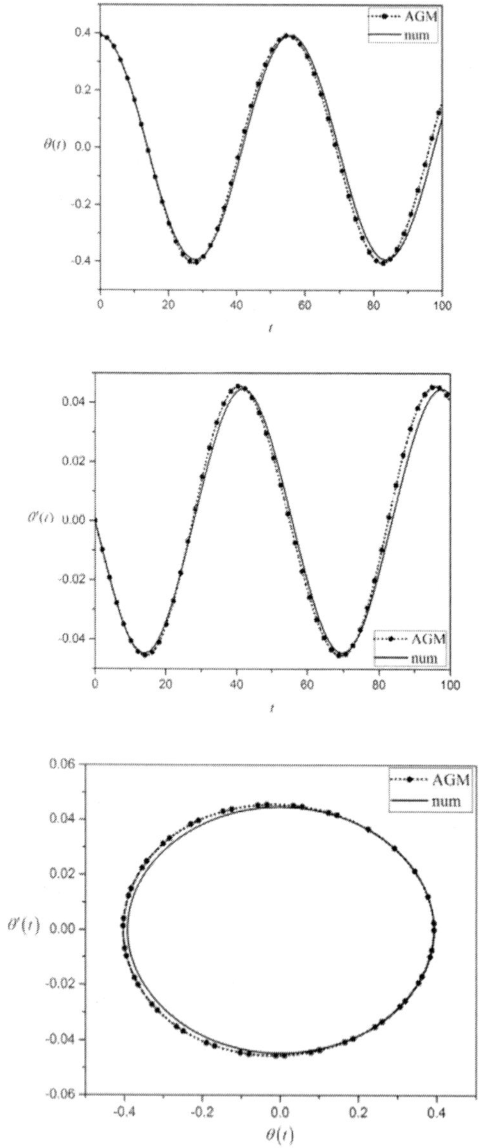

Fig. 2.1.3. (a, b, c) A comparison between the solutions obtained by AGM and the numerical method for $R = 15$, $\alpha = \pi/2$.

1) The initial conditions are applied to Eq. (2.1.5) in the form of

$$\theta = \theta(IC) \tag{2.1.6}$$

the initial condition (IC) is the abbreviation of the initial conditions. As a result, applying the initial conditions to Eq. (2.1.5) is performed as follows:

$$\theta(0) = A \rightarrow a_0 \cos(\varphi_0) + a_1 \cos(\varphi_1) = A \tag{2.1.7}$$

$$\dot{\theta}(0) = 0 \rightarrow -a_0\omega\sin(\varphi_0) - 2a_1\omega\sin(\varphi_1) = 0 \tag{2.1.8}$$

2) The initial conditions are applied to the main differential equation, which in the problem is Eq. (2.1.3), and on its derivatives in the following general forms:

$$G(\theta(t)) \rightarrow G(\theta(IC)) = 0,\ G'(\theta(IC)) = 0, \ldots \tag{2.1.9}$$

Therefore, after substituting Eq. (2.1.5), which has been considered as the solution to the main differential equation, into Eq. (2.1.3), the initial conditions are applied to the equation obtained and its derivative based on Eq. (2.1.9) as follows:

$$\begin{aligned} G(\dot{\theta}(0)) &: \left(\frac{3}{2}R^2 - \frac{4}{3}\frac{R\sin(\alpha)\cos(\psi)}{\alpha}\right)(\Gamma_2) \\ &+ \frac{2}{3}\frac{R^2\sin(\alpha)\sin(\psi)(\Gamma_1)^2}{\alpha} \\ &+ \frac{2}{3}\frac{g\sin(\alpha)\sin(\psi)}{\alpha} = 0 \end{aligned} \tag{2.1.10}$$

$$
\begin{aligned}
G'(\dot{\theta}(0)) &= \frac{4}{3}\frac{R\sin(\alpha)\sin(\psi)(\Gamma_1)(\Gamma_2)}{\alpha} \\
&+(\frac{3}{2}R^2 - \frac{4}{3}\frac{R\sin(\alpha)\cos(\psi)}{\alpha})(\Gamma_3) \\
&+\frac{2}{3}\frac{R^2\sin(\alpha)\cos(\psi)(\Gamma_1)^3}{\alpha} \\
&+\frac{4}{3}\frac{R^2\sin(\alpha)\sin(\psi)(\Gamma_1)(\Gamma_2)}{\alpha} \\
&+\frac{2}{3}\frac{g\sin(\alpha)\cos(\psi)(\Gamma_1)}{\alpha} = 0
\end{aligned}
\tag{2.1.11}
$$

$$
\begin{aligned}
G'(\dot{\theta}(0)) &: \frac{4}{3}\frac{R\sin(\alpha)\cos(\psi)(\Gamma_1)^2(\Gamma_2)}{\alpha} \\
&+\frac{4}{3}\frac{R\sin(\alpha)\sin(\psi)(\Gamma_2)^2}{\alpha} \\
&+\frac{8}{3}\frac{R\sin(\alpha)\sin(\psi)(\Gamma_1)(\Gamma_3)}{\alpha} \\
&+(\frac{3}{2}R^2 - \frac{4}{3}\frac{R\sin(\alpha)\cos(\psi)}{\alpha})(\Gamma_4) \\
&-\frac{2}{3}\frac{R^2\sin(\alpha)\sin(\psi)(\Gamma_1)^4}{\alpha} \\
&+\frac{10}{3}\frac{R^2\sin(\alpha)\cos(\psi)(\Gamma_1)^2(\Gamma_2)}{\alpha} \\
&+\frac{4}{3}\frac{R^2\sin(\alpha)\sin(\psi)(\Gamma_2)^2}{\alpha} \\
&+\frac{4}{3}\frac{R^2\sin(\alpha)\sin(\psi)(\Gamma_1)(\Gamma_3)}{\alpha} \\
&-\frac{2}{3}\frac{g\sin(\alpha)\sin(\psi)(\Gamma_1)^2}{\alpha} \\
&+\frac{2}{3}\frac{g\sin(\alpha)\cos(\psi)(\Gamma_2)}{\alpha}
\end{aligned}
\tag{2.1.12}
$$

To simplify, we use the following statements:

$$\begin{aligned}
\psi &= b_0\cos(\phi_0)+b_1\cos(\phi_1),\\
\Gamma_1 &= -b_0\omega\sin(\phi_0)-2b_1\omega\sin(\phi_1),\\
\Gamma_2 &= -b_0\omega^2\cos(\phi_0)-4b_1\omega^2\cos(\phi_1),\\
\Gamma_3 &= b_0\omega^3\sin(\phi_0)+8b_1\omega^3\sin(\phi_1),\\
\Gamma_4 &= b_0\omega^4\cos(\phi_0)+16b_1\omega^4\cos(\phi_1).
\end{aligned} \tag{2.1.13}$$

Solving a set of five algebraic equations with five unknowns from Eqs. (2.1.7) and (2.1.8) and Eqs. (2.1.10)-(2.1.12), with the assumption of $A=\frac{\pi}{8}\ R=15,\alpha=\frac{2\pi}{3},g=9.81,$ the constant coefficients b_0,b_1,ϕ_0,ϕ_1 and ω Eq. (2.1.5), can easily be yielded as follows:

$$\begin{aligned}
&b_0=-0.3972120591,\ \mathrm{b}_1{=}0.004512977299,\\
&\phi_0=-9.424777961,\ \phi_1=-9.424777961,\\
&\omega=-0.09096384741
\end{aligned} \tag{2.1.14}$$

After substituting the obtained values from Eq. (2.1.14) and into Eq. (2.1.5), the solution to the problem with the assumed physical constants will be obtained as follows:

$$\begin{aligned}
&\theta(t)=-0.3972120591\ \times\\
&cos\,(0.09096384741t+9.424777961)\\
&+0.004512977299\times\\
&cos\left(0.1819276948t+9.424777961\right)
\end{aligned} \tag{2.1.15}$$

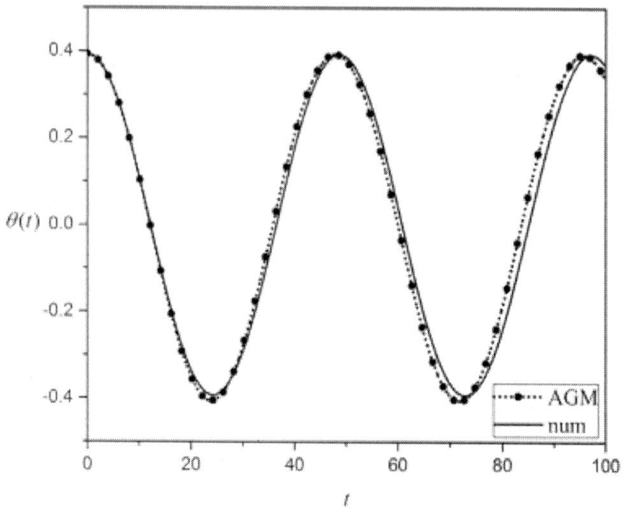
0.4
0.2
θ(t) 0.0
-0.2
-0.4
AGM
num
0
20
40
60
80
100
t

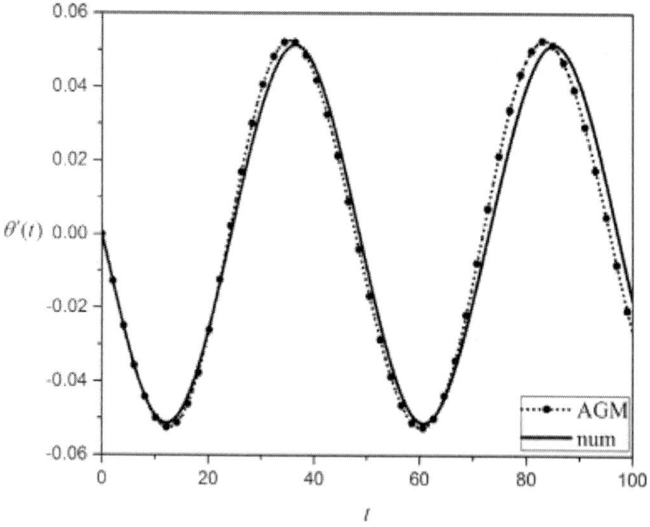
0.06
0.04
0.02
θ'(t) 0.00
-0.02
-0.04
-0.06
AGM
num
0
20
40
60
80
100
t

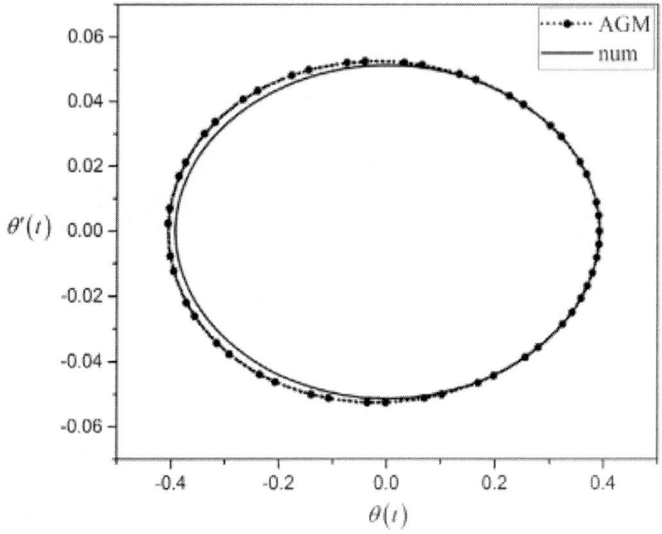

Fig. 2.1.4. (a, b, c) A comparison between the solutions obtained by AGM and the numerical method for R = 15, α = π/3.

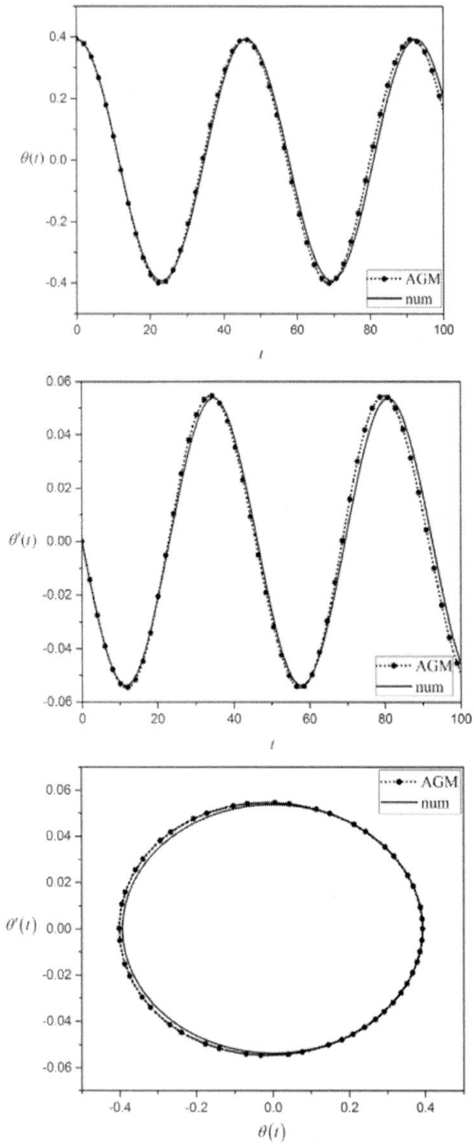

Fig. 2.1.5. (a, b, c) A comparison between the solutions obtained by AGM and the numerical method for R = 10, α = 2π/3.

In this problem, AGM has been utilized to solve the nonlinear differential equation of the swinging oscillation of a solid circular sector, and the results have been compared with the numerical solution. Although the processes of obtaining an analytical solution were accomplished in the previous section to make comprehensive comparisons and illustrate the accuracy and efficiency of the proposed method, we have investigated the effects of semicircular radius (R) and semicircular angle (α). Therefore, we have assumed the following appropriate amounts for our purposes:

$$1)R = 15 \rightarrow \alpha = \frac{2\pi}{3}, \alpha = \frac{\pi}{2}, \alpha = \frac{\pi}{3}$$

$$2)R = 10 \rightarrow \alpha = \frac{2\pi}{3}, \alpha = \frac{\pi}{2}, \alpha = \frac{\pi}{3}$$

$$3)R = 5 \rightarrow \alpha = \frac{2\pi}{3}, \alpha = \frac{\pi}{2}, \alpha = \frac{\pi}{3}$$

Figs. 2.1.2–2.1.4 demonstrate that the oscillation frequency will increase considerably by reducing the semicircular angle (α) with a constant semicircular radius (R). Notably, the frequency of the oscillation determined with AGM is generally higher than that determined with the numerical solution. Because the oscillation frequency is negatively correlated with the oscillation frequency, the oscillation period will increase. Apart from the mentioned results, these figures depict that the oscillation trends are somehow harmonic and stable. This is mainly because the vibration amplitude does not change at all. The same results are achieved in Figs. 2.1.5–2.1.10. A comparison of Fig. 2.1.2, Fig. 2.1.5, and Fig. 2.1.8 illustrates that the oscillation frequency will increase by reducing the semicircular radius (R). In addition, the velocities reach their peaks when the semicircular radius (R) or the semicircular angle (α) decreases (Tables 2.1.1–2.1.3).

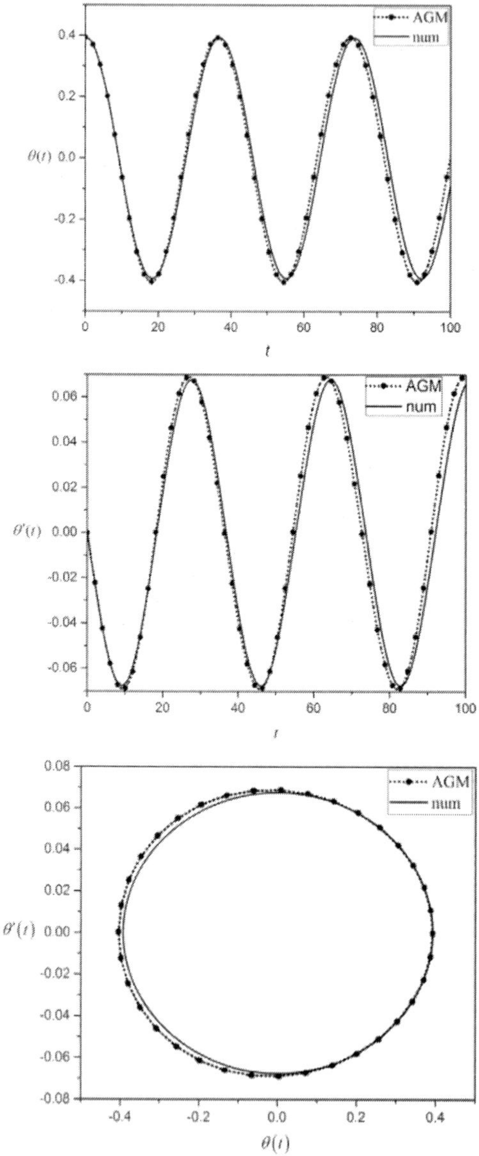

Fig. 2.1.6. (a, b, c) A comparison between the solutions obtained by AGM and the numerical method for R = 10, α = π/2.

Table 2.1.1. The results obtained by the Numerical Solution and AGM for R = 15 and α=2π/3

t	*Num*	*AGM*	*Error* = |*Num* − *AGM*|
0	0.39269908	0.39269908	0.00000000
10	0.24512253	0.24501045	0.00011208
20	−0.08993096	−0.09371990	0.00378893
30	−0.35555274	−0.36692738	0.01137464
40	−0.35254525	−0.35162485	0.00092039
50	−0.08307460	−0.06065639	0.02241821
60	0.25054818	0.26978783	0.01923964
70	0.39263866	0.39135302	0.00128564
80	0.23961867	0.21847002	0.02114865
90	−0.09675804	−0.12620233	0.02944428
100	−0.35844921	−0.37955984	0.02111063

Table 2.1.2. The results obtained by the numerical solution and AGM for R=15, α=π/2

t	*Num*	*AGM*	*Error* = |*Num* − *AGM*|
0	0.39269908	0.39269908	0.00000000
10	0.17012305	0.16964486	0.00047818
20	−0.24887845	−0.25944763	0.01056917
30	−0.38187454	−0.38749464	0.00562010
40	−0.08143699	−0.05286683	0.02857015
50	0.31332380	0.33172291	0.01839910
60	0.34991634	0.33408668	0.01582966
70	−0.01205181	−0.04827707	0.03622526
80	−0.36004875	−0.38611530	0.02606655
90	−0.29838583	−0.26304319	0.03534264
100	0.10482796	0.16555209	0.06072413

In this problem, AGM has been used to solve a nonlinear equation of circular sector oscillation systems. The plots and tables represent that AGM has acceptable accuracy compared with the numerical method. We can claim that AGM is a strong analytical method for solving linear and nonlinear equations, especially vibrational problems.

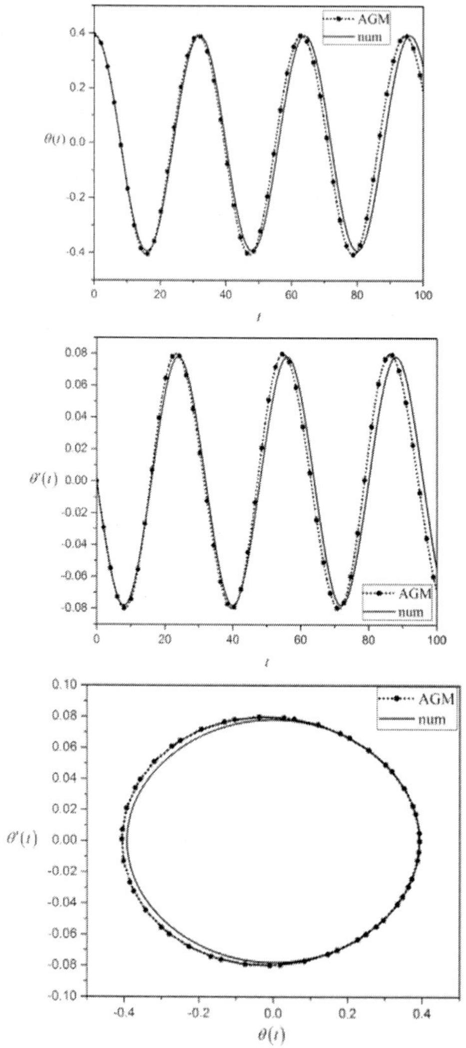

Fig. 2.1.7. (a, b, c) A comparison between the solutions obtained by AGM and the numerical method for R = 10, α = π/3.

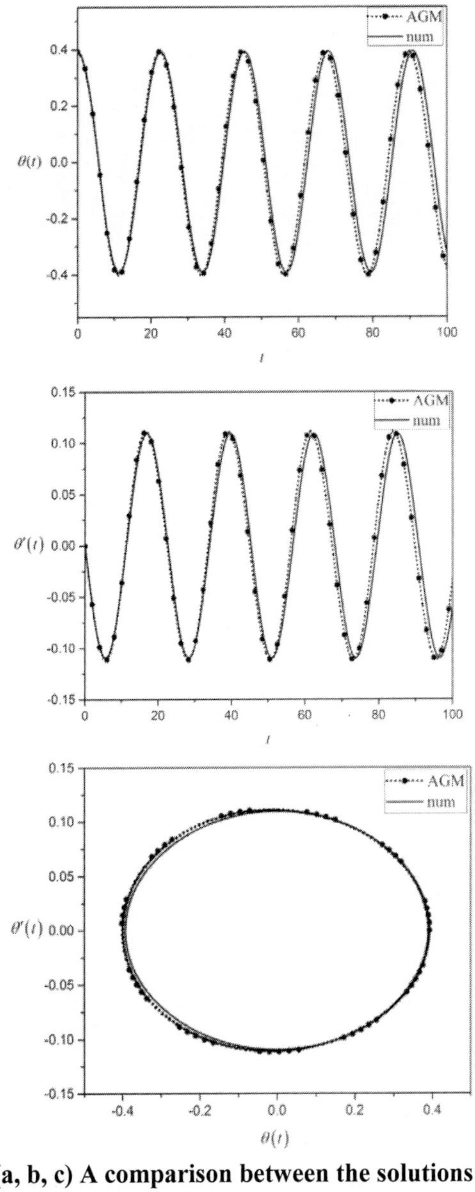

Fig. 2.1.8. (a, b, c) A comparison between the solutions obtained by AGM and the numerical method for R = 5, α = 2π/3.

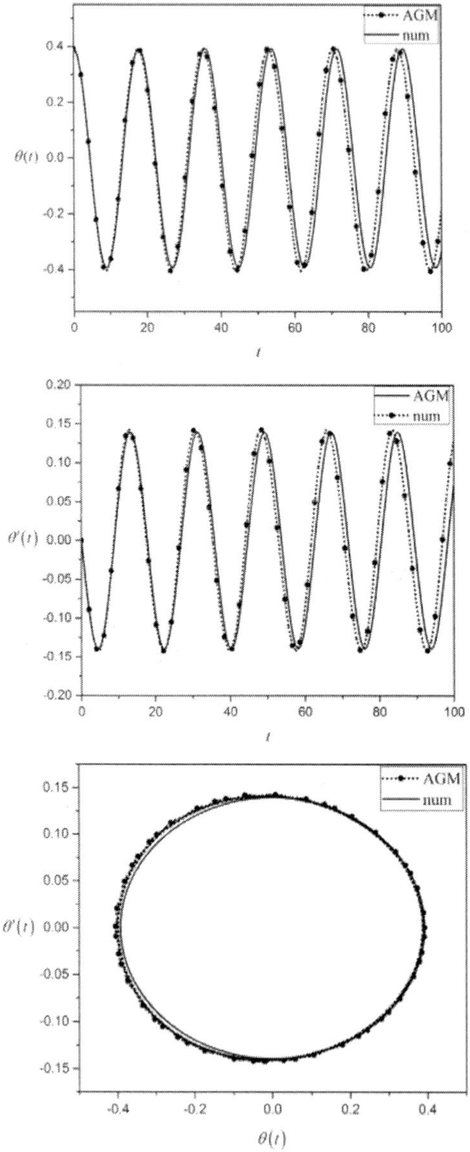

Fig. 2.1.9. (a, b, c) A comparison between the solutions obtained by AGM and the numerical method for R = 5, α = π/2.

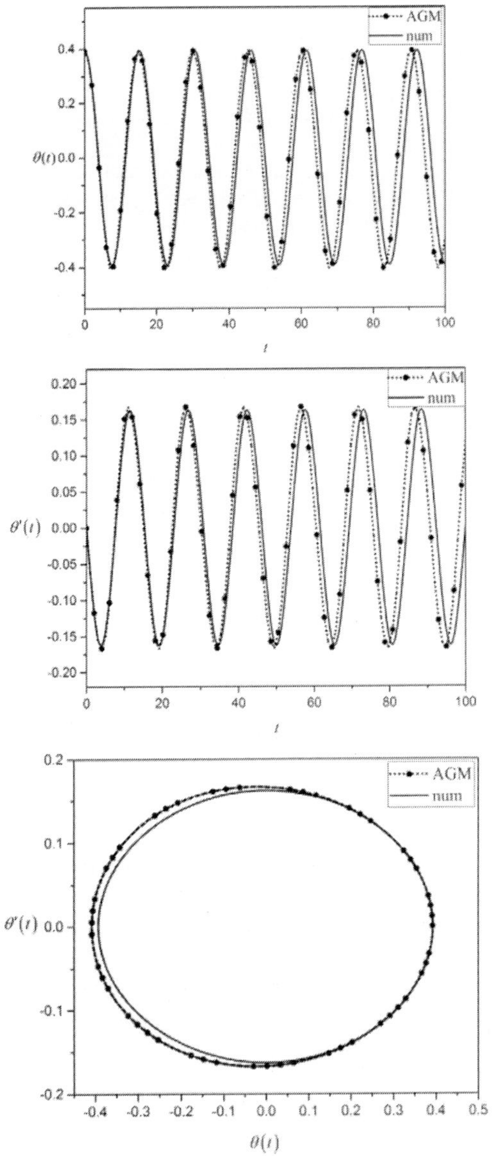

Fig. 2.1.10. (a, b, c) A comparison between the solutions obtained by AGM and the numerical method for R = 5, α = π/3.

Table 2.1.3. The results obtained by the numerical solution and AGM for R=15, α=π/3

t	*Num*	*AGM*	*Error* = |*Num* − *AGM*|
0	0.39269908	0.39269908	0.00000000
10	0.10875275	0.10762438	0.00112836
20	−0.33471506	−0.35104698	0.01633191
30	−0.29201803	−0.27870234	0.01331569
40	0.17597552	0.20864948	0.03267395
50	0.38631310	0.37789019	0.00842291
60	0.03769972	−0.00330218	0.04100190
70	−0.36626688	−0.39470691	0.02844003
80	−0.23945080	−0.18427538	0.05517541
90	0.23706674	0.29236833	0.05530158
100	0.36732664	0.33439609	0.03293055

In AGM, it is very convenient and easy to find the solution to the differential equation and the angular frequency (ω) simultaneously by merely selecting a function as the solution to the differential equation regarding the kind of operating system. In addition to the explanations above, after applying initial conditions to the considered answer, we exit from the field of differential equations into a set of algebraic equations. By solving a set of algebraic equations, a simple method, constant response coefficients are considered, and the angular frequency can be easily obtained.

2.2. Vibration in Arched Beam

Analysing and modelling the vibrational behaviour of arched bridges during an earthquake to decrease the damage to the structure is a very hard task. This has been performed analytically for the first time in the present study.

In general, vibrational equations and their initial conditions are defined for different systems as follows:

$$f\,(\ddot{u}, \dot{u}, u, F_0 \sin(\omega_0 t)) = 0 \tag{2.2.1}$$

Parameter (ω_0) is the angular frequency of the harmonic force exerted on the system, and (F_0) is its maximum amplitude. The initial conditions are as follows:

$$\{u(t)=u_0,\dot{u}(t)=0, \text{ at } t=0\} \tag{2.2.2}$$

AGM requires a total answer with constant coefficients to solve differential equations in various fields of study such as vibrations, structures, fluids, and heat transfer. Depending on the kind of vibration in vibrational systems, it is necessary to choose the answer mentioned in AGM. To clarify, we divide vibrational systems into two general forms:

Differential equations governing this kind of vibrational system are introduced in the following form:

$$f(\ddot{u},\dot{u},u)=0 \tag{2.2.3}$$

Now, the answer to this kind of vibrational system is chosen as follows:

$$u(t)=e^{-bt}\{Acos(\omega t)+Bsin(\omega t)\} \tag{2.2.4}$$

According to trigonometric relationships, Eq. (2.2.4) is rewritten as follows:

$$\mathrm{u}(\mathrm{t})=\mathrm{e}^{-bt}\ \{\mathrm{a}\cos(\omega \mathrm{t}+\varphi)\} \tag{2.2.5}$$

It is notable that in Eq. (2.2.5), $a=\sqrt{A^2+B^2}, \varphi=\arctan(\frac{B}{A})$.

Sometimes to increase the precision of the considered answer of Eq. (2.2.3), we can add another term in the form of the Fourier cosine series expansion as follows:

$$u(t)=e^{-bt}\{acos(\omega t+\varphi_1)+d\cos(2\omega t+\varphi_2)\} \tag{2.2.6}$$

In the above equation, we can omit the term (e^{-bt}) to facilitate the computational operations in AGM if the system is considered without any damping components.

Generally speaking, in AGM, Eq. (2.2.5) or Eq. (2.2.6) is assumed as the answer of the vibrational differential Eq. (2.2.3) that its constant coefficients, which are *a*, *b*, *d*, *ω* (angular frequency) and *φ* (initial vibrational phase) can easily be obtained by applying the given initial conditions in Eq. (2.2.2). The above procedure will be explained completely through the presented example in the preceding part of the problem. It is noteworthy that if there is no damping component in the vibrational system, the constant coefficient b in Eqs. (2.2.5) and (2.2.6) will automatically be computed as zero in the AGM solution procedure. On the contrary, the parameter b in Eqs. (2.2.5) and (2.2.6) for the other vibrational system with a damping component is obtained as a nonzero parameter in AGM.

In this step, it is assumed that the external forces exerting on the vibrational systems are defined as:

$$\mathrm{F(t\,)} = \mathrm{F_0 sin}\,(\omega_0 t) \tag{2.2.7}$$

As a result, the differential equation governing the vibrational system is expressed like Eq. (2.2.1) as follows:

$$f\,(\ddot{u}, \dot{u}, u\,, F_0\,\sin(\omega_0 t\,)) = 0 \tag{2.2.8}$$

The answer to the above equation is introduced as the sum of the particular solution (u_p) and the harmonic solution (u_h) as follows:

$$u_h(t) = e^{-bt}\,\{Acos\left(\omega t\right) + Bsin\left(\omega t\right)\}$$

$$u_p(t\,) = Mcos\,\left(\omega_0 t\,\right) + Nsin\,\left(\omega_0 t\,\right)\} \tag{2.2.9}$$

The result answer differential equation Eq. (2.2.8) as follows:

$$\mathrm{u\,(t\,)} = \mathrm{u}_p + \mathrm{u}_h \tag{2.2.10}$$

By utilizing trigonometric relationships $\{a = \sqrt{A^2 + B^2}\,, d = \sqrt{M^2 + N^2}\,\}$ and $\{\varphi = \arctan(\frac{B}{A}), \phi = \arctan(\frac{N}{M})\}\chi$ also by substituting the yielded equations into Eq. (2.2.10), the desired answer will be obtained in the form of:

$$u\left(t\right) = e^{-bt}\,\{acos\left(\omega t + \varphi\;\right)\} + d\cos\left(\omega t + \varphi\;\right) \tag{2.2.11}$$

To increase the precision of the equation, we can add another term in the form of the Fourier cosine series expansion as follows:

$$\begin{aligned} u(t) &= e^{-bt}\{a.cos(\omega t+\varphi_1) \\ &+c.\cos(2\omega t+\varphi_2)\}+d.\cos(\omega_0 t+\phi) \end{aligned} \tag{2.2.12}$$

And finally, by Eq. (2.2.12), the exact solution of all vibrational differential equations can be obtained in the following equation:

$$u(t)=e^{-bt}\{\sum_{k=1}^{\infty} a_k cos(k\omega t+\varphi_k)\}+dcos(\omega_0 t+\phi) \tag{2.2.13}$$

The constant coefficients of Eq. (2.2.13) $\{a_1,a_2,\ldots,\phi_1,\phi_2,\ldots,b,\omega,d,\varphi\}$ will easily be computed in AGM by applying the initial conditions of Eq. (2.2.2). To understand the above procedure thoroughly, reading the following lines is recommended.

Since the constant coefficient (b) in vibrational systems without damping components is always zero (b = 0), which in this case, decreases computational operations of Eq. (2.2.12) and (or Eq. (2.2.13)) in the following form:

$$\left\{\begin{aligned} &u(t)=\{a.cos(\omega t+\varphi_1) \\ &+c.cos(2\omega t+\varphi_2)+\ldots\}+d.cos(\omega_0 t+\phi) \\ &or\ u(t)=\{\sum_{k=1}^{\infty} a_k cos(k\,\omega t+\varphi_k)\}+dcos(\omega_0 t+\phi) \end{aligned}\right\} \tag{2.2.14}$$

Based on the above explanations, by applying initial conditions to a system without a damping component, parameter (b) is always zero for Eqs. (2.2.11)–(2.2.13). Therefore without the damping component, parameter (b)'s role in both Eqs. (2.2.11)–(2.2.13), which can be considered the answer to the vibrational problems, is individually considered as a catalyst for increasing the precision of the assumed answer. However, according to Eqs. (2.2.11)–(2.2.13) after applying initial conditions on the vibrational system in both states (with external force and without external force) by AGM, the value of the parameter (b) is computed as zero because the system has free vibration without any damping component.

Again, we mention that to decrease computational operations for systems without damping components, and since we know that (b) in the term (e^{-bt}) is zero, (e^{-bt}) can be omitted from Eqs. (2.2.11)–(2.2.13). Consequently, Eq. (2.2.11), which has been considered the answer to the systems without any damping component, can be rewritten as follows:

$$u(t) = a.cos\left(\omega t + \varphi\ \right) + d.cos(\omega_0 t + \phi) \tag{2.2.15}$$

In AGM, the application of initial conditions of Eq. (2.2.2) is done in the two following forms:

Regarding the kind of vibrational system (with external force and without external force) which was completely discussed in the previous part of this case study, a function is chosen as the answer to the differential equation from Eq. (2.2.5) or Eq. (2.2.6) for the systems without external forces and from Eqs. (2.2.11)–(2.2.15) for the defined systems with external forces, and then the initial conditions are applied to the selected function as follows:

$$\mathrm{u\ (t\ ) = u\ (IC)} \tag{2.2.16}$$

Notably, IC is the abbreviation of the introduced initial conditions of Eq. (2.2.2).

After choosing a function as the answer to the differential equation according to the kind of vibrational system, this is the best time to substitute the answer into the main differential equation instead of its dependent variable (u). Assume the general equation of the vibration such as Eq. (2.2.1) with time-independent parameter (t) and dependent function (u) as:

$$f\ (\ddot{u}, \dot{u}, u\ , F_0\ \sin(\omega_0 t\ )) = 0 \tag{2.2.17}$$

Therefore based on the kind of vibrational system, a function as the answer of the differential equation such as Eq. (2.2.5) or Eq. (2.2.6) and Eqs. (2.2.11)–(2.2.15) is considered as follows:

$$u\ = g\ (t\ ) \tag{2.2.18}$$

In this step, the equation above is substituted into Eq. (2.2.17) instead of (u) in the following form:

$$f(t) = f(g''(t), g'(t), g(t), F_0 sin(\omega_0 t)) \tag{2.2.19}$$

Eventually, the application of initial conditions on Eq. (2.2.19) and its derivatives is expressed as:

$$\begin{aligned} &f(IC) = f(g''(IC), g'(IC), g(IC), ...), \\ &f'(IC) = fj(g''(IC), g'(IC), g(IC), ...), \\ &f''(IC) = f''(g''(IC), g'(IC), g(IC), ...), ... \end{aligned} \tag{2.2.20}$$

It is better to say that in AGM, after applying the initial conditions on answer function Eq. (2.2.16), the differential function equation and its derivatives from Eqs. (2.2.20), according to the order of differential equation and utilizing the two given initial conditions of Eq. (2.2.2), a set of algebraic equations consisting of n equations with n unknowns is created. Therefore, the constant coefficients (a, b, c, d, angular frequency ω, and initial phase φ and ϕ) at Eqs. (2.2.11), (2.2.12) are easily achieved; this procedure will be explained as an example in the previous part of the problem. Notably, in Eq. (2.2.19), we can use the derivatives $f(t)$ with higher orders until the number of yielded equations equals the number of the mentioned constant coefficients of the assumed answer.

Consider an arched-shaped beam with hinged support and fixed end as a suitable model for a bridge with certain physical properties such as bending stiffness EI, cross-sectional area A, density ρ, length L, and a live harmonic load in the middle of the beam W_L (centre outlet e) in Fig. 2.2.1. It is citable that a concentrated sinusoidal force f (x, t) is exerted on the centre of the vibrational system. Based on the given information, the nonlinear partial differential equation governing the vibrational beam is achieved as follows:

$$\begin{aligned} &\mathrm{EI}\frac{\partial^4 w}{\partial x^4} + kw - \frac{EA}{L}\left(\frac{\partial^2 w}{\partial x^2} + \frac{\partial^2 y}{\partial x^2}\right) \\ &\times \int_0^L \left\{\frac{\partial w}{\partial x}\cdot\frac{\partial y}{\partial x} + \frac{1}{2}\left(\frac{\partial w}{\partial x}\right)^2\right\} dx + \rho A \frac{\partial^2 w}{\partial t^2} = -f(t) \end{aligned} \tag{2.2.21}$$

Where $f(t) = W_L.sin(\omega_0 t); y = e.sin(\frac{\pi x}{L})$

In Eq. (2.2.21), y is defined as a function for the length of the arched beam, and (e) is the centre outlet.

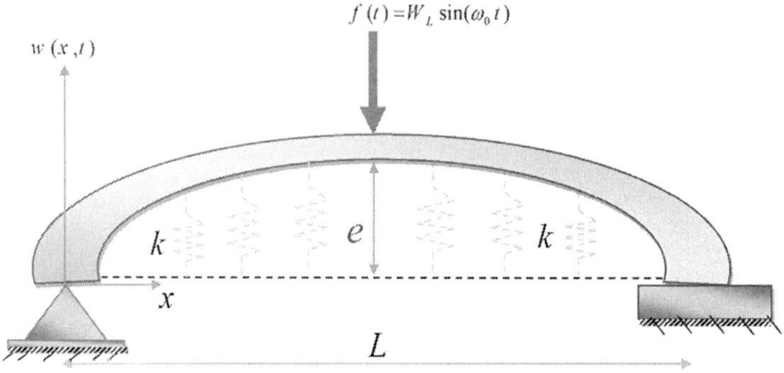

Fig. 2.2.1. The schematic diagram of the physical model.

By applying the method of separation of variables which indicates $w(x,t) = \varphi(x).z(t)$, Eq. (2.2.21) can be rewritten in the following form:

$$\ddot{Z} + \frac{1}{\rho A L^4 \phi} \left\{ \begin{array}{c} EIL^4 \phi'''' + kL^4 \phi \\ +\pi^3 e^2 \sin(\frac{\pi x}{L}) \int_0^L \phi' \cos(\frac{\pi x}{L}) dx \end{array} \right\}$$

$$Z + \frac{1}{2\rho L^4 \phi} \left\{ \begin{array}{l} \pi^2 eL \sin(\frac{\pi x}{L}) \int_0^L (\phi')^2 dx - \\ 2\pi e L^2 E \phi'' \int_0^L \phi' \cos(\frac{\pi x}{L}) dx \end{array} \right\} Z^2 \qquad (2.2.22)$$

$$+\frac{E\phi''\int_0^L(\phi')^2dx}{-2\rho L\phi}Z^3+\frac{W_L\delta(x-\frac{L}{2}).\sin(\omega_0 t)}{\rho A\phi}=0$$

In Eq. (2.2.22), parameter δ is the Dirac delta function defined as follows:

$$\int_0^\infty \phi(x).\delta(x-\frac{L}{2})dx=\phi(\frac{L}{2}) \qquad (2.2.23)$$

In the above equation, φ (x) is the modal shape function, and as a result, Eq. (2.2.22) can be expressed as follows:

$$\ddot{Z}+\alpha Z+\beta Z^2+\gamma Z^3+\lambda\phi(\frac{L}{2})\sin(\omega_0 t)=0 \qquad (2.2.24)$$

And the related initial conditions are introduced as follows:

$$Z(0)=Z_0\ ,\ \dot{Z}(0)=0 \qquad (2.2.25)$$

By utilizing the Galerkin Method and according to the orthogonality of the modes, the parameters α, β, γ, and η are defined as follows:

$$\alpha=\frac{1}{\rho AL^4\int_0^L(\phi)^2dx}$$

$$\int\left\{\begin{aligned}&EIL^4\phi'''' + kL^4\phi\\&+\pi^3e^2\sin\left(\frac{\pi x}{L}\right)\int_0^L\phi'\cos\left(\frac{\pi x}{L}\right)dx\end{aligned}\right\}\phi dx \qquad (2.2.26)$$

$$\beta = \frac{1}{2\rho L^4 \int_0^L (\phi)^2 dx}$$

$$\int_0^L \left\{ \pi^2 eL \sin\left(\frac{\pi x}{L}\right) \int_0^L (\phi')^2 dx - 2\pi eL^2 E \phi'' \int_0^L \phi' \cos\left(\frac{\pi x}{L}\right) dx \right\} \phi dx$$

$$\gamma = \frac{1}{-2\rho L \int_0^L (\phi)^2 dx}$$

$$\int_0^L \left\{ \phi'' \int_0^L (\phi')^2 dx \right\} \phi dx \; ; \lambda = \frac{W_L}{\rho A \int_0^L (\phi)^2 dx}$$

The modal shape function ϕ(x) for the hinged-fixed beam is obtained as follows:

$$\phi_n(x) = \frac{1}{\cos\left(\frac{kL}{2}\right)}$$

$$\left\{ \cos\left(\frac{kL}{2}\right).\sinh\left(\frac{Kx}{2}\right) - \cosh\left(\frac{kL}{2}\right).\sin\left(\frac{kx}{2}\right) \right\} \quad (2.2.27)$$

Where $\mathrm{k} = (4\,\mathrm{n} + 1)\frac{\pi}{2L}$ for $n = 1, 2, 3..$

Therefore, the modal shape function is depicted for different vibration modes in Figs. 2.2.2–2.2.5. These figures prove that the achieved modal shape functions are correct because it is clear that the left-hand side of all the graphs has negative slopes (hinged support), and the slope of their right-hand sides is zero (fixed end).

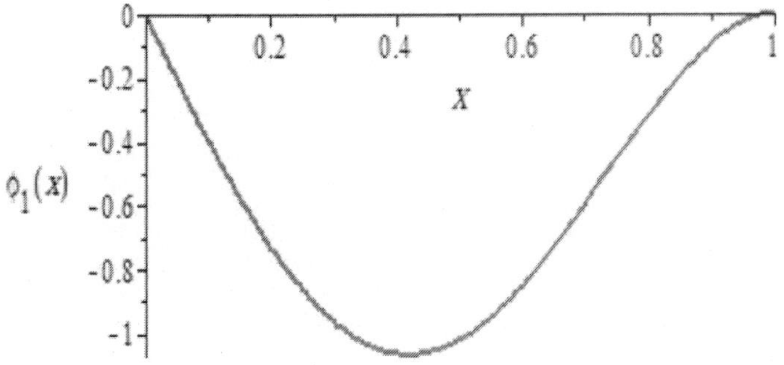

Fig. 2.2.2. The mode shape function for the first vibration mode (n = 1).

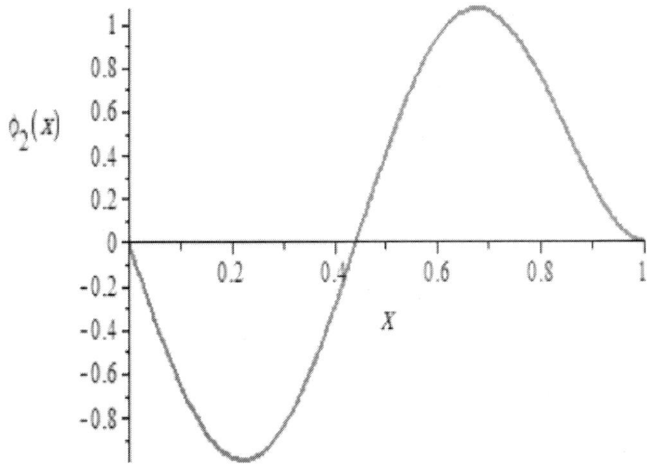

Fig. 2.2.3. The mode shape function for the second vibration mode (n = 2).

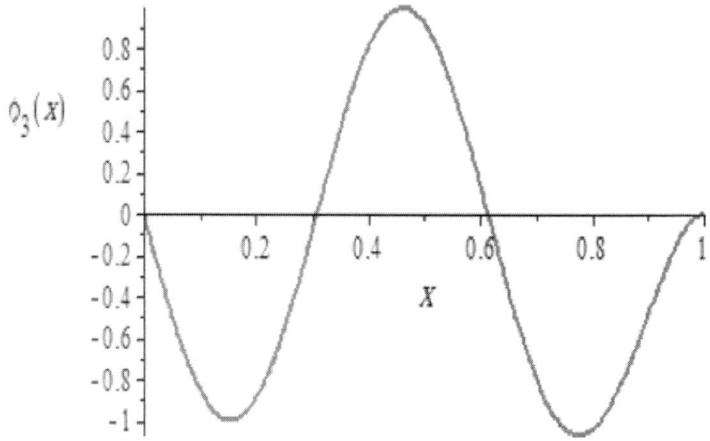

Fig. 2.2.4. The mode shape function for the third vibration mode (n = 3).

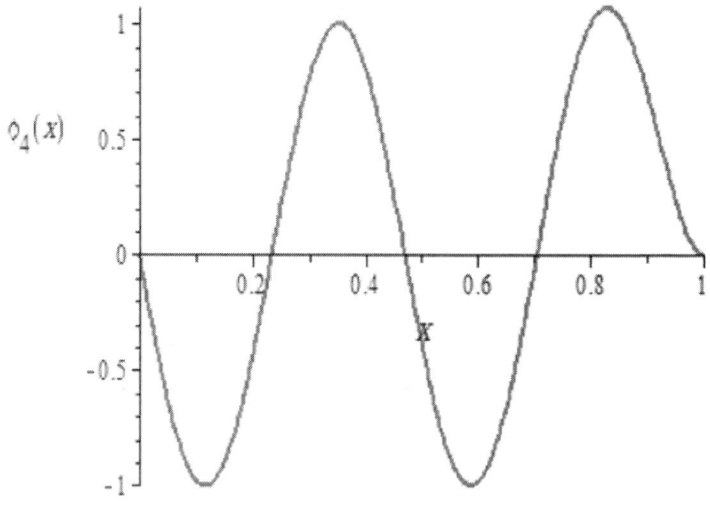

Fig. 2.2.5. The mod shape function for the fourth vibration mode (n = 4).

Consider the characteristic time function of Eq. (2.2.24). Then, to solve the obtained differential equations by AGM, we need a solution function which is a combination of a general solution and a private solution as follows:

$$z = a\cos\left(\omega t + \varphi\right) + b\sin\left(\omega_0 t + \phi\right) \tag{2.2.28}$$

The constant coefficients of Eq. (2.2.28), which are ω (angular frequency), ϕ (initial vibrational phase), a, and b, can easily be obtained by applying the initial conditions.

It is noteworthy that in the proposed method, initial or boundary conditions are applied in two ways as follows:

(a) In general, the initial conditions are applied on Eq. (2.2.28) in the form of:

$$\mathrm{z}(0) = \mathrm{z}(\,\mathrm{I\,C}\,) \tag{2.2.29}$$

For simplicity, IC is considered the abbreviation of initial conditions.

As a result, applying the initial conditions on Eq. (2.2.28) is done as follows:

$$\mathrm{z}(\mathrm{t} = 0) : \mathrm{a}\cos(\varphi) + \mathrm{b}\sin(\phi) = 0 \tag{2.2.30}$$

$$\dot{\mathrm{z}}(\mathrm{t} = 0) : -\mathrm{a}\omega\sin(\varphi) + \mathrm{b}\omega_0\cos(\phi) = 0 \tag{2.2.31}$$

The application of initial or boundary conditions on the main differential equation, which in this case is Eq. (2.2.24), and also on its derivatives is done in the following general form (for z (t)):

$$f(\,\mathrm{z}(\,\mathrm{IC}\,)\,) = 0\,, f'(\,\mathrm{z}(\,\mathrm{IC}\,)\,) = 0\,, \ldots. \tag{2.2.32}$$

Therefore, after substituting Eq. (2.2.28), which has been considered as the answer to the main differential equation, into Eq. (2.2.24), the initial conditions are applied on the equation obtained and also on its derivatives based on Eq. (2.2.32) as follows:

$$
\begin{aligned}
&f\ \ (\mathrm{z(t=0)}):\\
&a\left(\alpha-\omega^{2}\right)\cos\varphi+b\left(\alpha-\omega_{0}{}^{2}\right)\sin\phi\\
&+\left(a\cos\varphi+b\sin\phi\right)^{2}\left\{\beta+\gamma\left(a\cos\varphi+b\sin\phi\right)\right\}=0
\end{aligned}
\tag{2.2.33}
$$

On the first derivative of the equation:

$$
\begin{aligned}
&f\,'(z\,(t=0)):-a\omega\left(\alpha-\omega^{2}\right)\sin\varphi\\
&+b\,\omega_{0}\left(\alpha-\omega_{0}{}^{2}\right)\cos\phi\\
&+\left(a\cos\varphi+b\sin\phi\right)\left(3a\gamma\cos\varphi+3\gamma b\sin\phi\right)\\
&\left(b\,\omega_{0}\cos\phi-a\omega\sin\varphi\right)=0
\end{aligned}
\tag{2.2.34}
$$

On the second derivative of the equation:

$$
\begin{aligned}
&f\,''(z\,(t=0)):a\omega^{2}\left(\alpha-\omega^{2}\right)\cos\varphi\\
&-b\,\omega_{0}{}^{2}\left(\alpha-\omega_{0}{}^{2}\right)\sin\phi\\
&-\left(a\cos\varphi+b\sin\phi\right)\left(a\omega^{2}\gamma\cos\varphi+b\,\omega_{0}{}^{2}\sin\phi\right)\\
&\left(3a\gamma\cos\varphi+3\gamma b\sin\phi+2\beta\right)\\
&+2\left(a\omega\sin\varphi-b\,\omega_{0}\cos\phi\right)^{2}\\
&\left(\beta+3a\gamma\cos\varphi+3\gamma b\sin\phi\right)=0
\end{aligned}
\tag{2.2.35}
$$

Solving a set of algebraic equations consists of five equations with five unknowns from Eqs. (2.2.30), (2.2.31) and (2.2.33) to Eq. (2.2.35); the constant coefficients a, b, φ, ϕ, and ω from Eq. (2.2.28) can easily be yielded as follows: For simplicity, the following new variables are introduced as:

$$\psi_1 = \left(\alpha + \beta Z_0 + \gamma Z_0^2\right)\times$$
$$\left(\alpha + 3\gamma Z_0^2 + 2\beta Z_0 - \omega_0^2\right)$$
$$\psi_2 = \alpha + \beta Z_0 + \gamma Z_0^2$$
$$\Delta_1 = -Z_0^2\left(\alpha + \beta Z_0 + \gamma Z_0^2\right)\times$$
$$\left(\alpha + 3\gamma Z_0^2 + 2\beta Z_0 - \omega_0^2\right)\times$$
$$\left(\alpha + \beta Z_0 + \gamma Z_0^2 - \omega_0^2\right) - \lambda^2\omega_0^2$$

$$\Delta_2 = -\left(\alpha + \beta Z_0 + \gamma Z_0^2\right)\times$$
$$\left(\alpha + 3\gamma Z_0^2 + 2\beta Z_0 - \omega_0^2\right)\times$$
$$\left(\alpha + \beta Z_0 + \gamma Z_0^2 - \omega_0^2\right) \tag{2.2.36}$$

$$\Gamma_1 = \left(\alpha + \beta Z_0 + \gamma Z_0^2\right)$$
$$\left(\alpha + 3\gamma Z_0^2 + 2\beta Z_0 - 2\omega_0^2\right) + \omega_0^4$$
$$\Gamma_2 = \left(\alpha + \beta Z_0 + \gamma Z_0^2\right)\times$$
$$\left\{\begin{array}{l} \alpha\lambda^2 + 4\gamma^3 Z_0^8 + 8\beta\gamma^2 Z_0^7 + 4\alpha\gamma^2 Z_0^6 \\ +5\gamma\beta^2 Z_0^6 + 4\alpha\gamma\beta Z_0^5 + \\ \beta^3 Z_0^5 + \alpha\beta^2 Z_0^4 + \gamma\lambda^2 Z_0^2 \\ +\beta\lambda^2 Z_0 - 2\lambda^2\omega_0^2 \end{array}\right\} + \lambda^2\omega_0^4$$
$$\Lambda = \left(\alpha + \beta Z_0 + \gamma Z_0^2 - \omega_0^2\right)^2$$

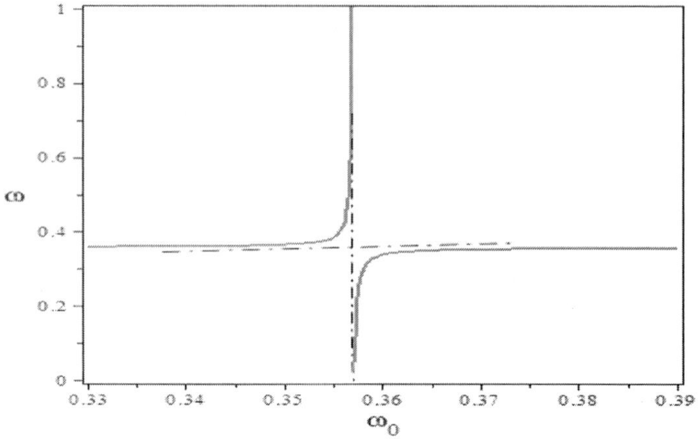

Fig. 2.2.6. The structural diagram of angular frequency in live load frequency (ω_0).

By the equations mentioned above, the constant coefficients of Eq. (2.2.28) can be obtained as follows:

$$a = \frac{\Lambda}{\Gamma_1}\sqrt{\frac{\Delta_1}{\Delta_2}} \ , b = \frac{\sqrt{\Gamma_2}}{\Gamma_1} \ ,$$
$$\varphi = tg^{-1}\left\{ Z_0 \sqrt{\frac{\Delta_2}{\Delta_1}} \right\} \ , \phi = -tg^{-1}\left\{ \frac{\lambda \psi_2}{\sqrt{\Gamma_2}} \right\} \tag{2.2.37}$$

The angular frequency is achieved as follows:

$$\omega = \sqrt{\frac{\psi_1}{\psi_2}} \Rightarrow$$
$$\omega = \sqrt{\frac{\left(\alpha + \beta Z_0 + \gamma Z_0^2\right)\left(\alpha + 3\gamma Z_0^2 + 2\beta Z_0 - \omega_0^2\right)}{\alpha + \beta Z_0 + \gamma Z_0^2 - \omega_0^2}} \tag{2.2.38}$$

And it is necessary to mention that $\omega_0 \neq \sqrt{\alpha + \beta Z_0 + \gamma Z_0^2}$

Notably, when $\omega_0 = \sqrt{\alpha + \beta Z_0 + \gamma Z_0^2}$ the vibrational system is resonant, the frequencies of the system and external harmonic load are identical. Therefore, this case should be avoided to keep the vibrational system safe. The vibrational frequency of the resonant modes tends toward infinity which this issue is shown graphically in Fig. 2.2.6 based on given physical values of Eqs. (2.2.40) and (2.2.41).

Based on Fig. 2.2.6 and the above explanations, the frequency of the external force (live load) cannot be selected arbitrarily; for instance, in this study, we know that 0.357 is related to the prevention of damaging the arched bridges by the frequency of the external force.

In this step, a variation of angular frequency in terms of spring stiffness has been illustrated graphically in Fig. 2.2.7. As regards the above diagram; it is clear that choosing some values for spring stiffness makes the vibrational system reach the resonance state where at this point, the system frequency will tend to infinity. Eventually, after substituting the obtained values from Eq. (2.2.37) and Eq. (2.2.38) into Eq. (2.2.28), the solution to the problem will be obtained as follows:

$$Z(t) = \frac{\Lambda}{\Gamma_1}\sqrt{\frac{\Delta_1}{\Delta_2}}\cos\left\{\sqrt{\frac{\psi_1}{\psi_2}}t + tg^{-1}\left[Z_0\sqrt{\frac{\Delta_2}{\Delta_1}}\right]\right\} + \frac{\sqrt{\Gamma_2}}{\Gamma_1}\sin\left[\omega_0 t - tg^{-1}\left[\frac{\lambda\psi_2}{\sqrt{\Gamma_2}}\right]\right] \tag{2.2.39}$$

By considering the following physical quantities for the system:

$$
\begin{aligned}
&W_L = 150(N), E = 3\times10^5\left(\frac{N}{m^2}\right),\\
&\rho = 2000\left(\frac{kg}{m^3}\right), L = 8(m), e = 2(mm)\\
&Z_0 = 0.02(m), \omega_0 = 2\left(\frac{Rad}{\sec}\right),\\
&A = 0.1(m^2), k = 8\left(\frac{N}{m}\right), I = 0.001(m^4)
\end{aligned}
\tag{2.2.40}
$$

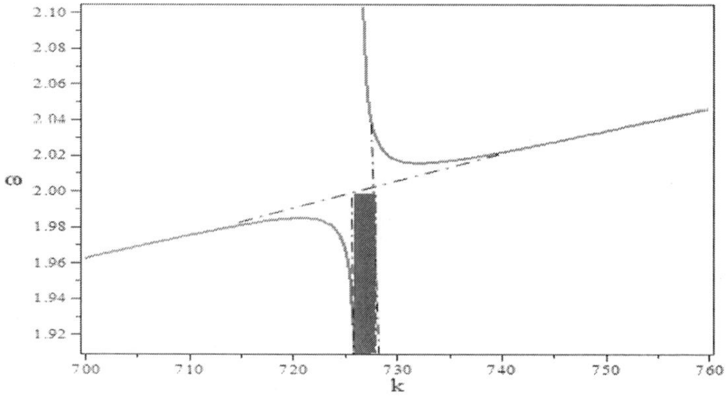

Fig. 2.2.7. Variation of angular frequency in terms of constant spring coefficients (k).

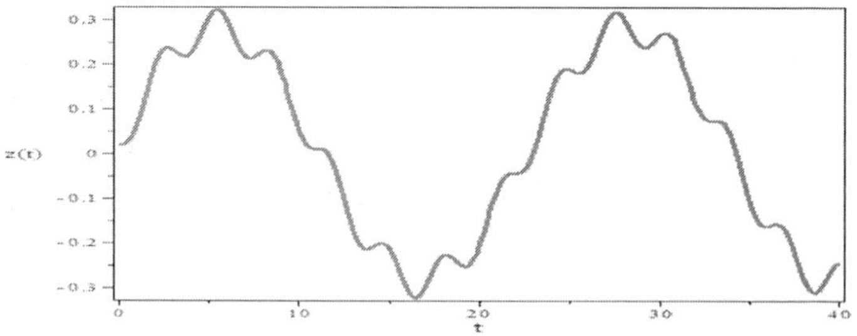

Fig. 2.2.8. The chart of the solution by AGM for the nonlinear characteristic time equation (Z (t)).

The coefficients of Eq. (2.2.24) are achieved as follows:

$$\alpha = 0.12716\left(\frac{1}{s^2}\right), \beta = 0.028513\left(\frac{1}{ms^2}\right)$$
$$\gamma = -0.001663\left(\frac{1}{m^2 s^2}\right), \lambda = -0.19167\left(\frac{m}{s^2}\right) \quad (2.2.41)$$

Therefore, the answer function of Eq. (2.2.28) is yielded in the following form:

$$Z(t) = 0.28432\cos(0.2755t - 1.5) + 0.03907\sin(2t + 3.1416) \quad (2.2.42)$$

And also, based on the given physical values, the shape function can be rewritten as follows:

$$\phi_n(x) = -0.02785\sinh(0.3927x) - \sin(0.3927x) \quad (2.2.43)$$

Then, the charts of the solution and its phase plane are drawn in Figs. 2.2.8 –2.2.10.

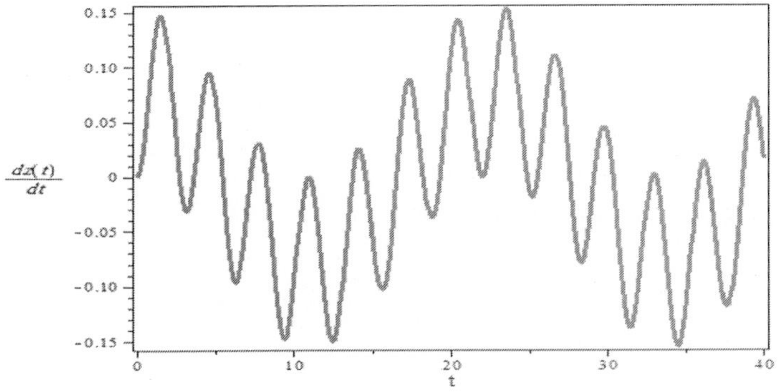

Fig. 2.2.9. The first derivative of the nonlinear characteristic time equation (Z (t)) obtained by AGM.

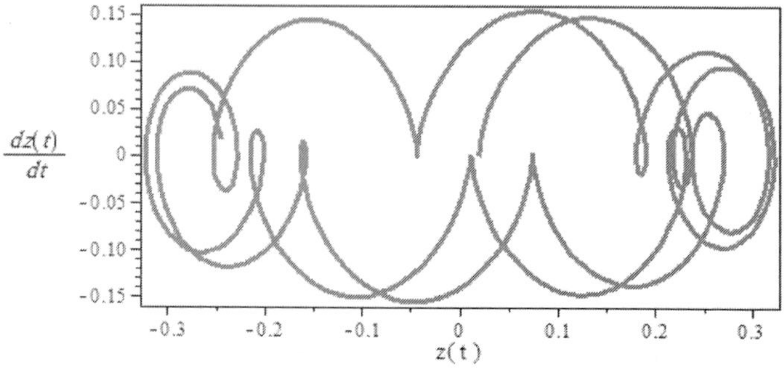

Fig. 2.2.10. The result from the phase plane for the nonlinear characteristic time equation (Z (t)) by AGM.

Table 2.2.1. The results for z (t) and its derivative are according to the given physical values.

t	0.0	8	16	24	32	40
$Z(t)_{Num}$	0.02	0.21777	−0.30443	0.136798	0.10038	−0.249136
$Z(t)_{AGM}$	0.02	0.22797	−0.2981	0.14086	0.10942	−0.24415
$\left\|\frac{Z(t)_{Num}-Z(t)_{AGM}}{Z(t)_{Num}}\right\|$	0.0	0.046	0.0205	0.029	0.090	0.020
$\dot{Z}(t)_{Num}$	0.02166	0.021658	−0.08127	0.122607	−0.0993	0.0161474

Concerning the given physical values from Eq. (2.2.40) and the determined domain t∈[0,40] defined in terms of a second (sec), the solution of the differential equations expressed numerically in Table 2.2.1 as follows:

Under the yielded solution by AGM from Eq. (2.2.39) and the obtained numerical solution whose results have been presented in Table 2.2.1, we will have the comparisons shown in Figs. 2.2.11–2.2.13.

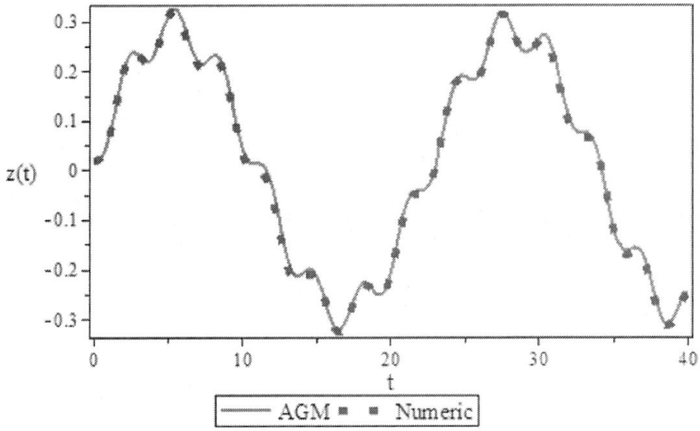

Fig. 2.2.11. A comparison between the achieved solutions by AGM and the numerical method.

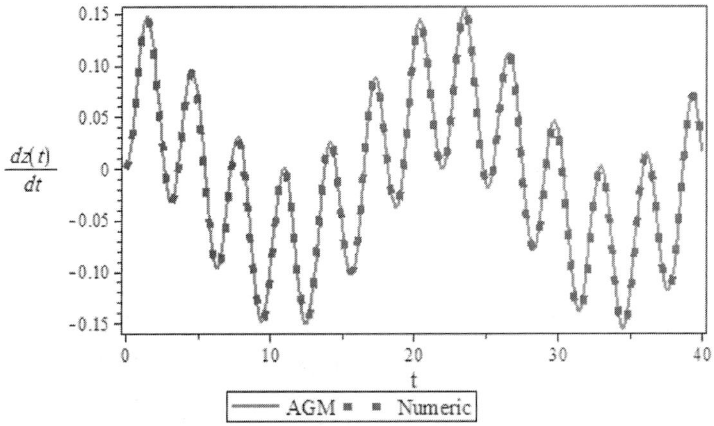

Fig. 2.2.12. A comparison between AGM's first derivative and the numerical method's solution.

According to these figures, it is obvious that AGM is an accurate and acceptable method for solving nonlinear differential equations, and so obviously, the charts of AGM and numerical solutions are overlapped; this overlap represents the precision of the method AGM at the solution of nonlinear equations.

The charts shown in Figs. 2.2.14 and 2.2.15 indicate the difference between the solutions obtained by AGM and the numerical method.

In Figs. 2.2.14 and 2.2.15, the difference between the solution obtained by AGM and the numerical method is negligible, proving this technique's reliability.

Concerning the method of separation of variables and the solution of the characteristic time equation and the shaped function $\phi_n(x)$ by AGM, the displacement of structural elements for any mode (n) of vibration can be obtained as follows:

$$w_n(x,t)=\phi_n(x).z(t) \tag{2.2.44}$$

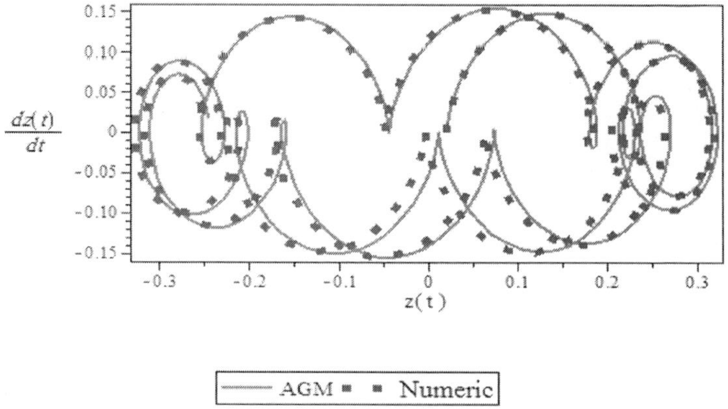

Fig. 2.2.13. Comparing the resulted phase planes by AGM and the numerical method.

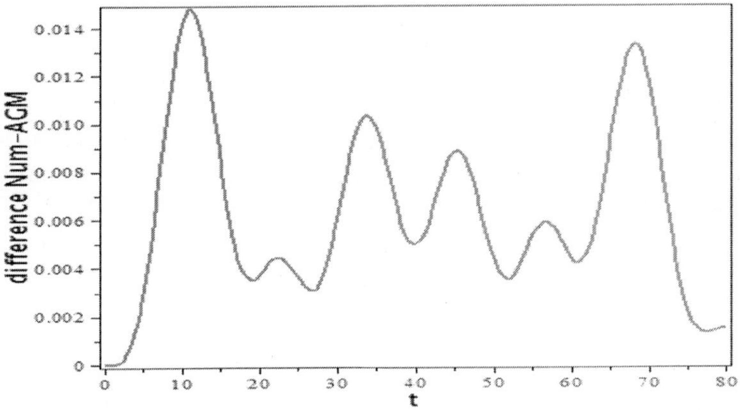

Fig. 2.2.14. The result of the difference between the solutions obtained by AGM and the numerical method for z (t).

And for all modes of vibration, we will have:

$$w(x,t) = \sum_{n=1}^{\infty} \phi_n(x).z(t) \qquad (2.2.45)$$

As it has been mentioned in this problem, we consider the first mode as an example, and the displacement for this mode about the given physical values of Eq. (2.2.40) can be achieved as follows:

$$\begin{aligned} &w(x,t)=\{-0.02785\sinh(0.3927x)-\sin(0.3927x)\} \\ &\times\{0.28432\cos(0.2755t-1.5) \qquad (2.2.46) \\ &+0.03907\sin(2t+3.\ 1416)\} \end{aligned}$$

The 3D displacement chart for structural elements in the first vibration mode (n=1) is depicted in Fig. 2.2.16.

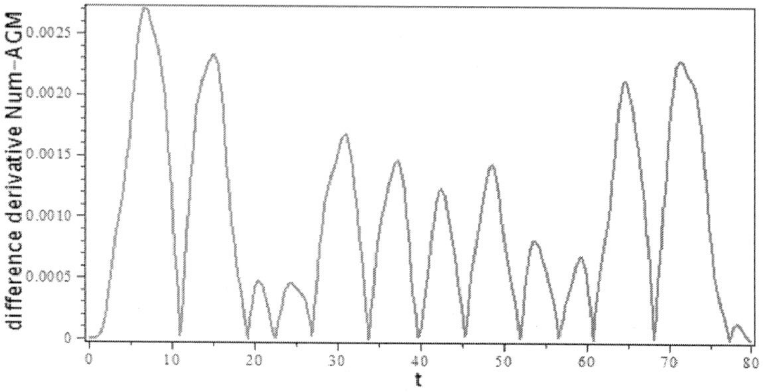

Fig. 2.2.15. The chart of the difference between the solutions obtained by AGM and the numerical method for $z'(t)$

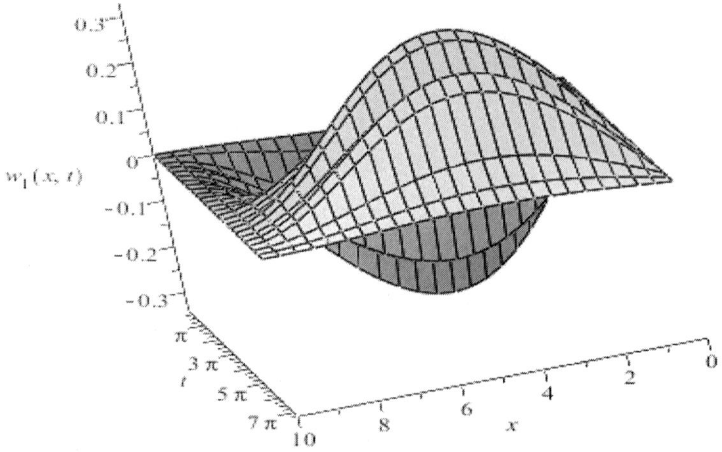

Fig. 2.2.16. Displacement of structural elements for the first mode (n = 1).

Based on the relation for angular frequency in Eq. (2.2.38), we can investigate the variation of vibrational frequencies in terms of the parameter e in Fig. 2.2.17.

Therefore, it shows that by increasing the initial amplitude of oscillation (Z_0), the angular frequency will be increased, and the beam bridge openings will be more curved as the vibration frequency (ω) increases. Moreover, the motion of external force and the arched beam length is investigated to analyze the variation of harmonic force amplitude $\left(b = \frac{\sqrt{\Gamma_2}}{\Gamma_1}\right)$ for bridge length (L) in Fig. 2.2.18. Concerning Fig. 2.2.18, the harmonic force's amplitude reaches its maximum amount in the middle of the beam length (at approximately $x \approx \frac{L}{2}$).

In this step, the effect of ω_0 on harmonic force amplitude has been investigated in Fig. 2.2.19.

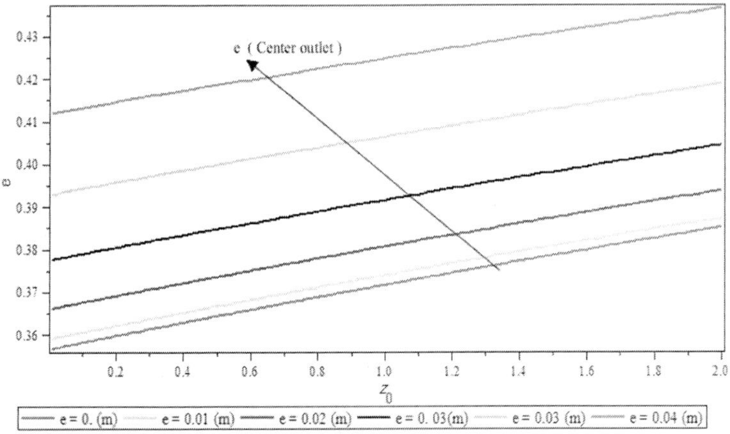

Fig. 2.2.17. The chart of angular frequency in terms of different beam centre outlets (e).

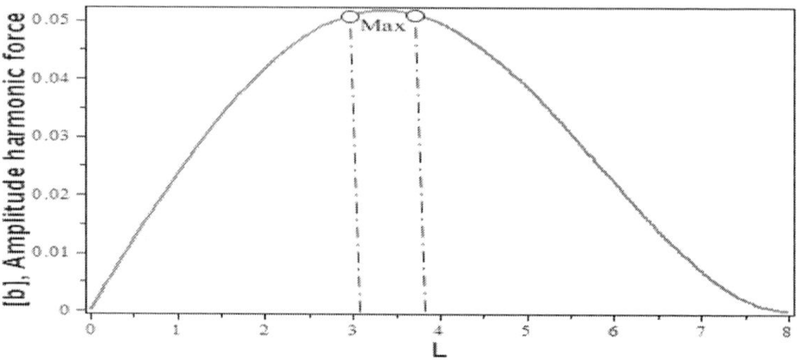

Fig. 2.2.18. Variation of the amplitude of harmonic force in terms of the beam length.

As regards Fig. 2.2.19, it is concluded that by increasing the parameter ω_0, the amount of amplitude of harmonic force is decreased remarkably. After that, the effect of the spring stiffness coefficient on the amplitude of harmonic force has been studied, and the results are presented graphically in Fig. 2.2.20. Eventually, based on the achieved observation from Fig. 2.2.20, it is logical to indicate that the amount of angular frequency of the vibrational system escalates significantly by increasing the size of spring stiffness coefficient.

Results of the difference between AGM and the numerical method (Figs. 2.2.14 and 2.2.15) depict that the difference between AGM and the numerical method is worthless, demonstrating this technique's accuracy. According to these figures, it is clear that AGM is an acceptable procedure for solving nonlinear differential equations. The charts of the AGM and the numerical solutions are overlapped, showing the AGM's precision at the solution of nonlinear equations. Therefore, using the AGM technique and considering the first mode and the displacement for this mode, it resulted that with increasing the initial amplitude of oscillation, (Z_0), the angular frequency increased, and the beam bridge openings were more curved as the vibration frequency (ω) increased. Also, the harmonic force's amplitude reaches its maximum in the middle of the beam length. According to Fig. 2.2.19, it has been found that by increasing the parameter ω_0, the amount of amplitude of harmonic force is decreased. As Fig. 2.2.20 depicted, the amount of angular frequency of the vibrational system increased rapidly by increasing the size of the spring stiffness coefficient.

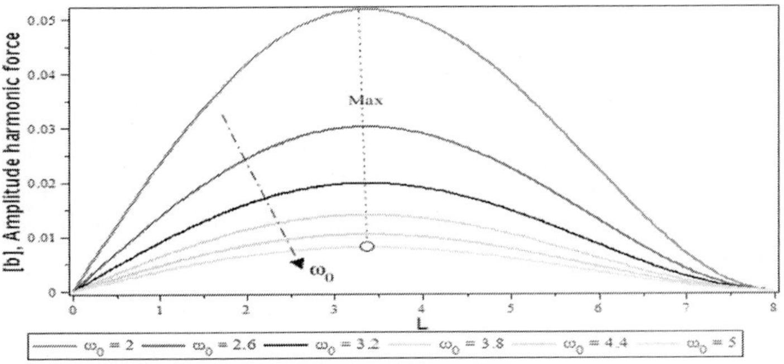

Fig. 2.2.19. The result of changing the amplitude of harmonic force and the beam length for different values of ω_0.

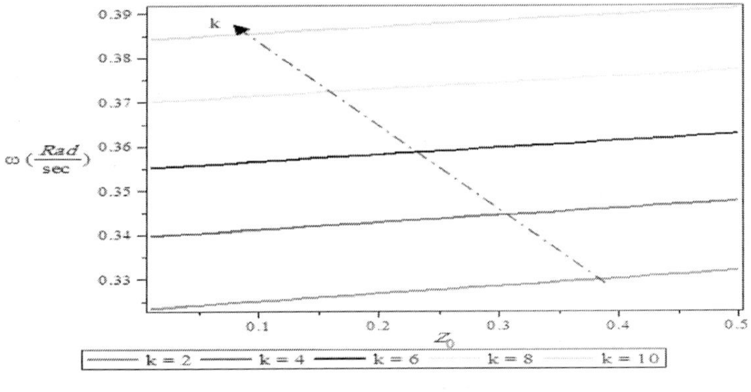

Fig. 2.2.20. The chart of angular frequency in terms of the initial vibration amplitude for different amounts of the spring stiffness (k).

The physical behaviour of arched bridges has been analyzed using a simplified model such as an arched beam. The proposed model's nonlinear differential equation has been assessed and analyzed analytically for the first time by combining AGM and the separation method. Then the solution was compared with the numerical solution, and the results indicate the precision of the achieved answers. It is concluded that the frequency of the external harmonic load or live load cannot be chosen arbitrarily because the resonance can occur. As a result, the vibrational system will be damaged.

Moreover, one should be cautious of the spring stiffness amount to prevent resonance phenomena. Also, the chart of angular frequency in terms of the parameter e has been achieved, and the observations prove that angular frequency is increased significantly by escalating the amount of parameter e. Notably, the displacement of the vertical load on the beam is heightened by increasing the frequency of the harmonic load. It is better to indicate that the amplitude of the harmonic load reaches its maximum size for $x \leq \frac{L}{2}$. Furthermore, the amplitude of harmonic force is decreased significantly by the rising ω_0. The last essential point yielded in this problem is related to the angular frequency of the arched beam, which rises remarkably by increasing the size of the spring stiffness coefficient.

2.3. Duffing-Type Nonlinear Oscillator

In this problem, attempts have been made to solve the nonlinear oscillator using AGM. Solving a nonlinear equation is difficult due to its high nonlinearity. This new approach emerged after comparing the achieved solutions with the numerical method and exact solution.

Consider a nonlinear oscillator modelled by the following governing nonlinear differential equation:

$$\frac{d^2}{dt^2}\theta(t)+\frac{\theta^3(t)}{1+\theta^2(t)}=0 \tag{2.3.1}$$

$$\theta(0)=A, \quad \frac{d}{dt}\theta(0)=0 \tag{2.3.2}$$

The system under consideration is shown in Fig. 2.3.1.

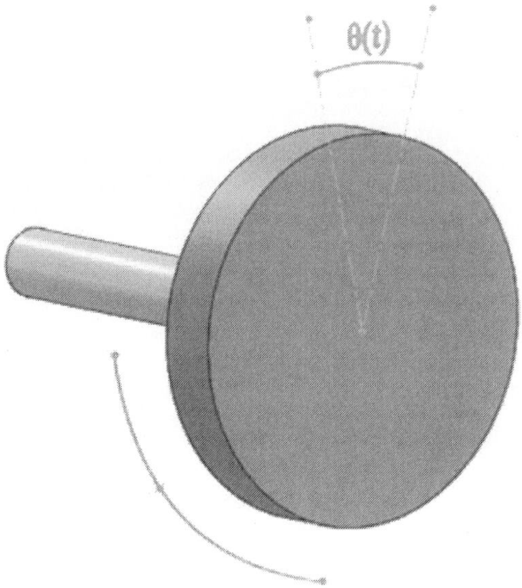

Fig. 2.3.1. The geometry of the oscillator.

The rigid disk is assumed to have mass M rotating at angular velocity $\dot{\theta}$, around the inerte Z-axis. Therefore, the torque τ effect on the hub causes it to rotate only. The process of formulation would be as follows:

$$M\ddot{\theta}(t) + k_t\theta(t) = \tau \tag{2.3.3}$$

By the assumption of the following values for the oscillator's equation, Eq. (2.3.1) would be obtained:

$$k_t = \frac{\theta^2(t)}{1+\theta^2(t)}, \quad M = 1, \quad \tau = 0 \tag{2.3.4}$$

$$\frac{d^2}{dt^2}\,\theta(t) + \frac{\theta^3(t)}{1+\theta^2(t)} = 0 \tag{2.3.5}$$

With the following initial conditions:

$$\theta(0) = A, \ \frac{d}{dt}\,\theta(0) = 0 \tag{2.3.6}$$

In general, vibrational equations and their initial conditions are defined for different systems as follows:

$$f(\ddot{u}.\dot{u}.u.F_0\sin(\omega_0 t)) = 0 \tag{2.3.7}$$

Parameter (ω_0) angular frequency of the harmonic force exerted on the system and (F_0) the maximum amplitude. And initial conditions are as follows:

$$\{u(t) = u_0.\quad \dot{u}(t) = 0.\quad at\ t = 0\} \tag{2.3.8}$$

In AGM, a total answer with constant coefficients is required to solve differential equations in various fields of study, such as vibrations, structures, fluids, and heat transfer. In vibrational systems concerning the kind of vibration, it is necessary to choose the answer mentioned in AGM. To clarify here, we divide vibrational systems into two general forms:

1) Vibrational systems without any external force. Differential equations governing this kind of vibrational system are introduced in the following form:

$$f(\ddot{u}.\dot{u}.u) = 0 \tag{2.3.9}$$

Now, the answer to this kind of vibrational system is chosen as follows:

$$u(t) = e^{-at}\{A\,cos(\omega t) + B\,sin(\omega t)\} \tag{2.3.10}$$

According to trigonometric relationships, Eq. (2.3.10) is rewritten as follows:

$$u(t) = e^{-at}\{b\,cos(\omega t + \varphi)\} \tag{2.3.11}$$

It is notable that in Eq. (2.3.7), $b = \sqrt{A^2 + B^2}$, $\varphi = arctan\,(\frac{B}{A})$.

Sometimes to increase the precision of the considered answer of Eq. (2.3.9), we can add another term in the form of the Fourier cosine series expansion as follows:

$$u(t) = e^{-at}\{b\,cos(\omega t + \varphi_1) + d\,cos(2\omega t + \varphi_2)\} \tag{2.3.12}$$

In the above equation, we can omit the term (e^{-at}) to facilitate the computational operations in AGM if the system is considered without damping components.

Generally speaking, in AGM, Eq. (2.3.11) or Eq. (2.3.12) is assumed as the answer of the vibrational differential Eq. (2.3.9) that its constant coefficients, which are a, b, d, ω (angular frequency) and φ (initial vibrational phase) can easily be obtained by applying the given initial conditions in Eq. (2.3.8). And also the above procedure will completely be explained through the presented example in the preceding part of the problem. It is noteworthy that if there is no damping component in the vibrational system, the constant coefficient (a) in Eqs. (2.3.11) and (2.3.12) will automatically be computed as zero in the AGM solution procedure.

On the contrary, the parameter b in Eqs. (2.3.11) and (2.3.12) for the other vibrational system with a damping component is obtained as a nonzero parameter in AGM.

2) Vibrational systems with external force. In this step, it is assumed that the external forces exerting on the vibrational systems are defined as follows:

$$F(t) = F_0 \sin(\omega_0 t) \tag{2.3.13}$$

As a result, the differential equation governing the vibrational system is expressed like Eq. (2.3.7) as follows:

$$f(\ddot{u}.\dot{u}.u.F_0 \sin(\omega_0 t)) = 0 \tag{2.3.14}$$

The answer to the above equation is introduced as the sum of the particular solution (u_p) and the harmonic solution (u_h) as follows:

$$u_h(t) = e^{-at}\{A\,cos(\omega t) + Bsin(\omega t)\} \tag{2.3.15}$$

$$u_p(t) = M\,cos(\omega_0 t) + N\,sin(\omega_0 t)$$

The resulting answer to differential equation Eq. (2.3.14) is as follows:

$$u(t) = u_h + u_p \tag{2.3.16}$$

By utilizing trigonometric relationships $\{\, b = \sqrt{A^2 + B^2}, d = \sqrt{M^2 + N^2}\,\}$ and $\{\varphi = arctan\,(\frac{B}{A}),\ \emptyset = arctan\,(\frac{N}{M})\}$ and by substituting the yielded equations into Eq. (2.3.16), the desired answer will be obtained in the form of

$$u(t) = e^{-at}\{b\,cos(\omega t + \varphi) + d\,cos(\omega_0 t + \o)\} \tag{2.3.17}$$

To increase the precision of the achieved equation, we can add another term in the form of the Fourier cosine series expansion as follows:

$$u(t) = e^{-at}\{b\,cos(\omega t + \varphi_1) + c\,cos(2\omega t + \varphi_2)\} + d\,cos(\omega_0 t + \o) \tag{2.3.18}$$

And finally, by Eq. (2.3.18), the exact solution of all vibrational differential equations can be obtained in the following equation:

$$u(t) = e^{-at}\left\{\sum_{k=1}^{\infty} b_k \cos(k\omega t + \varphi_k)\right\} + d\,cos(\omega_0 t + \o) \tag{2.3.19}$$

The constant coefficients of Eq. (2.3.19) which are $\{a_1, a_2. \dots, \varphi_1, \varphi_2, \dots, b, \omega, d, \o\}$ will easily be computed in AGM by applying the initial conditions of Eq. (2.3.8).

To understand the above procedure thoroughly, reading the following lines is recommended. Since the constant coefficient (a) in vibrational systems without damping components is always obtained zero $(a = 0)$, which in the case, to decrease computational operations of Eq. (2.3.18) and (or Eq. (2.3.19)) we would rewrite them in the following form:

$$\left[\begin{array}{l} u(t) = \{b\,cos(\omega t + \varphi_1) + c\,cos(2\omega t + \varphi_2)\} + d\,cos(\omega \\ \text{or } u(t) = \left\{\sum_{k=1}^{\infty} b_k \cos(k\omega t + \varphi_k)\right\} + d\,cos(\omega_0 t + \end{array}\right. \tag{2.3.20}$$

Based on the above explanations, by applying initial conditions on a system without a damping component, parameter (a) is always zero for Eqs. (2.3.17)–(2.3.19). Therefore, without the damping component, parameter (a)'s role in both Eqs. (2.3.17)–(2.3.19), each of which can be considered as the answer to the vibrational problems, is individually considered as a catalyst for increasing the precision of the assumed answer. However, according to Eqs. (2.3.17)–(2.3.19) after applying initial conditions on the vibrational system in both states (with external force and without external force) by AGM, the value of the parameter (a) is computed as zero because the mentioned system has a free vibration without any damping component.

Again, we mention that in order to decrease computational operations for systems without damping components, and since we know that (a) in the term (e^{-at}) is zero, so (e^{-at}) can be omitted from Eqs. (2.3.17)–(2.3.19). Consequently, Eq. (2.3.17), which has been considered the answer to the systems without any damping component, can be rewritten as follows:

$$u(t) = b\, cos(\omega t + \varphi) + d\, cos(\omega_0 t + \emptyset) \tag{2.3.21}$$

In AGM, the application of initial conditions of Eq. (2.3.8) is done in the two following forms:

a) Applying the initial conditions to the answer of the differential equation. With the kind of vibrational system (with external force and without external force) which was completely discussed in the previous part of this case study, a function is chosen as the answer to the differential equation from Eq. (2.3.11) or Eq. (2.3.12) for the systems without external forces and from Eqs. (2.3.17)–(2.3.19) for the defined systems with external forces, and then the initial conditions are applied to the selected function as follows:

$$u(t) = u(\mathrm{IC}) \tag{2.3.22}$$

Notably, IC is the abbreviation of the introduced initial conditions of Eq. (2.3.8).

b) Appling the initial conditions on the main differential equation and its derivatives

After choosing a function as the answer to the differential equation according to the kind of vibrational system, this is the best time to substitute the mentioned answer into the main differential equation instead of its dependent variable (u).

Assume the general equation of the vibration such as Eq. (2.3.7) with a time-independent parameter (t) and dependent function (u) as follows:

$$f(\ddot{u}.\dot{u}.u.F_0 \sin(\omega_0 t)) = 0 \tag{2.3.23}$$

Therefore, based on the kind of vibrational system, a function as the answer of the differential equation, such as Eq. (2.3.11) or Eq. (2.3.10) and Eqs. (2.3.17)–(2.3.19) is considered as follows:

$$u = g(t) \tag{2.3.24}$$

In this step, the equation above is substituted into Eq. (2.3.23) instead of (u) in the following form:

$$f(t) = f(g''(t), g'(t), g(t), F_0 \sin(\omega_0 t)) \tag{2.3.25}$$

Eventually, the application of initial conditions on Eq. (2.3.25) and its derivatives is expressed as follows:

$$\begin{aligned} &f(IC) = f(g''(IC), g'(IC), g(IC), \dots), \\ &f'(IC) = f'(g''(IC), g'(IC), g(IC), \dots), \\ &f''(IC) = f''(g''(IC), g'(IC), g(IC), \dots), \dots. \end{aligned} \tag{2.3.26}$$

It is better to say that in AGM, after applying the initial conditions on answer function Eq. (2.3.20), and also the functional differential equation and on its derivatives from Eq. (2.3.26) according to the order of differential equation and utilizing the two given initial conditions of Eq. (2.3.8), a set of algebraic equations which is consisted of n equations with n unknowns is created. Therefore, the constant coefficients (a, b, c, d, angular frequency ω and initial phase φ and ø) at Eqs. (2.3.17) and (2.3.18) are easily achieved. This procedure will be thoroughly explained as an example in the previous part of this problem.

It is noteworthy that in Eq. (2.3.25), we can use the derivatives of $f(t)$, with higher orders until the number of yielded equations equals the number of the mentioned constant coefficients of the assumed answer.

First of all, we rewrite problem Eq. (2.3.1) in the following order:

$$f(t) = \frac{d^2}{dt^2}\,\theta(t) + \frac{\theta^3(t)}{1+\theta^2(t)} = 0 \tag{2.3.27}$$

Based on the given explanations in the previous section, we consider the answer to Eq. (2.3.27) as follows:

$$\theta(t) = e^{-at}\left(b_1 \cos(\omega t + \varphi_1) + b_2 \cos(2\omega t + \varphi_2)\right) \tag{2.3.28}$$

Based on the given explanations, the existence of the term e^{-at} in Eq. (2.3.28) indicates a damping component in the oscillating system. Since no damping components exist in the example, the constant a in Eq. (2.3.28) will automatically be zero after applying the initial conditions in AGM. Moreover, the constant coefficient b, the initial vibrational phase φ and finally, the angular frequency ω can be computed by applying the initial conditions. According to the above theory, the answer to the problem is assumed as follows:

$$\theta(t) = b_1 \cos(\omega t + \varphi_1) + b_2 \cos(2\omega t + \varphi_2) \tag{2.3.29}$$

Concerning the proposed physical model, there are no boundary conditions, so the constant coefficients of Eq. (2.3.29) are just acquired for the given initial conditions, which have been presented in Eq. (2.3.2). Notably, initial or boundary conditions are applied in two ways in the following form:

(a) The initial conditions are applied on Eq. (2.3.29) in the form of:

$$\theta(t) = \theta(\mathrm{IC}) \tag{2.3.30}$$

To simplify, IC is considered the abbreviation of the initial conditions.

As a result, applying the initial conditions on Eq. (2.3.29) is done as follows:

$$\theta(0) = A \;\rightarrow\; b_1 \cos(\varphi_1) + b_2 \cos(\varphi_2) = A \tag{2.3.31}$$

And

$$\theta'(0) = 0 \;\rightarrow\; -b_1 \sin(\varphi_1)\omega - 2b_2 \sin(\varphi_2)\omega = 0 \tag{2.3.32}$$

(b) The application of initial or boundary conditions on the main differential equation, which in this case is Eq. (2.3.27) and its derivatives, is done in the following general forms:

$$f(\theta'(t)) \;\rightarrow\; f(\theta'(IC)) = 0,\; f'(\theta'(IC)) = 0,\; \ldots \tag{2.3.33}$$

Therefore, according to Eq. (2.3.33), the initial conditions are applied after substituting Eq. (2.3.29), which was considered as the answer to the main differential equation into Eq. (2.3.27).

As a result, the above procedure is done on the yielded equation and its derivative as follows:

$$f(\theta'(0)) : -b_1\cos(\varphi_1)\,\omega^2 - 4b_2\cos(\varphi_2)\,\omega^2 + \frac{(b_1\cos(\varphi_1)+b_2\cos(\varphi_2))^3}{1+(b_1\cos(\varphi_1)+b_2\cos(\varphi_2))^2} = 0 \tag{2.3.34}$$

$$f'(\theta'(0)) : b_1\sin(\varphi_1)\,\omega^3 + 8b_2\sin(\varphi_2)\,\omega^3 + \frac{3(b_1\cos(\varphi_1)+b_2\cos(\varphi_2))^2(-b_1\sin(\varphi_1)\omega - 2b_2\sin(\varphi_2)\omega)}{1+(b_1\cos(\varphi_1)+b_2\cos(\varphi_2))^2} - \frac{2(b_1\cos(\varphi_1)+b_2\cos(\varphi_2))^4(-b_1\sin(\varphi_1)\omega - 2b_2\sin(\varphi_2)\omega)}{(1+(b_1\cos(\varphi_1)+b_2\cos(\varphi_2))^2)^2} = 0 \tag{2.3.35}$$

$$f''(\theta'(0)) : b_1\cos(\varphi_1)\,\omega^4 + 16b_2\cos(\varphi_2)\,\omega^4 + \frac{6(b_1\cos(\varphi_1)+b_2\cos(\varphi_2))(-b_1\sin(\varphi_1)\omega - 2b_2\sin(\varphi_2)\omega)^2}{1+(b_1\cos(\varphi_1)+b_2\cos(\varphi_2))^2} - \frac{14(b_1\cos(\varphi_1)+b_2\cos(\varphi_2))^3(-b_1\sin(\varphi_1)\omega - 2b_2\sin(\varphi_2)\omega)^2}{(1+(b_1\cos(\varphi_1)+b_2\cos(\varphi_2))^2)^2} + \frac{3(b_1\cos(\varphi_1)+b_2\cos(\varphi_2))^2(-b_1\cos(\varphi_1)\omega^2 - 4b_2\cos(\varphi_2)\omega^2)}{1+(b_1\cos(\varphi_1)+b_2\cos(\varphi_2))^2} + \frac{8(b_1\cos(\varphi_1)+b_2\cos(\varphi_2))^5(-b_1\sin(\varphi_1)\omega - 2b_2\sin(\varphi_2)\omega)^2}{(1+(b_1\cos(\varphi_1)+b_2\cos(\varphi_2))^2)^3} - \frac{2(b_1\cos(\varphi_1)+b_2\cos(\varphi_2))^4(-b_1\cos(\varphi_1)\omega^2 - 4b_2\cos(\varphi_2)\omega^2)}{(1+(b_1\cos(\varphi_1)+b_2\cos(\varphi_2))^2)^2} = 0 \tag{2.3.36}$$

Solving a set of algebraic equations consists of five equations with five unknowns from Eqs. (2.3.31) and (2.3.32) and Eqs. (2.3.34)–(2.3.36); the constant coefficients b_1, b_2, φ_1, φ_2 and ω of Eq. (2.3.29) can easily be yielded as follows.

To simplify, the following new variable is introduced:

$$\Gamma = \frac{3A^2-\sqrt{9A^4-14A^2-23}-1}{A^2+3} \tag{2.3.37}$$

For Eq. (2.3.37), the constant coefficients of Eq. (2.3.29) can be rewritten as follows:

$$\omega = \frac{1}{4}\frac{\sqrt{-2A^2\Gamma+16A^2-6\Gamma+8A}}{A^2+1}, \quad b_1 = A-\frac{1}{6}\Gamma A, \quad b_2 = -\frac{1}{6}\Gamma A, \quad \varphi_1 = 0, \quad \varphi_2 = \pi \tag{2.3.38}$$

By considering the following physical quantities for the amplitude vibration (see Fig. 2.3.2),

$$A = 10 \tag{2.3.39}$$

After substituting the mentioned physical quantities into Eqs. (2.3.37) and (2.3.38), then by substituting the final obtained values in Eq. (2.3.29), the solution will be obtained as follows:

$$\theta(t) = 4.954484635 \cos(0.9674593759t) + 0.045515365 \cos(1.934918752t) \tag{2.3.40}$$

The above procedure can be applied to various values of A. However, to obtain better results, we continued our research for solving the mentioned problem with different values for A.

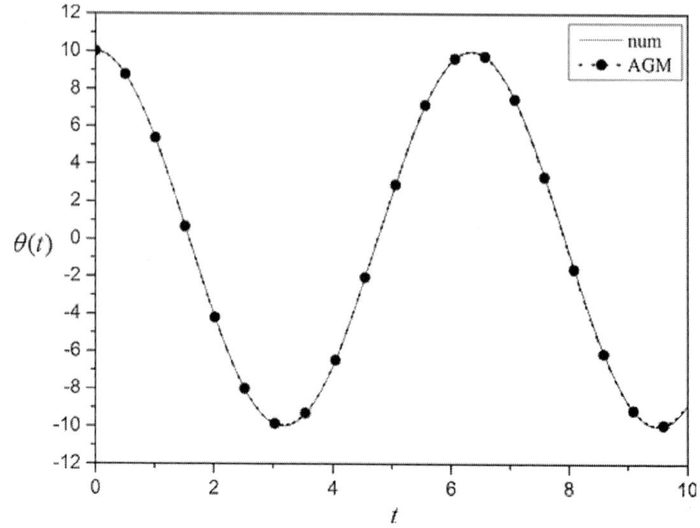

Fig. 2.3.2. Comparison between the results of $\theta\ (x)$, by AGM and the numerical method for $A = 10$.

For instance, solving the problem procedure for the minimal amount of A, such as 0.1 and 0.01, we assumed our trial function with two terms by solving the algebraic calculations (four equations and four unknowns). With the acceptance of the real part of the obtained result as the solution, the constant coefficients b_1, b_2, φ_1, φ_2 and ω of Eq. (2.3.29) can easily be yielded as follows:

For $A = 10$ we will have (see Figs. 2.3.3–2.3.7):

$$\omega = -0.08583205170,\quad b_1 = -0.1053709856,\quad b_2 = 0.005370985604,\quad \varphi_1 = -3.141592654,\quad \varphi_2 = -9.424777961 \tag{2.3.41}$$

So the solution to the problem we would have obtained is as follows:

$$\theta\ (t) = 0.005370985604\ \cos\left(0.1716641034t + 9.424777961\right) - 0.1053709856\ \cos\left(0.08583205170t + 3.141592654\right) \tag{2.3.42}$$

For $A = 0.01$ we will have (see Fig. 2.3.4):

$$\omega = -0.008633792486, \quad b_1 = -0.01055537038, \quad b_2 = 0.0005553703765$$
$$\varphi_1 = -3.141592654, \quad \varphi_2 = -9.424777961 \tag{2.3.43}$$

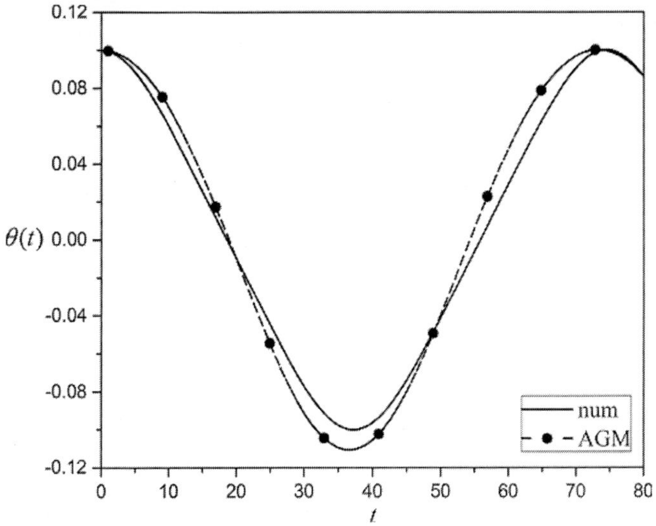

Fig. 2.3.3. Comparison between the results of $\theta\,(x)$, by AGM and the numerical method for $A = 0.1$.

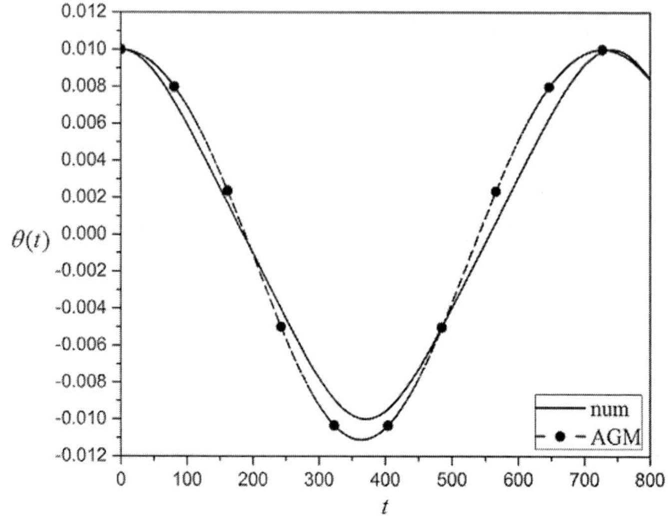

Fig. 2.3.4. Comparison between results obtained of $\theta\ (x)$, by AGM and the numerical method for $A = 0.01$.

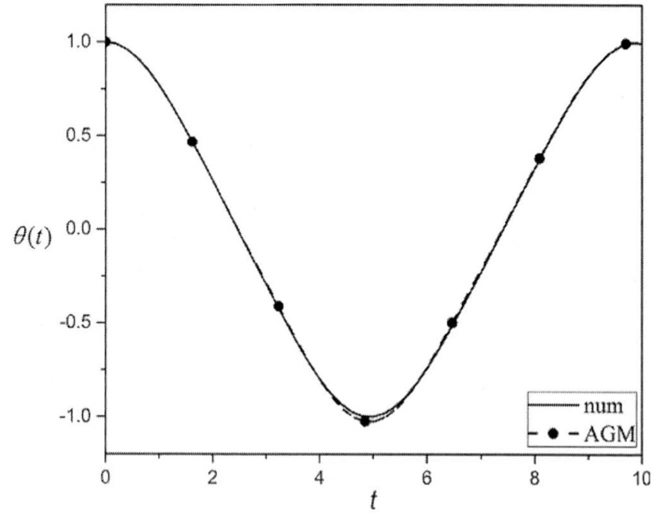

Fig. 2.3.5. Comparison between the results of $\theta\ (x)$ by AGM and the numerical method for $A = 1$

So the solution to the problem would have been obtained as follows:

$$\theta\,(t) = -0.01055537038 \times \cos(0.008633792486t + 3.141592654) + 0.0005553703765 \cos(0.01726758497t + 9.424777961) \quad (2.3.44)$$

These are accepted answers with low errors with the exact number of answers obtained by the Runge-Kutta method.

Another example, in this case, is solving Eq. (2.3.1) with $A = 1$, and this process could be applied for a different value of A, so in this case, we choose the procedure of solving due to our need to obtain an answer which depends on our opinion or accuracy that we are looking for. So, in this case, we apply the trial function with up to four terms as follows:

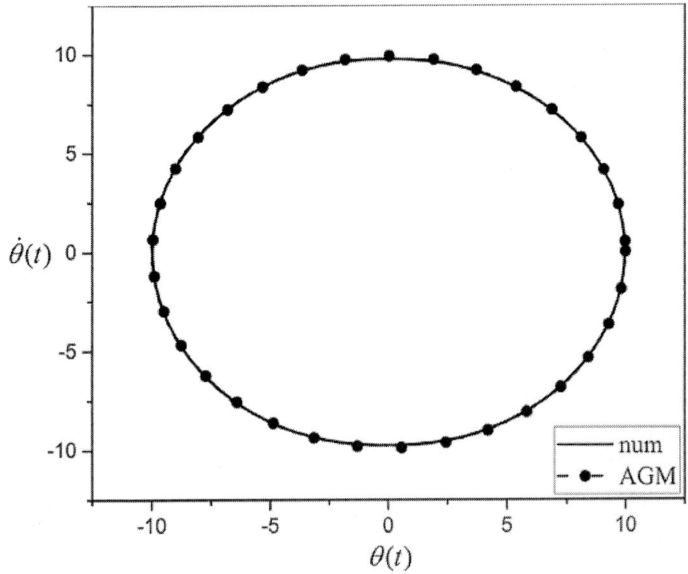

Fig. 2.3.6. Comparison between the results of $\theta\,(x)$ by AGM and the numerical method for $A = 10$

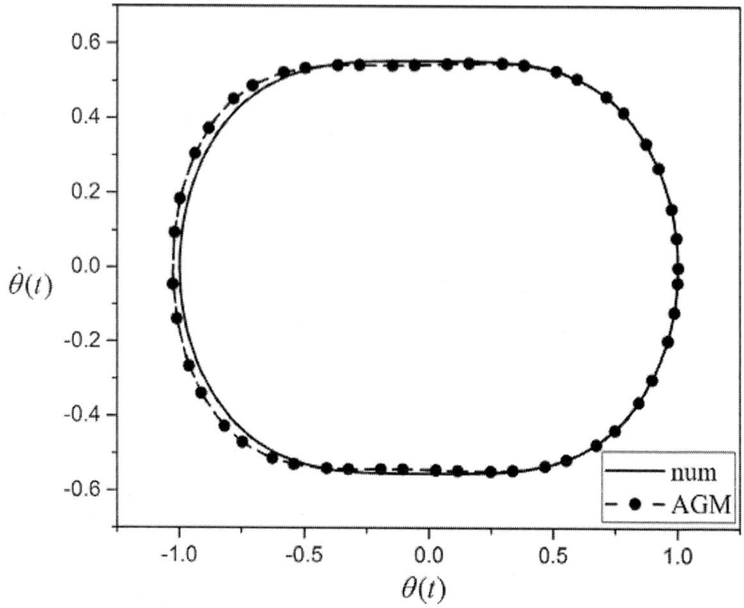

Fig. 2.3.7. Comparison between the results of $\theta\ (x)$ by AGM and the numerical method for $A = 1$

$$\theta\ (t) = b_1 \cos(\omega t + \varphi_1) + \ b_2 \cos(2\omega t + \varphi_2) + b_3 \cos(3\omega t + \varphi_3) + \ b_4 \cos(4\omega t + \varphi_4) \tag{2.3.45}$$

So by solving the set of algebraic calculations (nine equations and nine unknowns), the constant coefficients $b_1, b_2, b_3, b_4, \varphi_1, \varphi_2, \varphi_3, \varphi_4$ and ω of Eq. (2.3.45) would be obtained as follows:

$$\begin{aligned}
&\omega = -\ 0.6373562350 \\
&b_1 = -\ 0.9735802147, \quad b_2 = -\ 0.009515248435 \\
&b_3 = -\ 0.03994694798, \quad b_4 = -\ 0.004011914225 \\
&\varphi_1 = -\ 9.424777961, \quad \varphi_2 = -\ 12.56637061 \\
&\varphi_3 = -\ 9.424777961, \quad \varphi_4 = -\ 6.283185307
\end{aligned} \tag{2.3.46}$$

So the solution to the problem would have been obtained as follows:

Table 2.3.1. Resulting errors through AGM for $\theta(x)$ with A =10.

t	Error (A = 10)
0	0.0000000004
1	0.0018673829
2	0.0487324767
3	0.0379726475
4	0.0157708646
5	0.0376813682
6	0.0168360947
7	0.0313311902
8	0.0942359176
9	0.0127440686
10	0.0607523738

Table 2.3.2. Resulting errors through AGM for $\theta(x)$ with A = 0.1.

t	Error (A = 0.1)
0	0.0000000000
8	0.0079258488
16	0.0072122058
24	0.0086336631
32	0.0137262721
40	0.0091370221
48	0.0020672041
56	0.0140705200
64	0.0173457904
72	0.0024931915
80	0.0000038234

Table 2.3.3. Resulting errors through AGM for $\theta(x)$ with A = 0.01.

t	Error (A = 0.01)
0	0.0000000000
80	0.0008019579
160	0.0007062095
240	0.0009228495
320	0.0014249079
400	0.0009132327
480	0.0000999953
560	0.0015633441
640	0.0017874547
720	0.0002129877
800	0.0000389138

$$\begin{aligned}\theta(t) = &-0.9735802147\cos(0.6373562350t+9.424777961)\\ &-0.009515248435\cos(1.274712470t+12.56637061)\\ &-0.03994694798\cos(1.912068705t+9.424777961)\\ &-0.004011914225\cos(2.549424940t+6.283185307)\end{aligned} \tag{2.3.47}$$

These are accepted answers with low errors with the exact number of answers obtained by the Runge-Kutta method.

Table 2.3.4. Resulting errors through AGM for θ(x) with A = 0.01.

t	Error (A = 1)
0	0.0000000000
1	0.0323225843
2	0.0565293300
3	0.0694878386
4	0.1727424290
5	0.1847768542
6	0.0622806814
7	0.3681746222
8	0.5009673679
9	0.4464822849
10	0.1533444647

The current investigation has been done to solve the discussed nonlinear problem with the analytical method AGM in order to get a better solution in comparison with the numerical method (Runge-Kutta, R_4) and the mathematical procedure has been done differently by assuming various trial functions as the solution of the problem, which presents the simplicity and acceptable accuracy of this method among other analytical methods. According to figures and related tables, the results of this method would be noticeable (see Tables 2.3.1–2.3.4). For obtaining better solutions more simply during mathematical calculation with appropriate software (such as Maple), some other assumptions could be mentioned; for instance, for A= 0.01, the following terms in order of complex numbers could be considered:

$$\begin{gathered} z_1 = e^{I(\omega t+\varphi_1)},\ z_2 = e^{-I(\omega t+\varphi_1)}, \\ z_3 = e^{I(2\omega t+\varphi_2)},\ z_4 = e^{-I(2\omega t+\varphi_2)} \end{gathered} \tag{2.3.48}$$

Then the trial function would be assumed as follows:

$$\theta(t) = \sum_{i=1}^{2} \frac{b_i}{2}\left(z_i + z_{i+1}\right) \tag{2.3.49}$$

The procedure will be applied as described previously. The reason for using the above assumption is just to simplify the calculation.

In this problem, attempts have been made to solve the nonlinear oscillator equation with AGM. Its results compare with the numerical method (Runge-Kutta, R4) for different amplitude vibration values and by various solution processes to reveal that AGM could be applied for solving nonlinear equations with high nonlinearity. According to figures and related tables, it is logical to say that AGM is a very applicable and suitable approach for solving nonlinear differential equations and has acceptable efficiency and accuracy. After applying the initial conditions on the considered solution, we exit from the field of the differential equation into a set of algebraic equations. Then, by solving a set of algebraic equations, a simple procedure, the constant coefficients of the trial function would be obtained.

2.4. Euler-Bernoulli Beam

In this problem, solving the nonlinear differential equation governing the calculation of the large rotation deviation of the beam (or column) has been discussed. Boundary and initial conditions are required for analytical methods of each linear and nonlinear differential equation according to the physics of the problem. Therefore, we can solve every differential equation with any degree. By the boundary conditions, the general manner of a differential equation is as follows:

$$p_k\colon f\left(u, u', u'', \dots, u^{(m)}\right) = 0, u = u(x) \tag{2.4.1}$$

P_k is the nonlinear differential equation's name and m is the derivative order.

Boundary condition:

$$\begin{cases} u(0) = u_0, u'(0) = u_1, \dots, u^{(m-1)}(0) = u_{m-1} \\ u(L) = u_{L_0}, u'(L) = u_{L_1}, \dots, u^{(m-1)}(L) = u_{L_{m-1}} \end{cases} \tag{2.4.2}$$

To solve the first differential equation concerning the boundary conditions at x=L in Eq. (2.4.2), the series of letters in the nth order with constant coefficients, which is the answer of the first differential equation, is considered as follows:

$$u(x) = \sum_{i=0}^{n} a_i x^i = a_0 + a_1 x^1 + a_2 x^2 + \cdots + a_n x^n \tag{2.4.3}$$

The more precise answer to Eq. (2.4.1), the more choice of series sentences from Eq. (2.4.3). In applied problems, approximately five or six sentences from the series are enough to solve nonlinear differential equations. In answer to Eq. (2.4.3), regarding the series from degree (n), there are (n+1) unknown coefficients that need (n+1) equations to be specified. The boundary conditions of Eq. (2.4.2) are used to solve a set of equations consisting of (n+1) ones.

The boundary conditions are applied to the functions as follows:

1) The application of the boundary conditions for the answer of differential Eq. (2.4.3) is in the form of If x=0,

$$\begin{cases} u(0) = a_0 = u_0 \\ u'(0) = a_1 = u_1 \\ u''(0) = a_2 = u_2 \\ \quad\vdots \end{cases} \tag{2.4.4}$$

and when x=L,

$$\begin{cases} u(L) = a_0 + a_1 L + a_2 L^2 + \cdots + a_n L^n = u_{L_0} \\ u'(L) = a_1 + 2a_2 L + 3a_3 L^2 + \cdots + n a_n L^{n-1} = u_{L_1} \\ u''(0) = 2a_2 + 6a_3 L + 12a_4 L^2 + \cdots + n(n-1) a_n L^{n-2} = u \\ \quad\vdots \end{cases} \tag{2.4.5}$$

2) After substituting Eq. (2.4.3) into Eq. (2.4.1), the application of the boundary conditions on differential Eq. (2.4.1) is done according to the following procedure

$$\begin{cases} p_0 = f\left(u(0), u'(0), u''(0), \ldots, u^{(m)}(0)\right) \\ p_1 = f\left(u(L), u'(L), u''(L), \ldots, u^{(m)}(L)\right) \\ \quad\vdots \end{cases} \tag{2.4.6}$$

Concerning the choice of n (n<m), sentences from Eq. (2.4.3) and to make a set of equations which is consisted of (n + 1) equations and (n + 1) unknowns, we confront some additional unknowns, which are indeed the same coefficients of Eq. (2.4.3). So we should derive m times from Eq. (2.4.1) according to the additional unknowns in the aforementioned set of differential equations. It is time to apply the boundary conditions of Eq. (2.4.2) to them.

$$\begin{aligned} p_k' &= f\left(u', u'', u''', \ldots, u^{(m+1)}\right) \\ p_k'' &= f\left(u'', u''', u^{(IV)}, \ldots, u^{(m+2)}\right) \\ &\ldots\ldots \\ &\ldots\ldots \end{aligned} \tag{2.4.7}$$

3) Application of the boundary conditions on the derivatives of the differential equation p_k in Eq. (2.4.7) is done in the form of :

$$p_k' = \begin{Bmatrix} f\left(u'(0), u''(0), u'''(0), \ldots, u^{(m+1)}(0)\right) \\ f\left(u'(L), u''(L), u'''(L), \ldots, u^{(m+1)}(L)\right) \end{Bmatrix} \tag{2.4.8}$$

$$p_k'' = \begin{Bmatrix} f\left(u''(0), u'''(0), \ldots, u^{(m+2)}(0)\right) \\ f\left(u''(L), u'''(L), \ldots, u^{(m+1)}(L)\right) \end{Bmatrix} \tag{2.4.9}$$

(n + 1) equations can be made from Eqs. (2.4.4) to (2.4.9), so that (n + 1) unknown coefficients of Eq. (2.4.3), i.e., a_0, a_1, a_2,…, a_n, will be computed. The answer to the nonlinear differential Eq. (2.4.1) will be gained by determining the coefficients of Eq. (2.4.3).

This research examines a case of loading on beams and columns, which are entirely nonlinear differential equations. Consider an element of the beam, as shown in Fig. 2.4.1, on which the displacement and the slope of the element are marked. According to Fig. 2.4.1, an element of the beam is illustrated in three directions (x, y, and z). The slope of the element is expressed as dw_0/dx. Considering the mentioned displacement and slope, the beam is distracted. The original Euler-Bernoulli theory is valid only for infinitesimal strain and small rotation. The theory can be extended to problems involving a moderately large rotation, provided that the strain remains small using the von Kármán strains. The Euler-Bernoulli hypothesizes that plane sections remain plane and normal to the axis of the beam, which leads to displacements of the form.

$$v_1 = v_0\left(x\right) - z\frac{dw_0}{dx}, \quad v_2 = 0, \quad v_3 = w_0\left(x\right) \tag{2.4.10}$$

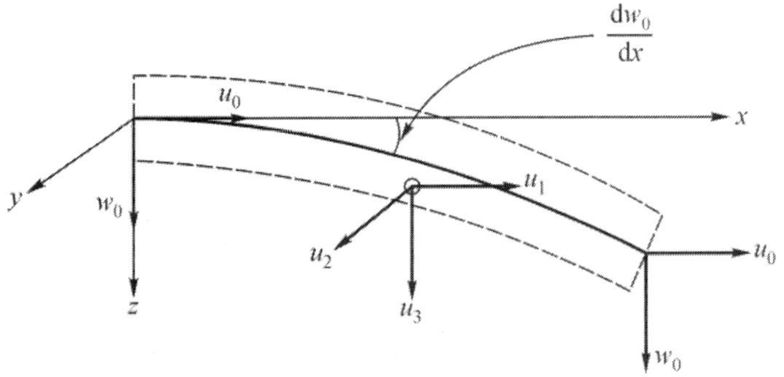

Fig. 2.4.1. Euler-Bernoulli beam

Using the definition of the Lagrangian Green strain from finite strain theory, we find that the von Kármán strains for the beam are valid for a large rotation but a small strain. These strains have the following forms:

$$\varepsilon_{11} = \frac{du_0}{dx_1} - x_3 \frac{d^2 w_0}{dx_1^2} + \frac{1}{2}\left[\left(\frac{du_0}{dx_1} - x_3 \frac{d^2 w_0}{dx_1^2}\right) + (\frac{dw_0}{dx_1})^2\right]$$

$$\varepsilon_{22} = 0,\ \varepsilon_{33} = \frac{1}{2}\left(\frac{dw_0}{dx_1}\right),\ \varepsilon_{23} = 0, \varepsilon_{12} = 0 \tag{2.4.11}$$

$$\varepsilon_{32} = \frac{1}{2}\left(\frac{du_0}{dx_1} - \frac{dw_0}{dx_1}\right) - \frac{1}{2}\left[\left(\frac{du_0}{dx_1} - x_3 \frac{d^2 w_0}{dx_1^2}\right)(\frac{dw_0}{dx_1})\right]$$

From the principle of virtual work, the balance of forces and moments in the beams gives us the equilibrium equations as follows:

$$\frac{dN_{xx}}{dx} + f(x) = 0$$

$$\frac{d^2 M_{xx}}{dx^2} + q(x) + \frac{d}{dx}\left(N_{xx}\frac{dw_0}{dx}\right) = 0 \tag{2.4.12}$$

where $f(x)$ is the axial load; $q(x)$ is the transverse load; and

$$N_{xx} = \int_A \sigma_{xx} dA\ , M = \int_A z\sigma_{xx} dA \tag{2.4.13}$$

The constitutive equations that relate stress to strains (and hence stress to displacements) are needed to close the system of equations. For a large rotation and small strain, these relations are

$$N_{xx} = A_{xx}\left[\frac{du_0}{dx} + \frac{1}{2}\left(\frac{dw_0}{dx}\right)^2\right] - B_{xx}\frac{d^2 w_0}{dx^2}$$

$$M_{xx} = B_{xx}\left[\frac{du_0}{dx} + \frac{1}{2}\left(\frac{dw_0}{dx}\right)^2\right] - D_{xx}\frac{d^2 w_0}{dx^2} \tag{2.4.14}$$

Where

$$A_{xx}\int_A EdA,\ B_{xx}=\int_A zEdA,\ D_{xx}=\int_A z^2 EdA \tag{2.4.15}$$

Where A_{xx} is the extensional stiffness,, B_{xx}, is the coupled extensional-bending stiffness, and D_{xx} is the bending stiffness. For the situation where the beam has a uniform cross-section and no axial load, the governing equation for a large-rotation Euler-Bernoulli beam is

$$f(x): EI\frac{dw^4}{dx^4}-\frac{3}{2}EI\left(\frac{dw}{dx}\right)^2\frac{d^2w}{dx^2}=q(x) \tag{2.4.16}$$

The beam with boundary conditions of fixed-hinged is shown in Fig. 2.4.2.

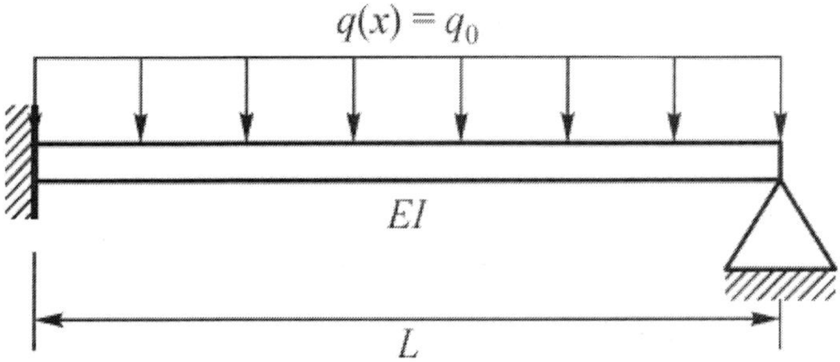

Fig. 2.4.2. Beam with fixed-hinged

Boundary conditions in fixed-hinged are as follows:

$$w(0)=0,\ w'(0)=0,\ w(L)=0,\ w''(L)=0 \tag{2.4.17}$$

As shown in Fig. 2.4.2, the element shown in Fig. 2.4.1 can be used for different boundary conditions for rotation, and also, a distributed load for this beam is considered. Solving the nonlinear differential equation by AGM based on the given explanations in the previous section, the answer of Eq. (2.4.16) in this method is considered by AGM as polynomials of Fourier series with constant coefficients as follows:

$$w(x)=\sum_{k=0}^{7}a_kx^k=a_0+a_1x+a_2x^2+a_3x^3+a_4x^4+a_5x^5+a_6x^6+a_7x^7 \tag{2.4.18}$$

In AGM, the constant coefficients $(a_0,a_1,\dots,a_7)$ of Eq. (2.4.18) can be easily computed by applying boundary conditions according to the physical aspects of the problem. Based on the explanations in the previous section of this problem, the constant coefficients $\{a_0,a_1,\dots,a_7\}$ in Eq. (2.4.18) are just achieved concerning the given initial conditions, which are applied in two manners in AGM. 1) In general, the initial conditions are applied on Eq. (2.4.18) in the form of

$$w(x=0)=0\rightarrow a_0=0 \tag{2.4.19}$$

$$\begin{aligned}
&w'(x=0)=0\rightarrow a_1=0\\
&w(x=L)=0\rightarrow a_0+a_1L+a_2L^2+a_3L^3\\
&+a_4L^4+a_5L^5+a_6L^6+a_7L^7=0\\
&w''(x=L)=0\rightarrow 2a_2+6a_3L+12a_4L^2\\
&+20a_5L^3+30a_6L^4+42a_7L^5=0
\end{aligned} \tag{2.4.20}$$

2) The application of initial or boundary conditions on the main differential equation, which in this case is Eq. (2.4.16), and also on its derivatives is done in the following general form:

$$\begin{aligned}
&f\left(w_{(x)}\right)\rightarrow f\left(w_{(BC)}\right)=0,\\
&f'\left(w_{(BC)}\right)=0,\ f''\left(w_{(BC)}\right)=0,\dots
\end{aligned} \tag{2.4.21}$$

It is notable that (B.C.) is the abbreviation of boundary conditions. Therefore, after substituting Eq. (2.4.18), which has been considered as the answer to the main differential equation, into Eq. (2.4.16), the initial conditions are applied on the equation obtained and also on its derivatives based on Eq. (2.4.22) as follows:

$$f\left(w_{(x=0)}\right):24a_4-3a_1^2a_2=-\frac{q_0}{EI} \tag{2.4.22}$$

$$\begin{aligned}&f\left(w_{(x=L)}\right)=24a_4+120\mathrm{a}_5\mathrm{L}+360\mathrm{a}_6\mathrm{L}^2+840\mathrm{a}_7L^3\\&-\frac{3}{2}\left(\mathrm{a}_1+2a_2L+3a_3L^2+4a_4L^3+5a_5L^5+6a_6L^6+7a_7L^6\right)^2\\&\left(2a_2+6a_3L+12a_4L^2+20a_5L^3+30a_6L^4+42a_7L^5\right)=-\frac{q_0}{EI}\end{aligned} \tag{2.4.23}$$

$$f'\left(w_{x=0}\right):120a_5-12a_1a_2^2-9a_1^2a_3=0 \tag{2.4.24}$$

$$\begin{aligned}&f'\left(w_{x=L}\right):120\mathrm{a}_5+720a_6L+2520a_7L^2\\&-3\left(a_1+2a_2L+3a_3L^2+4a_4L^3+5a_5L^5+6a_6L^6+7a_7L^6\right)\\&\times\left\{\left(2a_2+6a_3L+12a_4L^2+20a_5L^3+30a_6L^4+42a_7L^5\right)^2\right.\\&-\frac{1}{2}\left(a_1+2a_2L+3a_3L^2+4a_4L^3+5a_5L^5+6a_6L^6+7a_7L^6\right)\\&\left.\left(6a_3+24a_4L+60a_5L^2+120a_6L^3+210a_7L^4\right)\right\}=0\end{aligned} \tag{2.4.25}$$

Solving a set of algebraic equations consists of eight equations with eight unknowns from Eqs. (2.4.20) to (2.4.28), the constant coefficients $a_0, a_1, \ldots, a_7$ from Eq. (2.4.18) can be easily yielded as follows:

$$\begin{aligned}&\psi=q_0L^4\left(290304000E^2I^2-q_0^2L^4\right)+\\&2304\sqrt{105}EI\times\\&\sqrt{83235962880E^4I^4+q_0L^8\left(150877440E^2I^2-q_0^2L^8\right)}\end{aligned} \tag{2.4.26}$$

$$\Delta=\frac{\psi^{\frac{1}{3}}}{2304L^7}-\frac{7741440E^2I^2-q_0^2L^8}{2304\psi^{\frac{1}{3}}}-\frac{25q_0}{2304L^4}$$

Therefore, the constant coefficients of Eq. (2.4.25) are achieved as follows:

$$a_0 = 0,\ a_1 = 0,\ a_2 = -\frac{L^2\left(q_0 + 80\Delta L^3\right)}{16EI}$$
$$a_3 = -\frac{L\left(5q_0 + 304\Delta L^3\right)}{48EI},\ a_4 = -\frac{q_0}{24EI}, \tag{2.4.27}$$
$$a_5 = 0,\ a_6 = -\frac{7\Delta L}{EI},\ a_7 = \frac{\Delta}{EI}$$

After substituting the obtained values from Eqs. (2.4.26) and (2.4.27) into Eq. (2.4.18), the solution to the mentioned problem will be obtained as follows:

$$w(x) = -\frac{L^2\left(q_0 + 80\Delta L^3\right)}{16EI}x^2 - \frac{L\left(5q_0 + 304\Delta L^3\right)}{48EI}x^3 - \frac{q_0}{24EI}x^4 - \frac{7\Delta L}{EI}x^6 + \frac{\Delta}{EI}x^7 \tag{2.4.28}$$

By selecting the following physical values:

$$L = 1m,\ E = 10^7\,Pa,\ I = 10^{-5}\,m^4,\ q_0 = 2.5\,{}^{N}\!/_{m} \tag{2.4.29}$$

$$w(x) = -0.0015625x^2 + 0.002604x^3 - 0.00104x^4 + 8.426\times10^{-10}x^6 - 3.62\times10^{-10}x^7 \tag{2.4.30}$$

Then, the charts of the solution obtained and its derivative are depicted in Figs. 2.4.3 and 2.4.4.

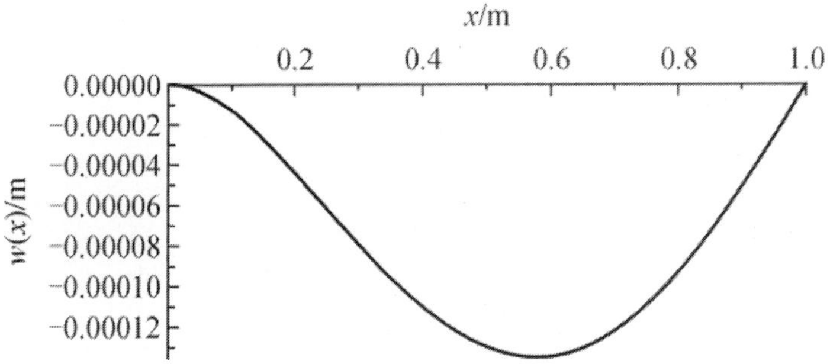

Fig. 2.4.3. The solution obtained by AGM

By the given physical values from Eq. (2.4.29) and the introduced domain $x \in \{0,1\}$, which is defined to simplify, the following new variables are introduced:

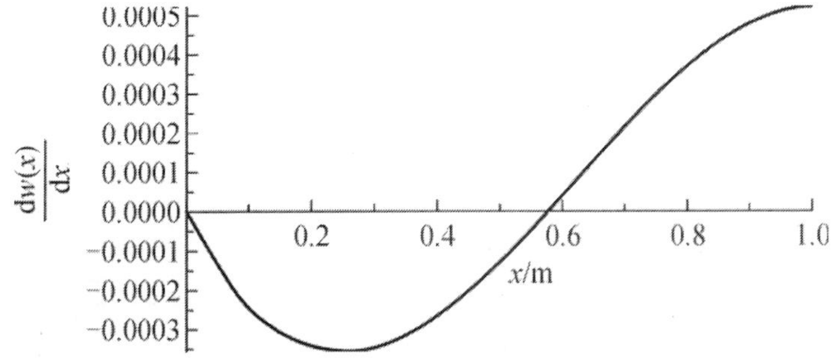

Fig. 2.4.4. The first derivative for the solution obtained by AGM

In terms of length (in "m"), the numerical solution to the problem is presented in Table 2.4.1. Due to the solution obtained from Eq. (2.4.28) by AGM, which is Eq. (2.4.30), and the obtained numerical solution, the results of which have been presented in Table 2.4.1, we have the comparisons as shown in Figs. 2-6 illustrate that the AGM is precise and reliable compared to the numerical method.

The following charts (Figs. 2.4.7 and 2.4.8) are the absolute value difference between the results based on the yielded solution by AGM and the results in Table 2.4.1 by the numerical method. According to the Figs. 2.4.7 and 2.4.8, it is notable that the difference between the numerical method and AGM is very small. As mentioned in the diagrams, AGM in solving complicated nonlinear differential equations is reliable, applicable, and close to the exact solution. According to Fig. 2.4.9, it is evident that the more normal load on the beam, the more displacement and deflection along the length of the beam. It shows its maximum deflection close to the middle of the beam. According to Fig. 2.4.2, because the beam is fixed-hinged, the maximum deflection is near the middle of the beam. Hence, it has been found that, for beams that used the Euler-Bernoulli theory, more load on the beam, more transposition alongside the beam.

In this study, one complicated nonlinear partial differential equation has been introduced and analyzed completely by AGM, the results of which have been compared with the numerical method (Runge-Kutta 4th).

Table 2.4.1. Results obtained for $w(x)$ and its derivative according to the given physical values

x/m	$w(x)_{Num}$, Runge-Kutte 4th	$w(x)_{AGM}$	$\left\|\frac{w(x)_{Num} - w(x)_{AGM}}{w(x)_{Num}}\right\| \times 100/\%$	$\frac{dw(x)}{dx}$ Runge-Kutte 4th
0.0	0.000000000	0.000000000	0.0000	0.000000000
0.2	-0.000043328	-0.000043333	0.0110	-0.000345840
0.4	-0.00010999	-0.000110000	0.0108	-0.000266680
0.6	-0.00013499	-0.000135000	0.0240	0.000037470
0.8	-0.000093333	-0.000093335	0.0021	0.000366637
1.0	0.000000000	0.000000000	0.0000	0.000520800

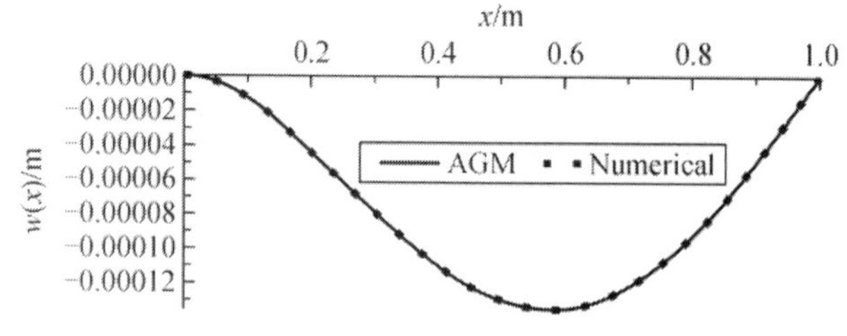

Fig. 2.4.5. Comparison between AGM and numerical solution

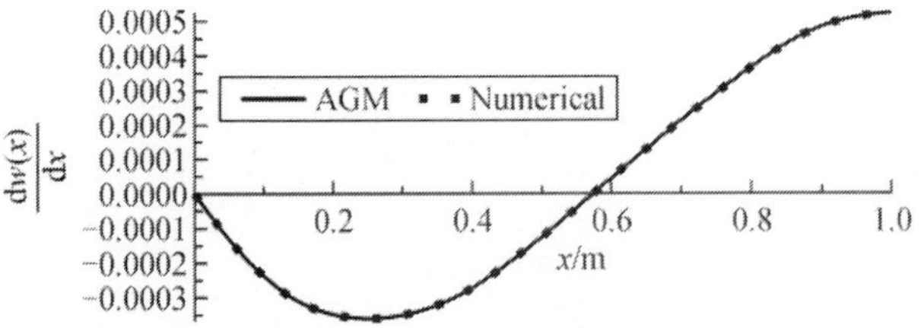

Fig. 2.4.6. Resulted from chart of the first derivative between AGM and the numerical method

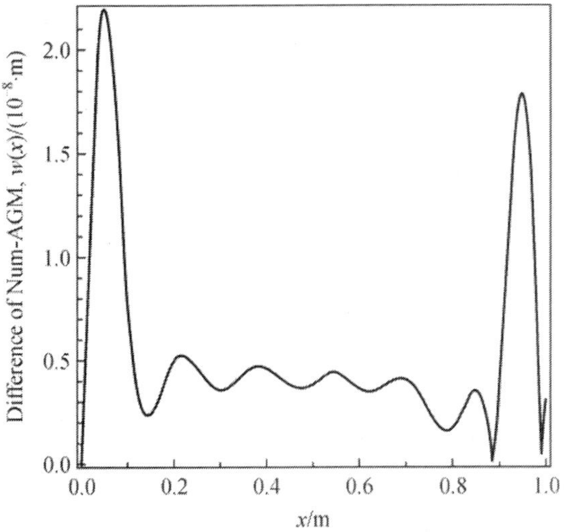

Fig. 2.4.7. Differences in the solutions obtained between AGM and the numerical method

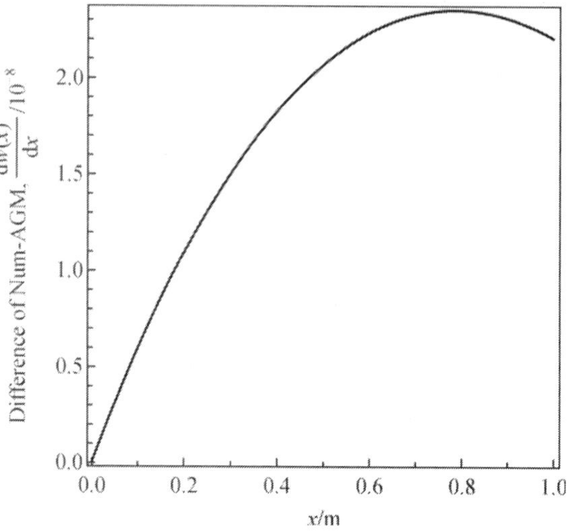

Fig. 2.4.8. Differences of the first derivative solutions obtained between AGM and the numerical method

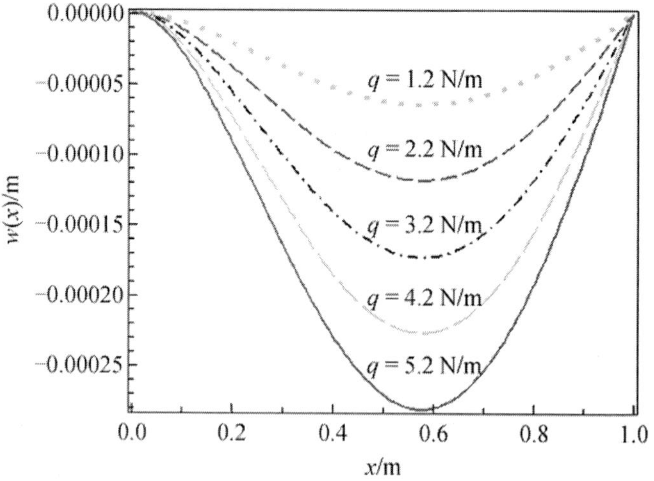

Fig. 2.4.9. Distribution load for different values of the beam

The process has shown AGM's ability to solve various differential equations in different fields of study, particularly vibrations. Consequently, it is concluded that AGM is a reliable and precise approach for solving miscellaneous differential equations. Moreover, a summary of the AGM's excellence and benefits is explained as follows:

1) By solving a set of algebraic equations with constant coefficients, we can easily obtain the nonlinear differential equation solution, which is possible even for researchers with intermediate mathematical knowledge.

2) Conversely, AGM can solve linear and nonlinear differential equations directly in most situations, which means the final solution can be obtained without any dimensionless procedure. Therefore, AGM can be considered a significant progress in nonlinear sciences. The shortage of boundary condition(s) in this method for solving differential equation(s) is completely terminated. AGM is operational for miscellaneous nonlinear differential equations, especially civil engineering and mechanical machinery vibrations. We are hopeful that enthusiastic young researchers will shortly apply this new approach.

2.5. Displacement of Structure and Heat Transfer Extended Surfaces

In this investigation, two complicated fluid and solid mechanics problems are solved with AGM, which is more convenient and precise than all of the well-known techniques for solving various nonlinear problems and is considered a revolution in nonlinear sciences.

Boundary and initial conditions are required for analytical methods of each linear and nonlinear differential equation according to the physics of the problem. This new method will solve two differential equations governing engineering processes. The nonlinear differential equation of p is a function of u, where the parameter u is a function of x. And their derivatives are considered as follows:

$$p_k = f\left(u, u', u'', \dots, u^{(m)}\right) = 0, u = u(x) \tag{2.5.1}$$

P_k is the name of the nonlinear differential equation, and m is the order of the derivatives.

Boundary conditions:

$$\begin{cases} u(0) = u_0, u'(0) = u_1, \dots, u^{(m-1)}(0) = u_{m-1} \\ u(L) = u_{L_0}, u'(L) = u_{L_1}, \dots, u^{(m-1)}(L) = u_{L_{m-1}} \end{cases} \tag{2.5.2}$$

To solve the first differential equation concerning the boundary conditions at x = L in Eq. (2.5.2), the series of letters in the *n*th order with constant coefficients, which is the answer to the first differential equation, is considered as follows:

$$u_x = \sum_{i=0}^{n} a_i x^i = a_0 + a_1 x^1 + a_2 x^2 + \dots + a_n x^n \tag{2.5.3}$$

The more precise answer to Eq. (2.5.1), the more choices of series sentences from Eq. (2.5.3) are available. In applied problems, approximately five or six sentences in the series are enough to solve nonlinear differential equations. In answer to differential Eq. (2.5.3), regarding the series from degree (n), there are (n + 1) unknown coefficients which need (n + 1) equations to satisfy. A total answer with constant coefficients is required to solve differential equations in AGM. The boundary conditions of Eq. (2.5.2) are used to solve a set of equations consisting of (n + 1).

The boundary conditions are applied to the functions as follows:

1) The application of the boundary conditions for the answer to Eq. (2.5.3) is in the form of

$$If\ x = 0, \begin{cases} u(0) = a_0 = u_0 \\ u'(0) = a_1 = u_1 \\ u''(0) = a_2 = u_2 \\ \quad\quad\quad \vdots \end{cases} \tag{2.5.4}$$

and when x = L,

$$\begin{cases} u(L) = a_0 + a_1 L + a_2 L^2 + \dots + a_n L^n = u_{L_0} \\ u'(L) = a_1 + 2a_2 L + 3a_3 L^2 + \dots + n a_n L^{n-1} = u_{L_1} \\ u''(L) = 2a_2 + 6a_3 L + 12 a_4 L^2 + \dots + n(n-1) a_n L^{n-2} = u_{L_{m-1}} \\ \vdots \quad\quad \vdots \quad\quad \vdots \quad\quad \vdots \quad\quad \vdots \quad\quad \vdots \end{cases} \tag{2.5.5}$$

2) After substituting Eq. (2.5.5) with Eq. (2.5.1), the application of the boundary conditions on Eq. (2.5.1) is done according to the following procedure:

$$\begin{aligned} &p_0 = f\left(u(0), u'(0), u''(0), \ldots, u^{(m)}(0)\right) \\ &p_1 = f\left(u(L), u'(L), u''(L), \ldots, u^{(m)}(L)\right) \\ &\vdots \qquad \vdots \qquad \vdots \qquad \vdots \qquad \vdots \end{aligned} \tag{2.5.6}$$

About the choice of n (n<m) sentences in Eq. (2.5.3) and to make a set of equations which consist of (n + 1) equations and (n + 1) unknowns, we confront several additional unknowns, which are certainly the same coefficients of Eq. (2.5.3). Therefore, to remove this problem, we should make m times derivation in Eq. (2.5.1) according to the additional unknowns in the aforementioned set of differential equations. This is the time to apply the boundary conditions of Eq. (2.5.2).

$$\begin{aligned} &p'_k = f\left(u', u'', u''', \ldots, u^{(m+1)}\right) \\ &p''_k = f\left(u'', u''', u^{(IV)}, \ldots, u^{(m+2)}\right) \\ &\vdots \qquad \vdots \quad \vdots \quad \vdots \qquad \vdots \end{aligned} \tag{2.5.7}$$

3) Application of the boundary conditions on the derivatives of the differential equation p_k in Eq. (2.5.7) is done in the form of

$$p'_k = \begin{cases} f\left(u'(0), u''(0), u'''(0), \ldots, u^{(m+1)}(0)\right) \\ f\left(u'(L), u''(L), u'''(L), \ldots, u^{(m+1)}(L)\right) \end{cases} \tag{2.5.8}$$

$$p''_k = \begin{cases} f\left(u''(0), u'''(0), \ldots, u^{(m+2)}(0)\right) \\ f\left(u''(L), u'''(L), \ldots, u^{(m+2)}(L)\right) \end{cases} \tag{2.5.9}$$

(n + 1) equations can be made from Eqs. (2.5.4) to (2.5.9), so that (n + 1) unknown coefficients of Eq. (2.5.3) will be calculated. For example, $a_0, a_1, a_2, ..., a_n$. The answer to the nonlinear differential Eq. (2.5.1) will be gained by determining the coefficients of Eq. (2.5.3). To comprehend the AGM benefits, two complicated fluid and solid mechanics problems have been presented in the following forms.

a) Example A

According to Fig. 2.5.1, the differential equation governing the structure along with displacement (Δ) and the end moment relevant to the two ends (M_{ab}, M_{ba}) can be acquired by writing the equilibrium equation.

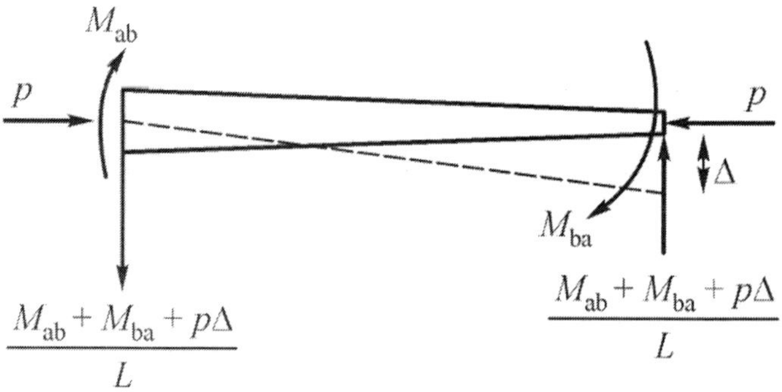

Fig. 2.5.1. Physical model

$$E(x)I(x)\frac{d^2y(x)}{dx^2}+py(x)$$
$$-\frac{M_{ab}+M_{ba}}{L}x+M_{ab}+\frac{p}{L}\Delta x=0 \tag{2.5.10}$$

In Eq. (2.5.10), the moment of inertia (I) and elastic modulus (E) are variable and are functions of the element along the length of the beam and achieved as follows:

$$I(x)=I_0(1-ax)^4 \tag{2.5.11}$$

$$E(x)=E_0(1-\varepsilon x) \tag{2.5.12}$$

In the mentioned equations, α and ε are constant values. They are defined as the reduction factor of the moment of inertia and elastic modulus along the beam or column. Also, I_0 E_0 are primary values for the structure. Notably, reduction factors (α and ε) can be easily computed by supposing the cross-section area is circular in the form as shown in Fig. 2.5.2. Where d_1 and d_2 are the diameters of cross-section 1, and cross-section 2, respectively; E_1 and E_2 are the elastic moduli of cross-section 1, and cross-section 2, respectively; and X indicates the direction of x coordinate.

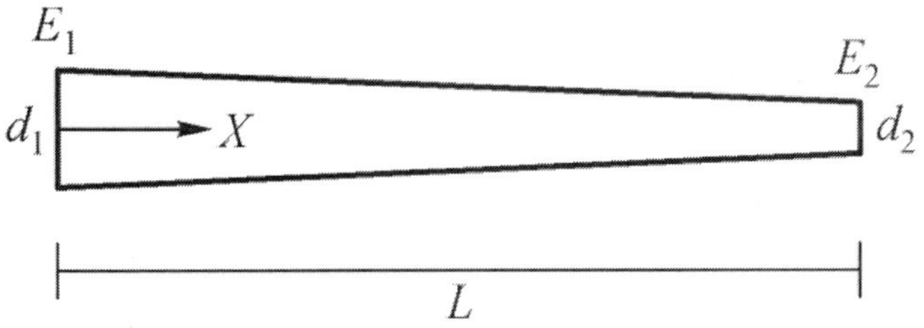

Fig. 2.5.2. Schematic diagram for computing the elastic modulus and diameter along the member length

$$\left|\begin{matrix}0\\E_1\end{matrix}\right.,\left|\begin{matrix}b\\E_2\end{matrix}\right.,\left|\begin{matrix}x\\E\end{matrix}\right.,E=E_1-\frac{E_1-E_2}{L}x$$

therefore (2.5.13)

$$E=E_0\left(1-\frac{E_1-E_2}{E_1L}x\right)\rightarrow\varepsilon=\frac{E_1-E_0}{E_1L}$$

and

$$\begin{vmatrix}0\\d_1\end{vmatrix},\begin{vmatrix}b\\d_2\end{vmatrix},\begin{vmatrix}x\\d\end{vmatrix},d=d_1-\frac{d_1-d_2}{L}x\rightarrow I=\frac{\pi}{64}d^4 \tag{2.5.14}$$

Finally, we have

$$I=\frac{\pi}{64}d_1^4\left(1-\frac{d_1-d_2}{Ld_1}x\right)^4\rightarrow$$

$$\begin{cases}I_0=\dfrac{\pi}{64}d_1^4\\ a=\dfrac{d_1-d_2}{Ld_1}\end{cases}\rightarrow I=I_0\left(1-ax\right)^4 \tag{2.5.15}$$

The reason for supposing elastic modulus (E) is variable for some structures, such as steel or concrete structures in bridges and dams, is that the structure material varies linearly in the length of the beam or column. By substituting Eqs. (2.5.11) and (2.5.12) into Eq. (2.5.10), the differential equation governing the structure is converted in the form of

$$\begin{aligned}f\left(x\right)&=E_0\left(1-\varepsilon x\right)I_0\left(1-ax\right)^4y''\\&+py-\frac{M_{ab}+M_{ba}}{L}x+M_{ab}+\frac{p\Delta}{L}x=0\end{aligned} \tag{2.5.16}$$

Equation (2.5.16) cannot be easily answered analytically because it is nonlinear. Let us consider solving an equation by AGM, the answer to which can be considered as a finite series of letters with constant coefficients in the form of

$$y(x)=\sum_{n=0}^{4}a_nx^n=a_0+a_1x+a_2x^2+a_3x^3+a_4x^4 \tag{2.5.17}$$

Constant coefficients (a_0 to a_4) in Eq. (2.5.17) are computed by applying boundary conditions which are applied as follows:

$$\begin{cases} y(0)=0 \rightarrow a_0=0 \\ y(L)=\Delta \rightarrow a_0+a_1L+a_2L^2+a_3L^3+a_4L^4=\Delta \end{cases} \tag{2.5.18}$$

Applying boundary conditions on Eq. (2.5.16) after substituting Eq. (2.5.17) into Eq. (2.5.16) is done as follows:

$$f(0)=2E_0I_0a_2+pa_0+M_{ab}=0 \tag{2.5.19}$$

And

$$\begin{aligned} &f(L)=E_0(1-\varepsilon L)I_0(1-aL)^4\times \\ &\left(2a_2+6a_3L+12a_4L^2\right) \\ &+p\left(a_0+a_1L+a_2L^2+a_3L^3+a_4L^4\right) \\ &-M_{ab}+p\Delta=0 \end{aligned} \tag{2.5.20}$$

To apply boundary conditions on the derivative of Eq. (2.5.16), first, its derivation is taken in terms of X; and after substituting Eq. (2.5.17) into Eq. (2.5.16), boundary conditions are applied in the form of

$$\begin{aligned} &f'(0)=-2E_0\varepsilon I_0a_2-8E_0I_0a_2a+ \\ &6E_0I_0a_3+pa_1-\frac{M_{ab}+M_{ba}}{L}+\frac{p\Delta}{L}=0 \end{aligned} \tag{2.5.21}$$

Constant coefficients (a_0 to a_4) of Eq. (2.5.17) can be gained by solving a set of equations (five equations with five unknowns) from Eqs. (2.5.18)–(2.5.21). To simplify, we define the following new variable:

$$\psi = 12E_0 I_0 L \begin{Bmatrix} \varepsilon(aL-1)^4 - a(aL-2)\times \\ (a^2L^2-2aL+2) \end{Bmatrix} \\ +pL^3a(aL-2)(a^2L^2-2aL+2) \\ -pL^3\varepsilon(La-1)^4 - 12E_0I_0 + pL^2 \tag{2.5.22}$$

Concerning the variable above, constant coefficients are yielded in the form of

$$a_0 = 0 \tag{2.5.23}$$

$$a_1 = \frac{1}{\psi L}\{\Delta L^2 \varepsilon a(aL-2)(a^2L^2-2aL+2)\times \\ \left(pL^2+12E_0L_0-\frac{pL}{3}\right)-L^4\varepsilon a M_{ba}(aL-2) \\ \times(a^2L^2-2aL+2)+L^4\varepsilon a M_{ab}\times \\ (\varepsilon a^3L^4+4a^4L^4-8+12a^2L^2-13a^3L^3-4\varepsilon L+6\varepsilon aL^2-4\varepsilon a^2L^3) \\ -12E_0L_0L\Delta(6La2-4a-4L^2a^3+L^3a^4-\varepsilon)- \\ L^2M_{ab}(8L^2a^2-12aL+8a^3L^3-12L^4a^4-3\varepsilon L+4L^5a^5-L^2\varepsilon^2) \\ +L^2M_{ba}(6L^2a^2-4La-4a^3L^3+L^4a^4-L\varepsilon+2) \\ +pL^2\Delta(L\varepsilon-2)+12E_0L_0\Delta\} \tag{2.5.24}$$

and a_2 is achieved in the form of

$$a_2 = -\frac{M_{ab}}{2E_0L_0} \tag{2.5.25}$$

The next terms are computed as follows:

$$
\begin{aligned}
a_3 = \frac{1}{6\psi E_0 L_0 L} & \{12 pL\,\varepsilon a E_0 L_0 \Delta (aL-2)\times(a^2L^2-2aL+2)+ \\
& 6E_0 L_0 L a \varepsilon M_{ab} \\
& \times\begin{pmatrix} 8-22La+28L^2a^2-17a^3L^3-4\varepsilon L \\ +6\varepsilon L^2 a-4\varepsilon L^3a^2+\varepsilon L^4a^3+4L^4a^4 \end{pmatrix} \\
& -6L\,\varepsilon a E_0 L_0 (aL-2)(a^2L^2-2aL+2) \\
& -12aE_0L_0 p\Delta(aL-2)(a^2L^2-2aL+2) \\
& -6E_0L_0M_{ab}\begin{pmatrix} 4q+\varepsilon-16L^3a^4+24L^2a^3 \\ -16La^2+4L^2a^5-L\varepsilon^2 \end{pmatrix} \\
& -12E_0L_0L\Delta(6La2-4a-4L^2a^3+L^3a^4-\varepsilon) \\
& -L^2M_{ab}\begin{pmatrix} 8L^2a^2-12aL+8a^3L^3-12L^4a^4 \\ -3\varepsilon L+4L^5a^5-L^2\varepsilon^2 \end{pmatrix} \\
& +6E_0L_0M_{ba}(6a^2L-4a-\varepsilon-4a^3L^2+a^4L^3) \\
& +2pL^2M_{ab}\begin{pmatrix} 4L^2a^3-L^3a^4+4a+\varepsilon-6La^2 \\ -4La\varepsilon+6L^2\varepsilon a^2-4L^3\varepsilon a^3+L^4\varepsilon a^4 \end{pmatrix} \\
& -pL(2p\Delta+2M_{ab}-M_{ba})+12E_0L_0\varepsilon p\,\Delta\}
\end{aligned}
\tag{2.5.26}
$$

Finally, a_4 is as below.

$$
\begin{aligned}
a_4 = &\frac{1}{6\psi E_0 I_0 L}\{12pL\varepsilon aE_0 I_0\Delta(aL-2)(a^2L^2-2aL+2) \\
&-6E_0I_0La\varepsilon M_{ab} \\
&\times\begin{pmatrix}-8+22La+28L^2a^2+17a^3L^3 \\ +4\varepsilon L-6\varepsilon L^2a+4\varepsilon L^3a^2\end{pmatrix} \\
&+6E_0I_0M_{ba}\begin{pmatrix}4La\varepsilon-6L^2\varepsilon a^2+4L^3\varepsilon a^3 \\ +6La^2-4L^2a^3+L^3a^4-4a-\varepsilon\end{pmatrix} \\
&+6E_0I_0M_{ab}\begin{pmatrix}L^5\varepsilon^2a^4+L^5\varepsilon a^5-4a-\varepsilon+16L^3a^4 \\ -24L^2a^3+16La^2-4L^4a^5+L\varepsilon^2\end{pmatrix} \\
&-12pLa^2E_0I_0\Delta(6-4aL+a^2L^2) \\
&+2pL^2M_{ab}\begin{pmatrix}4L^2a^3-L^3a^3+4a+\varepsilon-6La^2 \\ -4La\varepsilon+6L^2\varepsilon a^2-4L^3\varepsilon a^3+L^4\varepsilon a^4\end{pmatrix} \\
&+2p\Delta(24aE_0I_0+6\varepsilon E_0I_0-pL)-pL(2M_{ab}-M_{ba})\}
\end{aligned}
\tag{2.5.27}
$$

By substituting constant coefficients of Eqs. (2.5.23)–(2.5.27) with Eq. (2.5.17), nonlinear differential Eq. (2.5.16) is answered as

$$y(x) = a_0 + a_1x + a_2x^2 + a_3x^3 + a_4x^4 \tag{2.5.28}$$

By choosing the following physical values for the steel alloy structure

$$
\begin{aligned}
&\Delta = 1mm,\ M_{ab} = 60N.m,\ M_{ba} = \beta M_{ab},\ \beta=0.23 \\
&a = 0.26,\ \varepsilon = 0.6,\ E_0 = 3\times10^8\ N/m^2,\ p = k^2E_0I_0 \\
&k = 0.45,\ p = 6075N,\ L = 70cm,\ I_0 = 0.0001m^4
\end{aligned}
\tag{2.5.29}
$$

The displacement function of the structure element is obtained in the form of

$$
\begin{aligned}
y(x) = &\ 0.002x - 0.001x^2 \\
&-0.0000773x^3 + 0.000431x^4
\end{aligned}
\tag{2.5.30}
$$

According to the slope-deflection method, the structure displacement is demonstrated in Figs. 2.5.3 and 2.5.4.

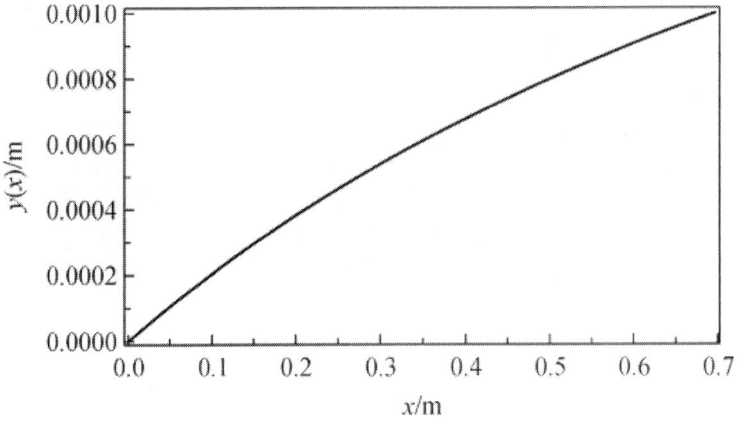

Fig. 2.5.3. Displacement of the structure element (y) versus length (x) obtained by AGM

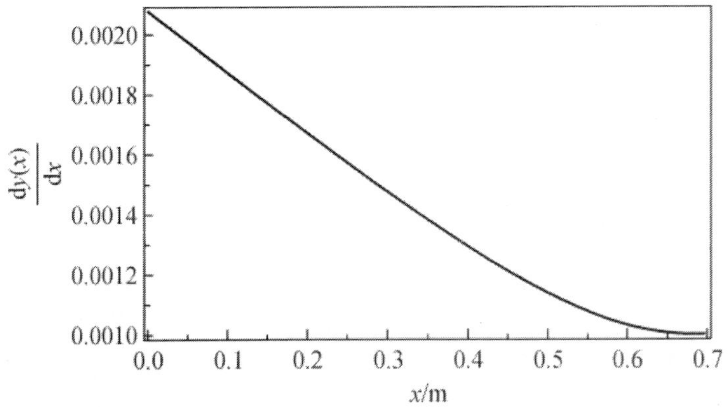

Fig. 2.5.4. Chart of the first derivative for the solution obtained by AGM

The number of series sentences in Eq. (2.5.17) can be increased for a more accurate answer. Therefore, by selecting degree 7, the answer is as follows:

$$y(x) = 0.00207x - 0.001x^2 - 0.0000779x^3$$
$$+0.000123x^4 + 0.000209x^5$$
$$-0.0001606x^6 + 0.000217x^7$$

The calculation of the bending moment in terms of rotation (end slopes) at both ends of the beam (structure) and displacement slopes at the ends of the beam (Fig. 2.5.5) by taking derivation of the displacement function of Eq. (2.5.28) is achieved as

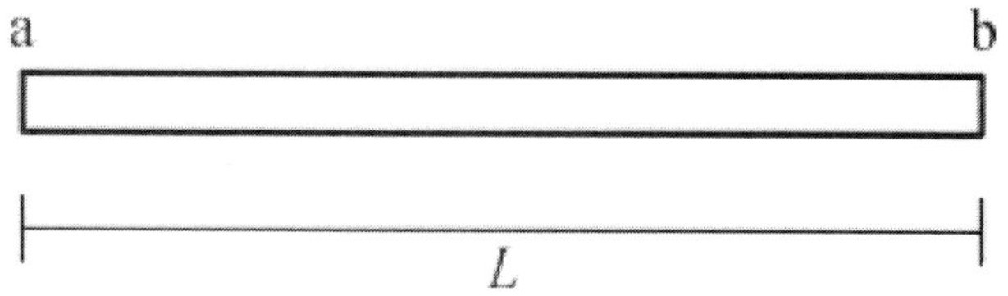

Fig. 2.5.5. Schematic of the structure for obtaining the bending moment at two ends

$$\theta_a = \frac{\partial y(x)}{\partial x}\Big|_{x=0} \quad \text{and} \quad \theta_b = \frac{\partial y(x)}{\partial x}\Big|_{x=L} \tag{2.5.31}$$

Where θ_a is the first derivate of displacement of the structure at x = 0, and θ_b is the first derivate of displacement at x = L. Since the displacement function $y(x)$ (Eq. (2.5.28)) is in terms of end moments (M_{ab}, M_{ba}) and displacement (Δ), bending moments at two ends are acquired by simultaneous solving Eq. (2.5.31) in the form of

$$\eta = \varepsilon L^2 a(aL - 2)(a^2L^2 - 2aL + 2)$$
$$-a^2L^2(a^2L^2 - 4aL + 6) - 6 \tag{2.5.32}$$

Based on the above variable, end moments are gained as follows:

$$M_{ab}=\frac{1}{\eta L^2}\{6E_0I_0L\begin{pmatrix}6L^3a^2\varepsilon-4L^4a^3\varepsilon\\-4L^2\varepsilon a-6L^2a^2\\+4La+4L^3a^3-L^4a^4\\+L\varepsilon+L^5\varepsilon a^4-4\end{pmatrix}\theta_a$$
$$-6E_0I_0L^2\left(a^4L^3-4a^3L^2+6a^2L-4a-\varepsilon\right)\theta_b$$
$$-12E_0I_0\Delta\begin{pmatrix}\varepsilon L^5a^4-a^4L^4-4\varepsilon L^2a-4\varepsilon L^4a^3\\-3-6a^2L^2+4aL+4a^3L^3+\varepsilon L+6\varepsilon L^3a^2\end{pmatrix}$$
$$+pL^2\left(L\theta_a-\Delta\right)\} \tag{2.5.33}$$

Then

$$M_{ba}=\frac{1}{\eta L^2}\{12E_0I_0L\Delta\left(\varepsilon L-1\right)$$
$$\times\left(aL-1\right)^4\left(4aL+\varepsilon L+3\right)$$
$$+6E_0I_0L\Delta\left(L\varepsilon-1\right)\left(La-1\right)^4\left(4La+\varepsilon L+2\right)$$
$$+6E_0I_0L\theta_b\left(L\varepsilon-1\right)\left(La-1\right)^4\left(4La+\varepsilon L+4\right)$$
$$+pL^3\theta_a\left(La-1\right)^4\left(L\varepsilon-1\right)$$
$$+L^2p\Delta\big(6L^3\varepsilon a^2-4L^2\varepsilon a-4L^4\varepsilon a^3+L^5\varepsilon a^4$$
$$-4La-6L^2a^2+4L^3a^3-L^4a^4-L\varepsilon-11)\} \tag{2.5.34}$$

By considering the constant cross-section area and material for the structure ($\alpha = 0$ and $\varepsilon = 0$), we will gain the well-known function of end moments with the slope-deflection method as follows:

$$M_{ab}=\frac{2E_0I_0}{L}\left(2\theta_a+\theta_b-\frac{3\Delta}{L}\right)-\frac{p}{6}\left(L\theta_a-\Delta\right) \tag{2.5.35}$$

$$M_{ba} = \frac{2E_0 I_0}{L}\left(2\theta_b + \theta_q - \frac{3\Delta}{L}\right) - \frac{p}{6}\left(L\theta_b - \Delta\right) \tag{2.5.36}$$

By taking derivation from the bending moment, we can calculate the maximum bending moment in the following procedure:

$$M_x = EI\frac{\partial^2 y(x)}{\partial x^2} \tag{2.5.37}$$

$$\frac{\partial M_x}{\partial x} = 0 \text{ or } \frac{\partial^3 y(x)}{\partial x^3} = 0 \tag{2.5.38}$$

The maximum bending moment can be computed after acquiring the value of x from Eq. (2.5.38) and substituting it in Eq. (2.5.37). When the material of the cross-section area is constant, the maximum bending moment is obtained as follows:

To simplify, we define the following new variable

$$\beta = \frac{M_{ab}}{M_{ba}} \text{ and } \beta < 1 \tag{2.5.39}$$

And

$$\begin{aligned}\Omega = &\left\{-12E_0 I_0\left(\beta+1\right) - pL^2(\beta-5)\right\} M_{ab} \\ &+2p\Delta\left(pL^2 + 12E_0 I_0\right)\end{aligned} \tag{2.5.40}$$

Then

$$\begin{aligned}&\lambda_1 = \left(2-\beta\right) M_{ab} + 2p\Delta,\ \lambda_2 = pL^2 - 12E_0 I_0, \\ &\lambda_3 = pL\left(\beta - 2\right) M_{ab} - 2p^2 L\Delta\end{aligned} \tag{2.5.41}$$

Finally, the maximum bending moment is acquired as follows:

$$M_{\max} = -M_{ab} + \frac{\Omega^2}{4pL^2\lambda_1\lambda_2} + \frac{\lambda_3\Omega^2}{8p^2L^3\lambda_1^2\lambda_2} \quad (2.5.42)$$

Regarding the length of the structure ($0 \le x \le 0.68m$), the numerical solution of Eq. (2.5.16) is achieved with the physical values of Eq. (2.5.29), as shown in Table 2.5.1.

Table 2.5.1. Obtained values based on the numerical solution of Eq. (2.5.16) along the length of the structure

x/m	$y(x)$/m	x/m	$y(x)$/m
0.00	0.000000	0.36	0.000620
0.04	0.000081	0.40	0.000670
0.08	0.000160	0.44	0.000720
0.12	0.000230	0.48	0.000770
0.16	0.000300	0.52	0.000810
0.20	0.000370	0.56	0.000850
0.24	0.000440	0.60	0.000900
0.28	0.000500	0.64	0.009400
0.32	0.000560	0.68	0.009800

Fig. 2.5.6 demonstrates the displacements of the structure (y (x)) in terms of the length (x) obtained by AGM and the numerical method. As the diagram shows, a comparison of the AGM and the numerical method reveals that AGM is very effective and precise.

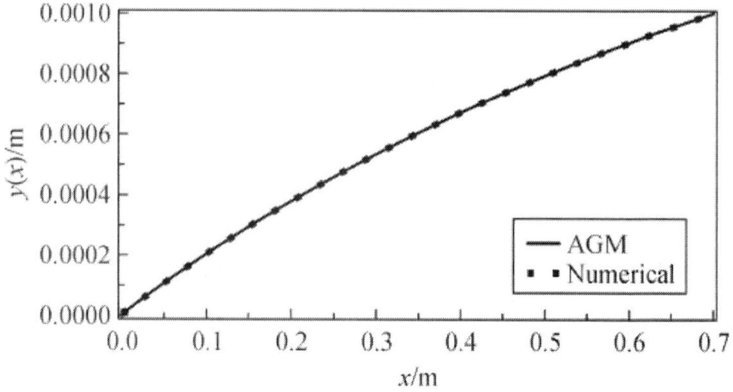

Fig. 2.5.6. Comparison between AGM and the numerical method for computing displacement of the structure

Also, Fig. 2.5.7 illustrates the comparison of the first derivative of the displacement structure versus length, which shows that AGM is reliable.

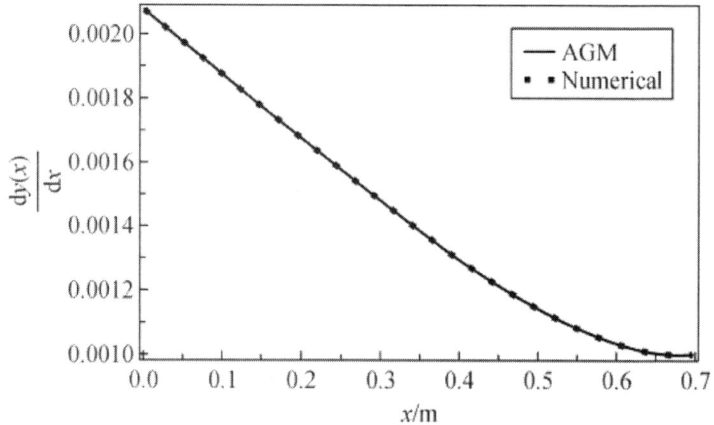

Fig. 2.5.7. Comparison between the first derivative of the solutions obtained by AGM and the numerical method

Equation (2.5.16) is solved by the analytical method as follows:

$$\xi_1 = \frac{i\varepsilon}{2(1-\varepsilon)}\sqrt{\frac{p}{aE_0I_0(a-\varepsilon)}} \text{ and}$$
$$\xi_2 = \frac{2i}{(1-\varepsilon)}\sqrt{\frac{p}{E_0I_0(1-\varepsilon)}} \tag{2.5.43}$$

i denotes the imaginary part of the complex number. Then

$$\xi_3 = \frac{\xi_2(L\varepsilon-1)}{La-1},\ \mathrm{X}=\frac{\xi_2(\varepsilon x-1)}{ax-1} \tag{2.5.44}$$

For mathematical functions, we have the Whittaker equation as follows:

$$W(\mu,\upsilon,z) = e^{-\frac{1}{2}z} z^{\frac{1}{2}+\upsilon} \times$$
$$hypergeom\left\{\left(\frac{1}{2}+\upsilon-\mu\right),(1+2\upsilon),z\right\} \tag{2.5.45}$$

$$hypergeom(n,d,z) = \sum_{k=0}^{\infty}\frac{z^k\prod_{i=1}^{p}\frac{\Gamma(n_i+k)}{\Gamma(n_i)}}{k!\prod_{j=1}^{q}\frac{\Gamma(d_j+k)}{\Gamma(d_j)}} \tag{2.5.46}$$

Where Γ is the Gamma function. To make it easier, we consider the following abbreviations:

$$W \equiv WW,\ \mathrm{M} \equiv \mathrm{WM} \tag{2.5.47}$$

So we have:

$$\phi = WW\left(\xi_1,\frac{1}{2},\xi_3\right)WM\left(\xi_1,\frac{1}{2},\xi_2\right) - WM\left(\xi_1,\frac{1}{2},\xi_3\right)WW\left(\xi_1,\frac{1}{2},\xi_2\right) \tag{2.5.48}$$

Eventually, the analytical answer of Eq. (2.5.16) is gained in the form of

$$\begin{aligned} y(x) = {} & \frac{ax-1}{p\phi(La-1)}\Bigg\{WM\left(\xi_1,\frac{1}{2},X\right) \\ & \Big\{-WW\left(\xi_1,\frac{1}{2},\xi_3\right)\times M_{ab} + WW\left(\xi_1,\frac{1}{2},\xi_2\right)LaM_{ab} \\ & +2WW\left(\xi_1,\frac{1}{2},\xi_2\right)p\Delta - WW\left(\xi_1,\frac{1}{2},\xi_2\right) \\ & +WW\left(\xi_1,\frac{1}{2},X\right)\times\Big\{2WM\left(\xi_1,\frac{1}{2},\xi_2\right)p\Delta \\ & -WM\left(\xi_1,\frac{1}{2},\xi_2\right)M_{ba} - WM\left(\xi_1,\frac{1}{2},\xi_3\right)M_{ab} \\ & +WM\left(\xi_1,\frac{1}{2},\xi_3\right)LaM_{ab}\Big\}\Bigg\} \\ & + \frac{(M_{ab}+M_{ba}-p\Delta)x - LM_{ab}x}{pL} \end{aligned} \tag{2.5.49}$$

Comparisons between exact solutions and AGM charts are shown in Figs. 2.5.8 and 2.5.9. As can be seen, Figs. 2.5.8 and 2.5.9 compare the solution between AGM and the exact solution, while Figs. 2.5.6 and 2.5.7 compare the solution between AGM and the numerical method.

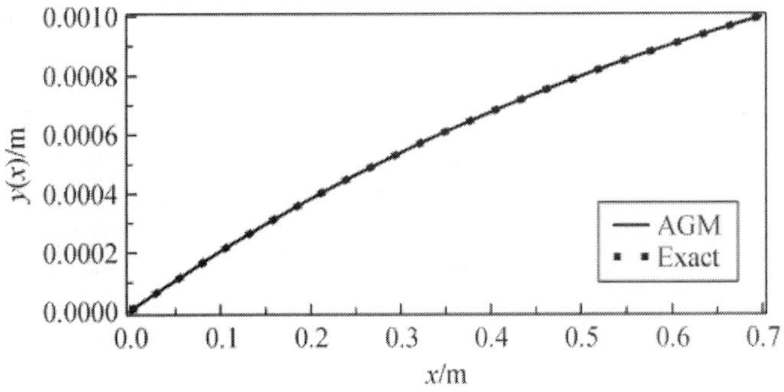

Fig. 2.5.8. Comparison between AGM and exact solution for computing the displacement of the structure

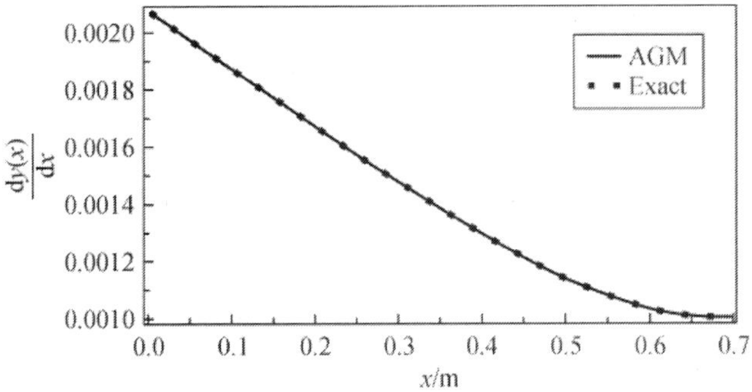

Fig. 2.5.9. Comparison between the first derivative of the solutions obtained by AGM and the exact solution

By comparing the answers between AGM, the numerical method, and the exact solution, it has been found that AGM is efficient and precise in solving linear and nonlinear equations. Also, since AGM is straightforward, it can be used to solve complex nonlinear equations in engineering sciences. Figs. 2.5.10 and 2.5.11 show the differences in the absolute value between the results based on the yielded solution by AGM and the results in Table 2.5.1 (the numerical method).

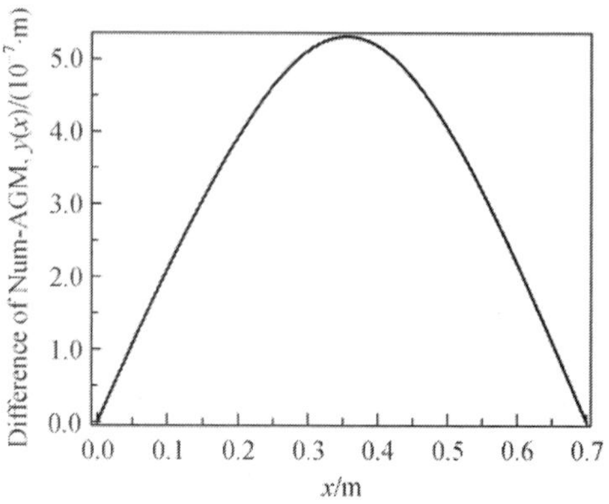

Fig. 2.5.10. Differences between the solutions obtained by AGM and the numerical method

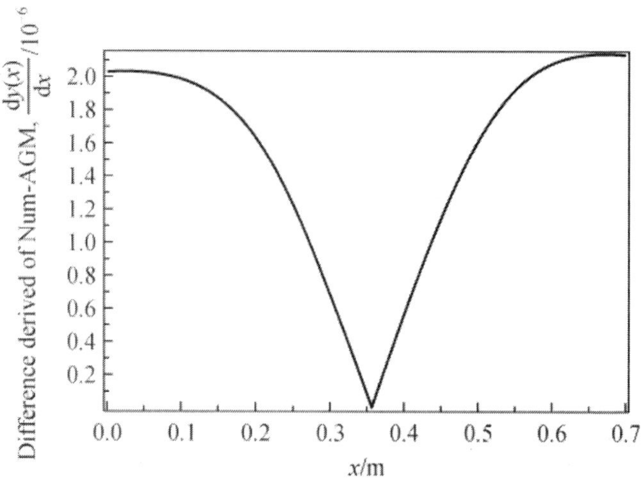

Fig. 2.5.11. Differences between the first derivative of the solution obtained between AGM and the numerical method

Since displacement in the middle of the structure is more than that at the two ends, the difference is more apparent at this position (Fig. 2.5.10). It is obvious that the first derivative is vice versa (Fig. 2.5.11). According to the Figs. 2.5.10 and 2.5.11, it could be seen that differences between AGM and the numerical method are negligible, concluding that solving the equations (linear and nonlinear) using AGM, in addition to simplicity, is more accurate.

b) Example B

Consider the one-dimensional heat transfer on the following extended surface (Fig. 2.5.12) with various logarithmic surface and heat generation, which are $A(x)$ and $G(x)$, respectively. It is also assumed that conductivity K is a function of temperature.

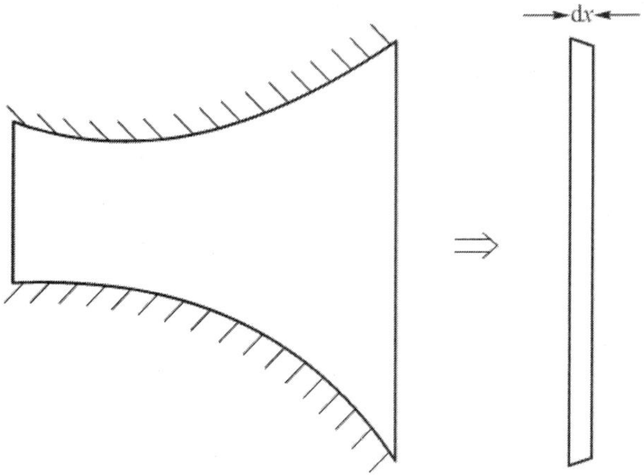

Fig. 2.5.12. Schematic of the extended surface

The following ordinary differential equations governing the system are in the form of

$$f(x)=\frac{d}{dx}\left(K(T).A(x)\frac{dT}{dx}\right)+G(x)=0 \qquad (2.5.50)$$

Where

$$A(x) = A_0 e^{\alpha x}, \; G(x) = G_0 e^{-\alpha x}, \\ K(T) = k_0(1+\beta T) \tag{2.5.51}$$

where, A_0, G_0 and k_0 are the values of $A(x)$, $G(x)$ and $K(T)$ at x = 0, T = 0, respectively. The boundary conditions are

$$\begin{cases} at\ x = 0 \rightarrow T = T_0 \\ at\ x = L \rightarrow T = T_L \end{cases} \tag{2.5.52}$$

For solving the governing differential Eq. (2.5.50) by AGM, the answer to parameter T can be considered as a finite series with constant coefficients in the form of

$$T(x) = \sum_{n=0}^{3} a_n x^n = a_0 + a_1 x + a_2 x^2 + a_3 x^3 \tag{2.5.53}$$

Constant coefficients $a_0\ to\ a_3$ are easily obtained by applying boundary conditions which are described in Eq. (2.5.52) in the following procedure:

$$\begin{cases} T(0) = T_0 \;\; then\ a_0 = T_0 \\ T(L) = T_L \;\; \text{so the result is } a_0 + a_1 L + a_2 L^2 + a_3 L^3 = T_L \end{cases} \tag{2.5.54}$$

After substituting Eq. (2.5.53) with Eq. (2.5.50), this is the best time to apply boundary conditions on the equation obtained, which is specified in AGM as follows:

$$f(0) = k_0 \beta A_0 a_1 + k_0 A_0 \alpha a_1 + 2k_0 A_0 a_2 + G_0 = 0 \tag{2.5.55}$$

and then for x = L, we have

$$f\left(L\right)=k_0\beta A_0\left(1+aL\right)\left(a_1+2a_2L+3a_3L^2\right)$$
$$+k_0\left(1+\beta L\right)A_0a\left(a_1+2a_2L+3a_3L^2\right)$$
$$+k_0\left(1+\beta L\right)A_0\left(1+aL\right)\left(2a_2+6a_3\right)$$
$$+G_0\left(1-aL\right)=0 \tag{2.5.56}$$

Constant coefficients a_0 *to* a_3 can be evaluated by solving sets of ordinary equations, Eqs. (2.5.53) to (2.5.56), which consists of four equations with four unknowns. To make it simple and convenient, we define the following new variable:

$$\xi=\beta L^2\left(6\beta La+6La^2+5\beta-10a\right)$$
$$-12L\left(\beta+a\right)-12 \tag{2.5.57}$$

Concerning the above procedure, constant coefficients a_0 *to* a_3 are obtained as follows:

$$a_0=T_0 \tag{2.5.58}$$

$$a_1=\frac{-1}{\xi k_0A_0L}\left\{L^2G_0\left(6K^2\beta\alpha+3L\alpha+5L\beta+6\right)\right\}$$
$$+6k_0A_0\left(4\beta\alpha L^2+\beta L+3\alpha L+2\right)\left(T_L-T_0\right) \tag{2.5.59}$$

Finally, for a_2 and a_3 are:

$$a_2=\frac{1}{\xi k_0A_0L}\left\{3A_0T_0k_0\begin{pmatrix}4\alpha^2L^2\beta+2\beta+6\beta\alpha L\\+4\beta^2L^2\alpha+3\beta^2L+3\alpha^2L\end{pmatrix}\right.$$
$$+3k_0A_0T_L\left(\beta+\alpha\right)\left(4\beta\alpha L^2+3\alpha L+3\beta L+2\right)$$
$$\left.-LG_0\left(L^2\alpha^2-9L^2\beta\alpha-9L\beta-9L\alpha-6\right)\right\} \tag{2.5.60}$$

$$a_3 = \frac{1}{\xi k_0 A_0 L}\{-A_0 T_0 k_0 \begin{pmatrix} 3\alpha^2 L\beta + 3\beta^2\alpha L \\ +2\alpha^2 + 2\beta\alpha + 2\beta^2 \end{pmatrix}$$
$$+2k_0 A_0 T_L \left(3\alpha^2 L\beta + 2\alpha^2 + 2\beta\alpha + 3\alpha\beta^2 L + 2\beta^2\right) \quad (2.5.61)$$
$$-LG_0\left(L\alpha^2 - 6\alpha - 3L\alpha\beta - 4\beta\right)\}$$

By substituting the constant coefficients from Eqs. (2.5.58)–(2.5.61) into Eq. (2.5.53), the temperature distribution in this special case is obtained in the form of

$$T(x) = a_0 + a_1 x + a_2 x^2 + a_3 x^3 \quad (2.5.62)$$

By choosing the following physical values, we will have

$$\alpha = 0.2, \beta = 0.12, T_L = 50^\circ C, T_0 = 20^\circ C, L = 0.4m$$
$$K_0 = 10\, W/(m.^\circ C), G_0 = 8\, W/m \quad (2.5.63)$$

Therefore, Eq. (2.5.62) is yielded as follows:

$$T(x) = 20 + 156.727x - 225.07x^2 + 51.894x^3 \quad (2.5.64)$$

The physical values are changeable according to the characteristics and conditions of the system. Figs. 2.5.13 and 2.5.14 explain the behaviour $T(x)$ in the specified domain.

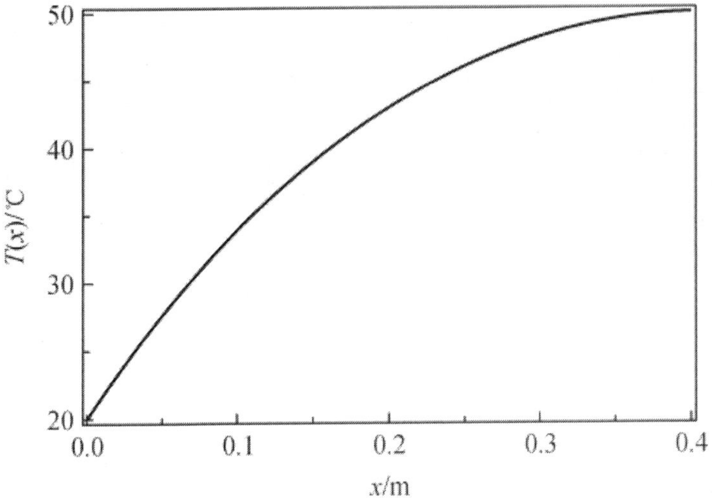

Fig. 2.5.13. Temperature distribution along the length of the extended surface

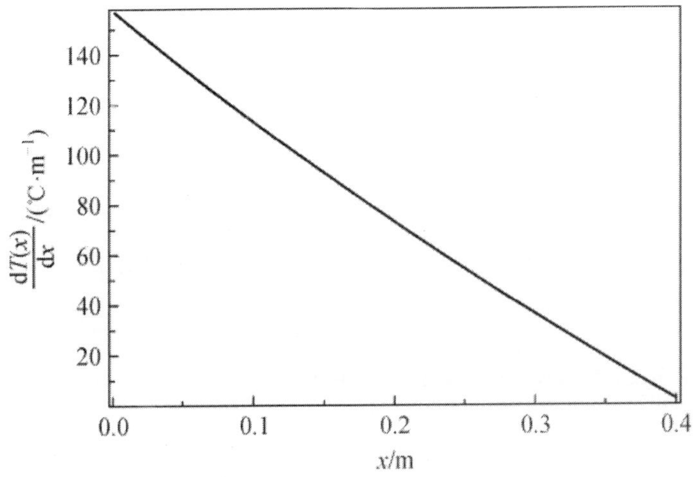

Fig. 2.5.14. The first derivative of the solution obtained by AGM

Conduction heat transfer $q = kA\frac{\partial T(x)}{\partial x}$ according to the physical values of Eq. (2.5.63) is as follows:

$$q(x) = (10 + 1.2x)(0.002 + 0.0004x) \times (146.727 - 450.153x + 155.683x^2) \quad (2.5.65)$$

And related diagram is shown in Fig. 2.5.15.

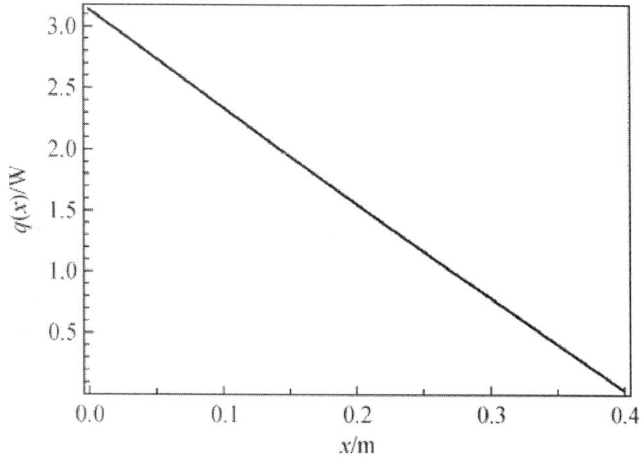

Fig. 2.5.15. Heat transfer rate for the solution obtained by AGM

The numerical solution of nonlinear differential Eq. (2.5.50) in the range of $x \in (0, 0.4)m$ is shown in Table 2.5.2.

Table 2.5.2. Results obtained according to the numerical solution of Eq. (2.5.50) in the specified domain

x/cm	0	5	10	15	20	25	30	35	40
$T(x)$/°C	20.00	27.25	33.42	38.54	42.67	45.85	48.10	49.48	50.0
$\frac{\partial T}{\partial x}$/(°C·m^{-1})	156.07	134.00	112.83	92.48	72.93	54.16	36.13	18.81	2.187

Fig. 2.5.16 compares AGM and numerical solutions for temperature distribution in the specified domain. As shown in Figs. 2.5.16 and 2.5.17 (temperature distribution) and Eq. (2.5.17) (first derivative of the temperature), differences between AGM and the numerical method are negligible. Indeed, it has been concluded that solving the equations using AGM is more accurate and reliable.

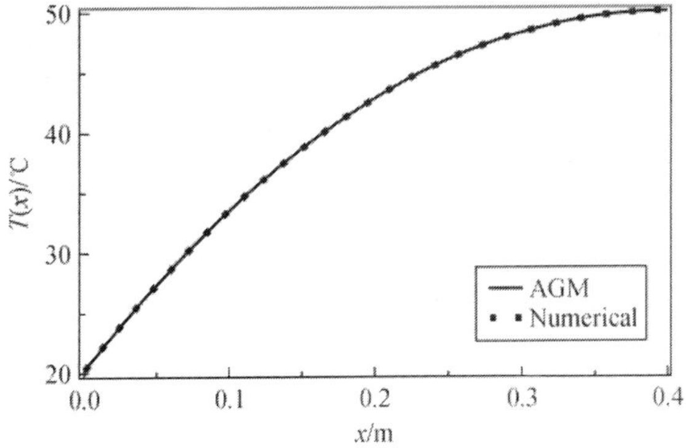

Fig. 2.5.16. Comparison between AGM and numerical solution for temperature distribution in the specified domain

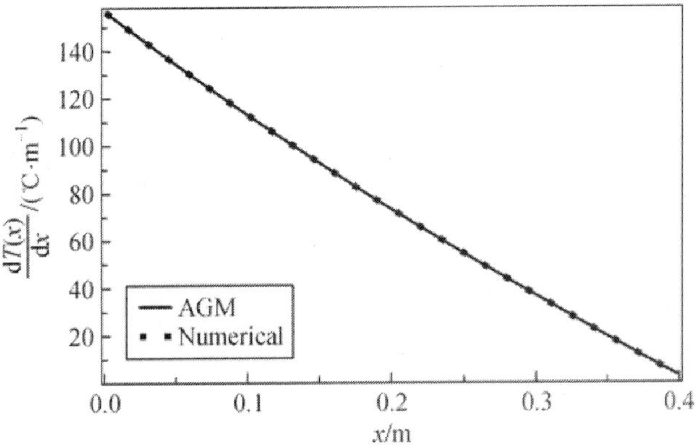

Fig. 2.5.17. Comparison between AGM and numerical solution for u(t)

Fig. 2.5.18 is the absolute value difference between the results based on the yielded solution by AGM and the results of the numerical method (Table 2.5.2). Also, Fig. 2.5.18 depicts that the absolute value difference between AGM and the numerical method is inconsiderable. Indeed, it has been found that solving equations using AGM is accurate and reliable.

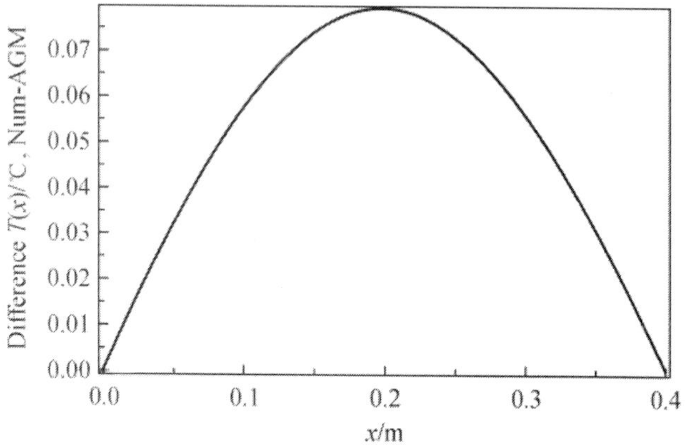

Fig. 2.5.18. Absolute value difference of the solutions obtained between AGM and the numerical method

This study proposes a new method (AGM) for solving two nonlinear differential equations. Two complex fluid and solid mechanical differential equations are:

1) Slope-deflection method used for a beam or column with variable cross-section areas; 2) the mechanism of conduction heat transfer on a distributed surface has been solved by this method.

The proceedings above have been done to indicate the ability of AGM to solve miscellaneous differential equations. Therefore, this claim is proved according to the investigation. Boundary conditions are needed to solve linear and nonlinear differential equations concerning the order of differential equations. If the amount of boundary conditions is fewer than the ideal limitation, solving differential equations is impossible. It is noteworthy that in AGM, by taking derivation from the own differential equation(s) n times, where n denotes the number of the absence of boundary condition(s), the difficulties above are eliminated in this problem in the form of examples, which has been explained thoroughly. There are explicit advantages in this approach in comparison to the other semi-analytic methods, such as:

1) Differential equations are directly solvable by this method;

2) Concerning the accretion of series sentences, the precision of the solution increases significantly;

3) It is not necessary to convert variables into new ones; and

4) We can solve equations without any dimensionless procedures.

This method's boundary conditions shortage for solving differential equations is terminated completely. AGM is operational for miscellaneous nonlinear differential equations, which we hope will be applied by enthusiastic young researchers soon.

2.6. Rigid Beams on Viscoelastic Foundation

The present study investigates a vibrational differential equation governing a rigid beam on a viscoelastic foundation.

In general, vibrational equations and their initial conditions are defined for different systems as follows:

$$f(\ddot{u}, \dot{u}, u, F_0 \sin\,(\omega_0 t)) = 0 \qquad (2.6.1)$$

$$\{u(0) = A, u'(0) = 0\} \qquad (2.6.2)$$

In AGM, a total answer with constant coefficients is required to solve differential equations in various fields of study, such as vibrations, structures, fluids, and heat transfer. In vibrational systems concerning the kind of vibration, it is necessary to choose the answer mentioned in AGM. To clarify here, we divide vibrational systems into two general forms.

Differential equations governing this kind of vibrational system are introduced in the following form:

$$f(\ddot{u}, \dot{u}, u) = 0 \qquad (2.6.3)$$

Now, the answer to this kind of vibrational system is chosen as follows:

$$u(t) = e^{-bt}[Acos(\omega t) + Bsin(\omega t)] \qquad (2.6.4)$$

According to trigonometric relationships, Eq. (2.6.4) is rewritten as follows:

$$u(t) = e^{-bt}[acos(\omega t + \varphi)] \tag{2.6.5}$$

It is notable that in the above equation $a = \sqrt{A^2 + B^2}$ and $\varphi = \arctan\left(\frac{B}{A}\right)$.

Sometimes to increase the precision of the considered answer of Eq. (2.6.3), another term in the form of the Fourier cosine series expansion can be added as follows:

$$u(t) = e^{-bt}[acos(\omega t + \varphi) + dcos(2\omega t + \varphi)] \tag{2.6.6}$$

In the above equation, the term (e^{-bt}) is omitted to facilitate the computational operations in AGM if the system is considered without any damping components.

Generally speaking, in AGM, Eq. (2.6.5) or Eq. (2.6.6) is assumed to be the answer to the vibrational differential Eq. (2.6.3) and its constant coefficients, which are a, b, d, ω (angular frequency) and φ (initial vibrational phase) can easily be obtained by applying the given initial conditions in Eq. (2.6.2). Also, the above procedure will completely be explained through the example in the previous part of this study.

Notably, if there is no damping component in the vibrational system, the constant coefficient of b in Eq. (2.6.5) and Eq. (2.6.6) will automatically be computed as zero in the AGM solution procedure. On the contrary, the parameter b in Eq. (2.6.5) and Eq. (2.6.6) for the other vibrational system with a damping component is obtained as a nonzero parameter in AGM.

In this step, it is assumed that the external forces exerting on the vibrational systems are defined as:

$$F(t) = F_0 \sin(\omega_0 t) \tag{2.6.7}$$

As a result, the differential equation governing the vibrational system is expressed like Eq. (2.6.1) as follows:

$$f(\ddot{u}, \dot{u}, u, F_0 \sin(\omega_0 t)) = 0 \tag{2.6.8}$$

The answer to the above equation is introduced as the sum of the particular solution (u_p) and the harmonic solution (u_h) as follows:

$$u_h(t) = e^{-bt}[Acos(\omega t) + Bsin(\omega t)] \tag{2.6.9}$$

$$u_p(t) = Mcos(\omega_0 t) + Nsin(\omega_0 t)$$

Then

$$u(t) = u_p + u_h \quad (2.6.10)$$

By utilizing trigonometric relationships and substituting the yielded equations into Eq. (2.6.10), the desired answer will be obtained in the form of:

$$u(t) = e^{-bt}[acos(\omega t + \varphi)] + dcos(\omega_0 t + \emptyset) \quad (2.6.11)$$

For increasing the precision of the achieved equation, another term in the form of the Fourier cosine series expansion is added as follows:

$$u(t) = e^{-bt}[acos(\omega t + \varphi) + ccos(2\omega t + \varphi)] + dcos(\omega_0 t + \emptyset) \quad (2.6.12)$$

Finally, the exact solution of all vibrational differential equations can be obtained in accordance with the following equation:

$$u(t) = e^{-at}\left[\sum_{k=1}^{\infty} b_k\, cos(k\omega t + \varphi_k)\right] + dcos(\omega_0 t + \emptyset) \quad (2.6.13)$$

To understand the above procedure thoroughly, reading the following lines is recommended:

Since the constant coefficient (b) in vibrational systems without damping components is always obtained at zero, we can add the term (bt) instead of (e^{-bt}) in Eq. (2.6.12) to decrease computational operations in the following form:

$$u(t) = bt + acos(\omega t + \varphi) + dcos(2\omega t + \varphi) \quad (2.6.14)$$

Based on the above explanations, by applying initial conditions on a system without a damping component, the value of the parameter (b) is always zero for Eq. (2.6.12) and Eq. (2.6.13). Therefore, the role of the parameter (b) in both Eq. (2.6.12) and Eq. (2.6.13), where each of them can be considered as the answer to the vibrational problems is individually considered as a catalyst for increasing the precision of the assumed answer. However, according to Eq. (2.6.15) and Eq. (2.6.19), after applying initial conditions on the vibrational system in both states (with external force and without external force) by AGM, the value of the parameter (b) is computed at zero because the system has a free vibration without any damping component.

The investigation showed that to decrease computational operations for systems without damping components that (b) in the term (e^{-bt}) is zero, so (e^{-bt}) can be omitted from Eq. (2.6.11). Consequently, Eq. (2.6.11), which has been considered the answer to the systems without any damping component, can be rewritten as follows:

$$u(t) = acos(\omega t + \varphi) + dcos(\omega_0 t + \emptyset) \qquad (2.6.15)$$

The constant coefficients of Eq. (2.6.11) or Eq. (2.6.12), which are $\{a, b, \omega, \varphi, c, d, \emptyset\}$, will easily be computed in AGM by applying the initial conditions of Eq. (2.6.2).

In AGM, the initial conditions of Eq. (2.6.2) are applied in the following forms.

a) Applying the initial conditions on the answer of the differential equation.

Regarding the kind of vibrational system (with external force and without external force), a function is chosen as the answer to the differential equation from Eq. (2.6.5) or Eq. (2.6.6) for the systems without external forces and from Eq. (2.6.11) or Eq. (2.6.12) for the defined systems with external forces. Then the initial conditions are applied to the selected function as follows:

$$u(t) = u(IC) \qquad (2.6.16)$$

Notably, IC is the abbreviation of the introduced initial conditions of Eq. (2.6.2).

b) Appling the initial conditions on the main differential equation and its derivatives.

The best time to substitute the answer into the main differential equation instead of its dependent variable (u) is after choosing a function as the answer of the differential equation according to the kind of vibrational system.

Assume the general equation of the vibration such as Eq. (2.6.1) with time-independent parameter (t) and dependent function (u) as:

$$f(\ddot{u},\dot{u},u,F_0\sin(\omega_0 t)) = 0 \tag{2.6.17}$$

Therefore, based on the kind of vibrational system, a function as the answer of the differential equation such as Eq. (2.6.5) or Eq. (2.6.6) and Eq. (2.6.11) or Eq. (2.6.12) is considered as follows:

$$u = g(t) \tag{2.6.18}$$

In this step, the aforementioned equation is substituted with Eq. (2.6.16) instead of (u) in the following form:

$$f(t) = f(g''(t), g'(t), g(t), F_0\sin(\omega_0 t)) \tag{2.6.19}$$

Eventually, the application of the initial conditions on Eq. (2.6.18) and its derivatives is expressed as:

$$\begin{aligned} f(IC) &= f(g''(IC), g'(IC), g(IC), \ldots) \\ f'(IC) &= f'(g''(IC), g'(IC), g(IC), \ldots) \\ f''(IC) &= f''(g''(IC), g'(IC), g(IC), \ldots) \\ &. \\ &. \\ &. \end{aligned} \tag{2.6.20}$$

Eq. (2.6.20) means that the differential equation's initial conditions (or boundary conditions) can be developed. In other words, to calculate the constants of the solution function (Eq. (2.6.11) or Eq. (2.6.12)), a set of algebraic equations need to be solved first (with an equal number of algebraic equations and the number of unknowns). After substituting Eq. (2.6.18) in the differential equation and successive differentiation (if needed), make the number of algebraic equations equal to the number of unknowns of the solution function. The coefficients of the solution function $(a, b, c, \omega, \emptyset)$ can be easily achieved by solving the algebraic equations.

In AGM, after applying the initial conditions of Eq. (2.6.17), Eq. (2.6.18), and also on its derivatives from Eq. (2.6.19) according to the order of differential equation and utilizing the own two given initial conditions of Eq. (2.6.2), a set of algebraic equations of n equations with n unknowns is created. Therefore, the constant coefficients (a, b, c, d, angular frequency

"ω" and initial phase) are easily achieved. This procedure will be explained as an example in the preceding part of this problem.

Notably, in Eq. (2.6.19), derivatives of f(t) with higher orders until the number of yielded equations is equal to the number of the constant coefficients of the assumed answer can be used.

The rigid beam shown in Fig. 2.6.1 with pin support in the middle on the viscoelastic foundation is modelled with a spring stiffness of (k) and a damping coefficient of (c) per unit length. The vibrational equation and angular frequency for great angles are obtained as shown in Fig. 2.6.1. Fig. 2.6.2 shows the forces acting on the vibrating rigid beam shown in Fig. 2.6.1.

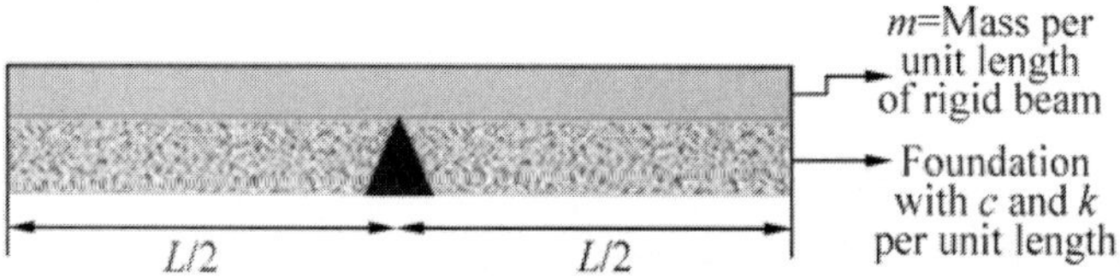

Fig. 2.6.1. The schematic of the rigid beam with pin support on the viscoelastic foundation

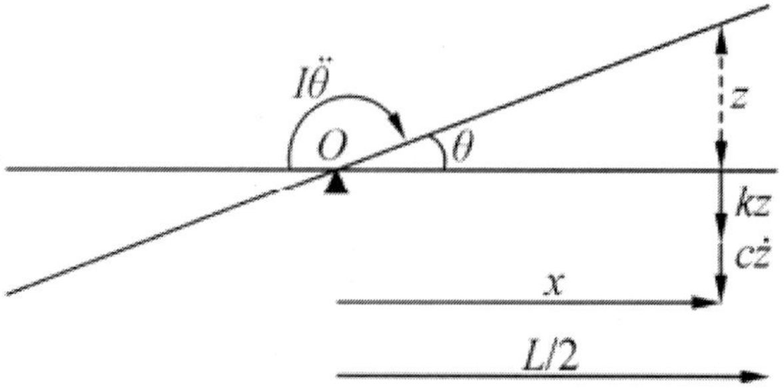

Fig. 2.6.2. The displacement diagram of the mentioned system

Point O, which is the pin support as follows:

$$\sum M_0 = 0 \rightarrow I_0\ddot{\theta} + 2\int_0^{L/2} kzx\,dx + 2\int_0^{L/2} c\dot{z}x\,dx = 0 \tag{2.6.21}$$

The relation between the parameters x, z, and the θ is as follows:

$$z = x\tan\theta;\ \dot{z} = x\,\dot{\theta}\left(1+\tan^2\theta\right) \tag{2.6.22}$$

The mass moment of inertia of the mass per unit length of the rigid beam is as follows:

$$I_0 = \left(mL\right).\left(\frac{L^2}{12}\right) = \frac{mL^3}{12} \tag{2.6.23}$$

By substituting Eq. (2.6.2) and Eq. (2.6.3) into Eq. (2.6.1), the following is assumed:

$$\frac{1}{12}mL^3\ddot{\theta} + 2k\tan\theta\int_0^{L/2} x^2dx + 2c\left(1+\tan^2\theta\right)\dot{\theta}\int_0^{L/2} x^2dx = 0 \tag{2.6.24}$$

The result of Eq. (2.6.4) is as follows:

$$\frac{1}{12}mL^3\ddot{\theta} + \frac{1}{12}cL^3\dot{\theta}\left(1+\tan^2\theta\right) + \frac{1}{12}kL^3\tan\theta = 0 \tag{2.6.25}$$

Rewrite the nonlinear differential Eq. (2.6.5) as follows:

$$\ddot{\theta} + \frac{c}{m}\left(1+\tan^2\theta\right)\dot{\theta} + \frac{k}{m}\tan\theta = 0 \tag{2.6.26}$$

The initial conditions for Fig. 2.6.1 according to Eq. (2.6.6) are expressed in the forms of:

$$\theta(0) = A;\ \dot{\theta}(0) = 0 \tag{2.6.27}$$

Before starting the solution procedure, it is necessary to mention that to solve a nonlinear differential equation by AGM; there is no need to use a Taylor series expansion for the term (sin θ), whereas other methods such as EBM, VIM and PMM, convert the trigonometric terms into their Taylor series expansions. As a result, the solution obtained by AGM is more accurate and precise than the methods above. For the solution of Eq. (2.6.26) in AGM, a polynomial and trigonometric function is as follows:

$$\theta = e^{-at}\left[\lim_{x\to\infty}\sum_{k=1}^{n} b_k \cos\left(k\omega t + \varphi_k\right)\right] + d\cos\left(\omega_0 t + \phi\right) \tag{2.6.28}$$

Eq. (2.6.8) is a more general form of the solution function of the AGM. Term $d\cos\left(\omega_0 t + \phi\right)$ in Eq. (2.6.8) is related to external forces applied to the system that does not exist in this problem. For the answer to the differential equation with acceptable accuracy, a term of answer function (n=1) can be chosen as follows:

$$\theta = e^{-at}\left[b\cos\left(\omega t + \varphi\right)\right] \tag{2.6.29}$$

The constant coefficients of Eq. (2.6.9), ω, φ, a, and b can easily be obtained by applying the initial conditions.

It is noteworthy that in the proposed method, initial or boundary conditions are applied in two ways as follows: 1) In general, the initial conditions are applied on Eq. (2.6.8) in the form of:

$$\theta(0) = \theta(IC) \tag{2.6.30}$$

For simplicity, IC is considered the abbreviation of the initial conditions. As a result, applying the initial conditions on Eq. (2.6.10) is done as follows:

$$\theta(0) = A \rightarrow b\cos(\varphi) = A \tag{2.6.31a}$$

$$\dot{\theta}(0)=0 \rightarrow b\left(a\cos(\varphi)+\omega\sin(\varphi)\right)=0 \tag{2.6.31b}$$

2) The application of initial or boundary conditions on the main differential equation, which in this case is Eq. (2.6.6), and also on its derivatives is done in the following general form:

$$f\left(\theta(t)\right) \rightarrow f\left(\theta(IC)\right)=0,\ f'\left(\theta(IC)\right)=0,\ldots \tag{2.6.32}$$

Therefore, after substituting Eq. (2.6.9), which has been considered as the answer to the main differential equation, into Eq. (2.6.6), the initial conditions are applied to the equation obtained and also on its derivatives based on Eq. (2.6.12) as follows:

$$\begin{aligned} & f\left(\theta(t=0)\right): \frac{b}{m}\left(ma^2-m\omega^2-ca\right)\cos(\varphi) \\ & +\frac{b\omega}{m}(2am-c)\sin(\varphi) \\ & +\frac{1}{m}\{k+c\tan\left[b\cos(\varphi)\right]\}\tan\left[b\cos(\varphi)\right]+\frac{c}{m}=0 \end{aligned} \tag{2.6.33a}$$

Then the first derivative of Eq. (2.6.6) is as follows:

$$\begin{aligned} & f'\left(\theta(t=0)\right): ab\left(3\omega^2-a^2\right)\cos(\varphi)-b\omega\left(3a^2-\omega^2\right)\sin(\varphi) \\ & -\frac{1}{m}\{k+2c\left[\left(ab\cos(\varphi)+b\omega\sin(\varphi)\right)\right]\} \\ & \times\tan\left[b\cos(\varphi)\right]\{1+\tan^2\left[b\cos(\varphi)\right]\} \\ & +\frac{cb}{m}\{a+\tan^2\left[b\cos(\varphi)\right]\} \\ & \times\left[2a\omega\sin(\varphi)+\left(a^2-\omega^2\right)\cos(\varphi)\right]=0 \end{aligned} \tag{2.6.33b}$$

Solving a set of algebraic equations that consist of four equations, four unknowns (from Eq. (2.6.11a), Eq. (2.6.11b), Eq. (2.6.13a), and Eq. (2.6.13b)) and constant coefficients a, b, φ and ω (from Eq. (2.6.9)) the following can be yielded: For simplicity, the following new variable is introduced as:

$$\psi = 4km\tan(A) - c^2 A\left[1 + \tan^4(A) + 2\tan^2(A)\right]$$
$$\Delta_1 = \frac{mA}{\tan(A)}, \qquad \Delta_2 = m^2 A^2 + \Delta_1^2 \tag{2.6.34}$$

According to Eq. (2.6.34), variables of response function coefficients are obtained as follows:

$$a = \frac{c\Delta_2 \tan^2(A)}{2m^3 A^3}, \qquad b = \frac{2mA\sqrt{kA}}{\sqrt{\Delta_1 \psi}}$$
$$\varphi = -\tan^{-1}\left(\frac{c\Delta_2}{2m\Delta_1\sqrt{k\Delta_1}}\right), \qquad \omega = \frac{1}{2m}\sqrt{\frac{\psi}{A}} \tag{2.6.35}$$

After substituting the obtained values from Eq. (2.6.15) into Eq. (2.6.9), the solution of nonlinear differential Eq. (2.6.6) by AGM will be obtained as follows:

$$\theta(t)2mA\sqrt{\frac{kA}{\Delta_1\psi}}\exp\left[-\left(\frac{c\Delta_2\tan^2 A}{2m^3A^2}\right)t\right]\times$$
$$\cos\left\{\frac{1}{2m}\sqrt{\frac{\psi}{A}}t - \tan^{-1}\left(\frac{c\Delta_2}{2m\Delta_1\sqrt{k\Delta_1}}\right)\right\} \tag{2.6.36}$$

By selecting the following physical quantities for Fig. 2.6.1:

$$m = 10,\ k = 2,\ c = 0.5,\ A = 0.2 \tag{2.6.37}$$

The answer function of Eq. (2.6.9) and the angular frequency are obtained as follows:

$$\omega = 0.4495\left(\frac{rad}{\sec}\right) \tag{2.6.38a}$$

$$\theta(t) = 0.200335e^{-0.026t}\left(0.4495t - 0.05784\right) \tag{2.6.38b}$$

Then, the charts of the solution obtained function, derivative, and phase plane are drawn in Figs. 2.6.3–2.6.5.

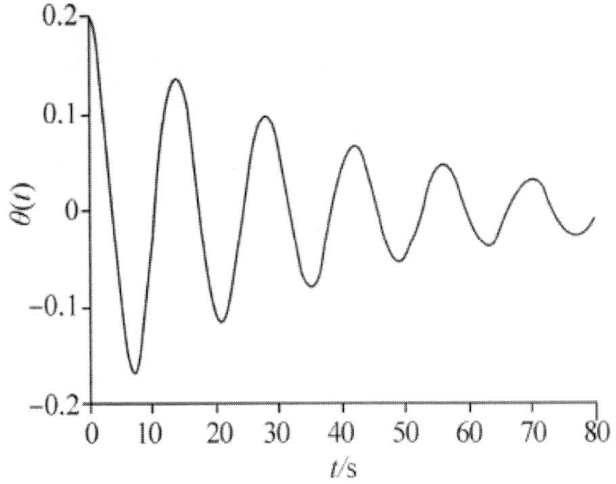

Fig. 2.6.3. The solution obtained by AGM

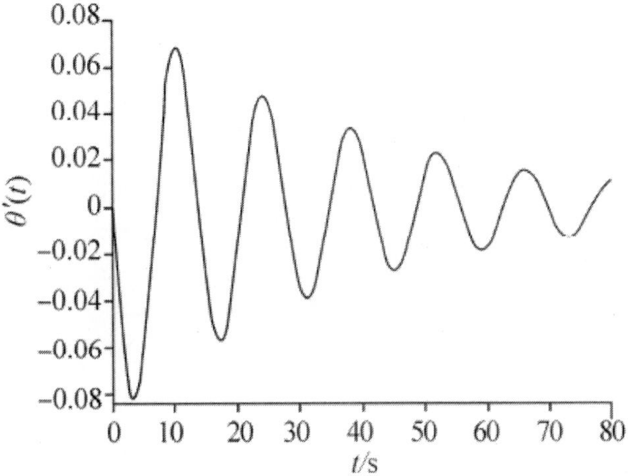

Fig. 2.6.4. The derivative of the solution obtained by AGM

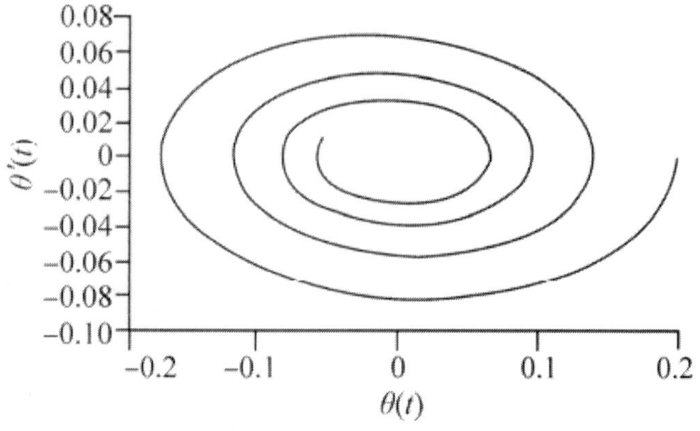

Fig. 2.6.5. The result in phase plane by AGM

Concerning the given physical values from Eq. (2.6.17) and the determined domain $t \in [0,80]$, which is defined in terms of second (s), the solution of the differential equation is expressed numerically (Runk45) in Table 2.6.1.

Table 2.6.1. Results obtained for θ(*t*) and its derivative according to the given physical values

t/s	$\theta(t)_{\text{Num}}$	$\theta(t)_{\text{AGM}}$	Error /%	$\dot{\theta}(t)_{\text{Num}}$	t/s	$\theta(t)_{\text{Num}}$	$\theta(t)_{\text{AGM}}$	Error /%	$\dot{\theta}(t)_{\text{Num}}$
0	0.2	0.2	0.00	0.0	48	−0.050 518 11	−0.050 635 71	0.23	−0.013 339 7
16	0.090 485 15	0.090 753 42	0.29	−0.046 517 2	64	−0.038 655 24	−0.038 728 01	0.18	0.006 100 332
32	−0.011 689 17	−0.011 547 80	1.20	−0.039 488 1	80	−0.011016078	−0.011 240 90	2.04	0.011 298 127

Due to the solution obtained from Eq. (2.6.16) by AGM and the obtained numerical solution, its results are presented in Table 2.6.1; the following comparisons are provided in Figs. 2.6.6–2.6.8.

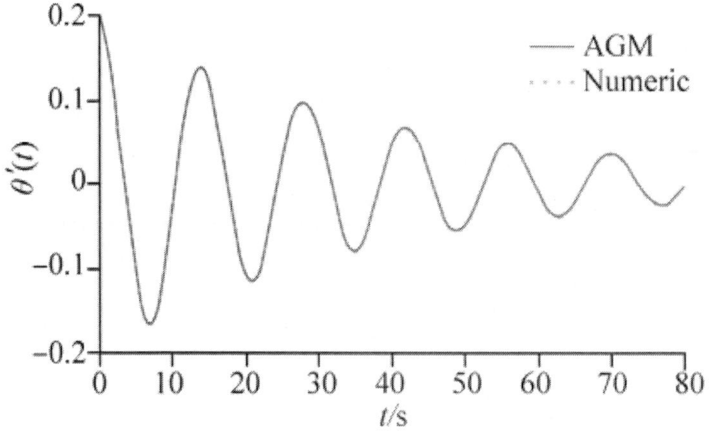

Fig. 2.6.6. A comparison between the achieved solutions by AGM and the numerical method

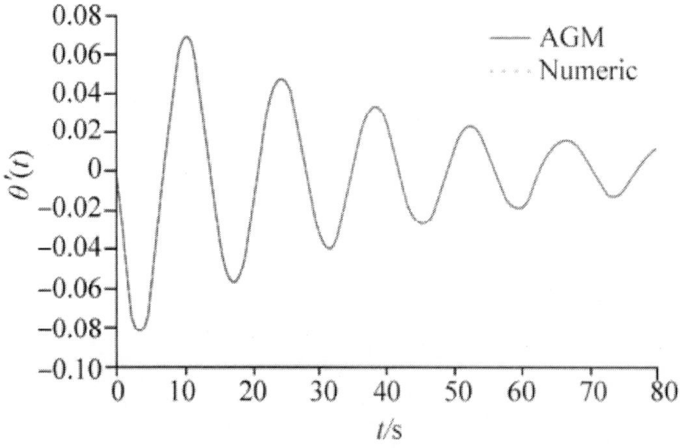

Fig. 2.6.7. A comparison between the derivative of the achieved solutions by AGM and the numerical method

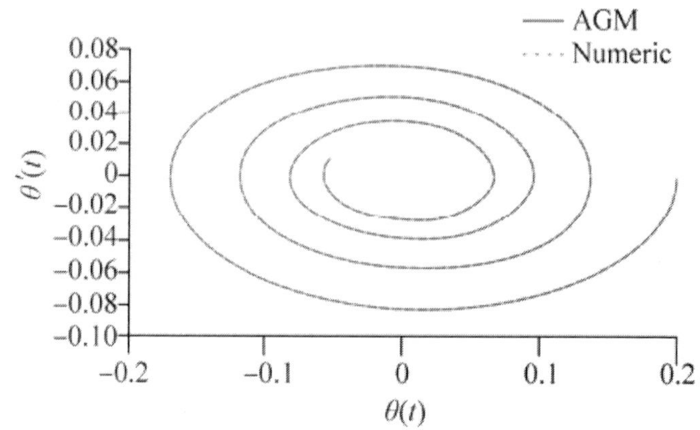

Fig. 2.6.8. Comparing the resulted phase planes by AGM and the numerical method

The following Figs. 2.6.9 and 2.6.10 are the results based on AGM's yielded solution from Eq. (2.6.16) and Table 2.6.1 by the numerical method.

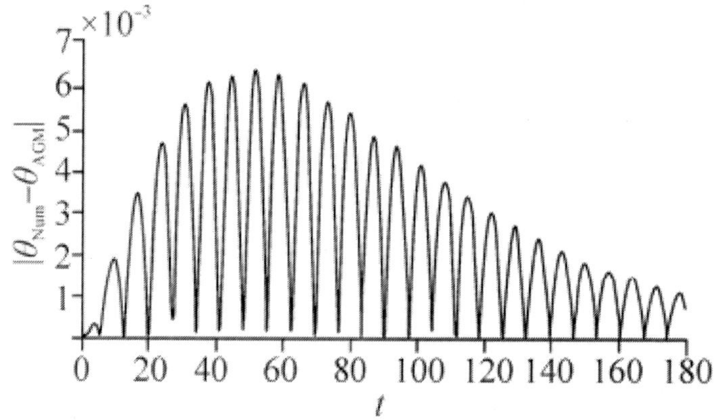

Fig. 2.6.9. Comparing the absolute differences between displacement functions which are predicted by AGM and the numeric method

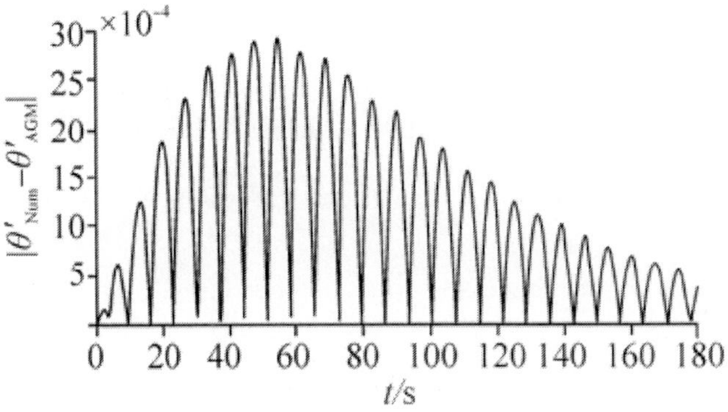

Fig. 2.6.10. Comparing the absolute differences between velocity functions which are predicted by AGM and the numeric method

Concerning Figs. 2.6.9 and 2.6.10, it is revealed that the errors increase rapidly until t = 50 s and then decrease steadily.

The angular frequency (ω) from Eq. (2.6.15) in the initial amplitude of vibration is seen in Fig. 2.6.11. It is clear that the more amplitude of vibration in the IC, the more the angular frequency increases.

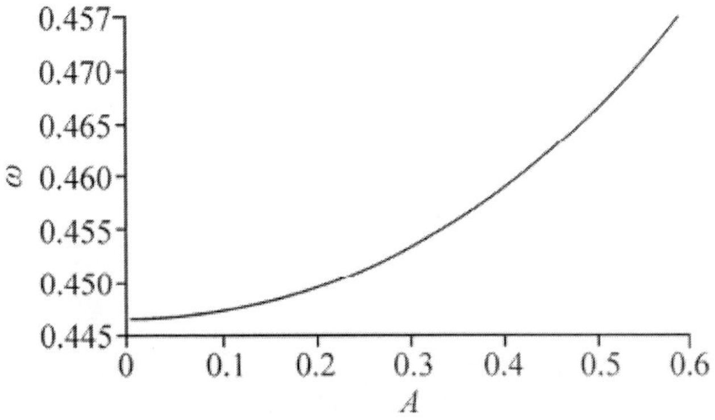

Fig. 2.6.11. The chart of angular frequency in terms of initial vibrational amplitude

The damping factor in the vibrational solution $b\left\{e^{-at}\cos\left(\omega t+\phi\right)\right\}$ is e^{-at} since in the vibrational differential equation, the term $e^{-\xi\omega t}$ is the main factor for damping, (ξ) can be obtained as follows:

$$e^{-at} \equiv e^{-\xi\omega t} \rightarrow \xi = \frac{a}{\omega} \tag{2.6.39}$$

The damping ratio (ξ) in terms of the initial amplitude of vibration (A) from Eq. (2.6.19) is depicted in Fig. 2.6.12.

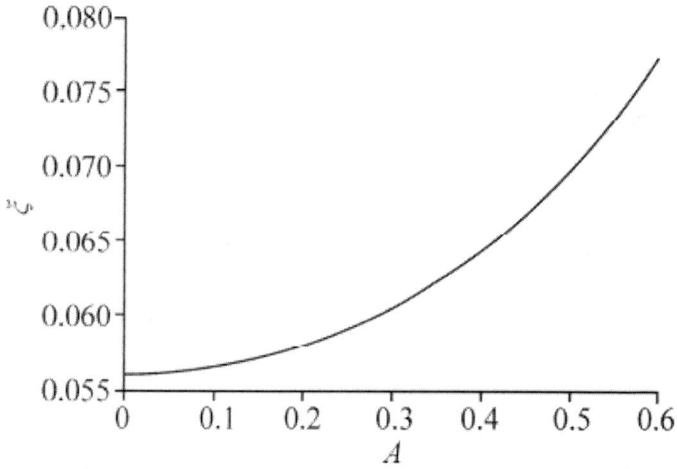

Fig. 2.6.12. The damping ratio variation in the initial amplitude of vibration (A).

Concerning Fig. 2.6.12, it is revealed that by increasing the amount of initial amplitude of vibration (A), the value of the damping ratio will be increased. Therefore, in this step, it is logical to indicate that since the formula of the damping coefficient is obtained, there is a direct relationship between the initial amplitude of vibration and the damping coefficient.

The energy dissipated per cycle by the damping force (F_d) can be shown with a graphical display. Terms of damping force in the differential Eq. (2.6.6) are obtained as follows.

$$F_d = \alpha\dot{\theta} \tag{2.6.40}$$

After insertion of the derivative of Eq. (2.6.16) into Eq. (2.6.20), the following result is obtained:

$$F_d = \alpha b e^{-at}\left\{\cos\left(\omega t+\varphi\right)+\omega\sin\left(\omega t+\varphi\right)\right\} \tag{2.6.41}$$

According to Eq. (2.6.9) as follows:

$$\theta = b e^{-at}\left\{\cos\left(\omega t+\varphi\right)\right\} \tag{2.6.42}$$

For each cycle of vibration, at times $\left(t=\dfrac{2k\pi}{\omega}, k=1,\ 2,\ldots\right)$

with the merger of the two equations Eqs. (2.6.21) and (2.6.22), the trigonometric relations' damping force can be obtained as follows:

$$F_d = \alpha\left[\theta \pm \sqrt{b^2\exp\left(-\frac{4ak\pi}{\omega}\right)-\theta^2}\right] \tag{2.6.43}$$

As well as the energy lost per cycle of vibration, the equation is as follows:

$$w_d = \oint F_d\,d\theta \rightarrow w_d = \oint \alpha\dot{\theta}^2\,dt \rightarrow$$
$$w_d = \int_0^{\frac{2k\pi}{\omega}} \alpha\dot{\theta}^2 dt,\ k=1,2,\ldots \tag{2.6.44}$$

The comparison of the energy loss for three cycles is depicted in Figs. 2.6.13–2.6.16. They show that the area enclosed by the ellipse determines the energy lost per cycle.

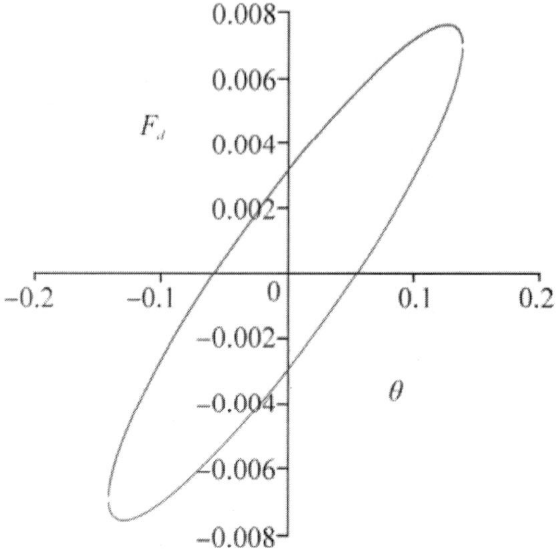

Fig. 2.6.13. The energy dissipated in the first cycle (*k*=1)

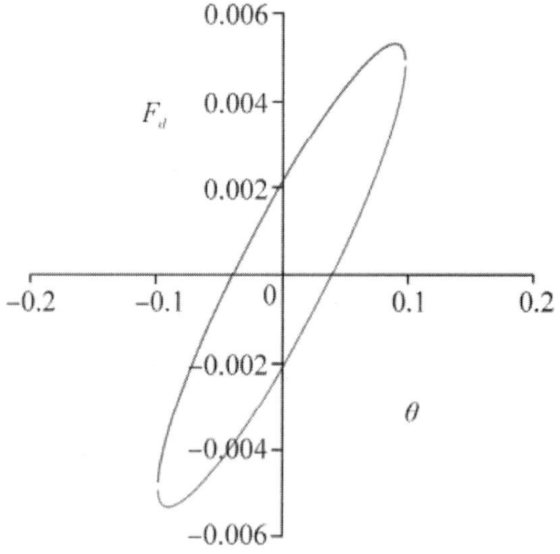

Fig. 2.6.14. The energy dissipated in the second cycle (k=2)

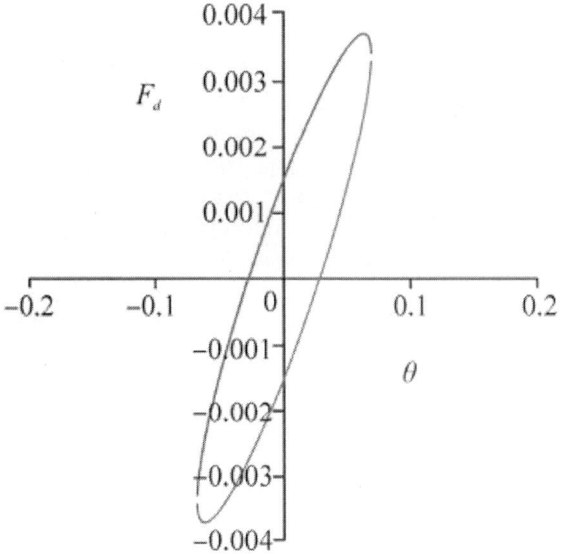

Fig. 2.6.15. The energy dissipated in the third cycle (*k*=3)

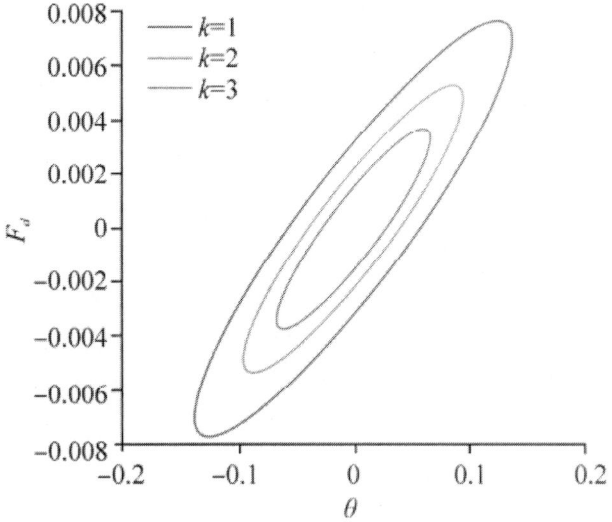

Fig. 2.6.16. The energy dissipated in the third cycle (*k*=1, 2, 3)

A rigid beam with pin support on the viscoelastic foundation has been modelled for this problem. The complicated nonlinear vibrational differential equations governing it have been introduced and analyzed by AGM, and the results have been compared with the numerical method (Runk45). This problem was solved analytically, and the solution was obtained for different values of frequencies and amplitudes. This showed that increasing the amplitude of vibration in the initial condition increases the angular frequency and damping ratio. The dissipated energy in different cycles was also investigated. The above process has shown that AGM can solve a broad range of differential equations in different fields of study, particularly in vibrations.

Consequently, it is concluded that AGM is a reliable and precise approach for solving differential equations. Moreover, a summary of the AGM's excellence and benefits is explained. Solving a set of algebraic equations with constant coefficients makes it possible to easily obtain the solution of a nonlinear differential equation along with the related angular frequency. Applying this procedure is possible even for students with intermediate mathematical knowledge. On the other hand, it is better to say that AGM can solve linear and nonlinear differential equations directly in most situations. This means that the final solution can be obtained without any dimensionless procedures. Therefore, AGM can be considered a significant progress in nonlinear sciences. The shortage of boundary condition (s) in this method for solving differential equation (s) is completely eradicated. AGM is operational for nonlinear differential equations, especially for civil and mechanical engineering vibrations.

List of Symbols

AGM	Akbari-Ganji's Method
R	Semicircular radius
G	Gravity acceleration
A	Vibration amplitude
α	Semicircular angle
$\theta(t)$	Angular displacement
$\dot{\theta}(t)$	Angular velocity

E	Modulus elasticity ($N.m^{-2}$)
L	Length of the beam (The length of each structure member)
W_L	Live harmonic load
EI	Bending stiffness
δ	Dirac delta function
A	Cross-sectional area
ρ	Density
$f(x, t)$	The sinusoidal load exerted on the arched beam (N/m^2)
ω_n	Modal frequencies (Rad/s)
D	Bending rigidity (N/m^2)
$\varphi(x)$	Mode shape function
ω	Angular frequency of the external harmonic load
z_0	The initial amplitude of oscillation
n	Vibrational mode
e	Centre outlet
k	Spring stiffness coefficient
τ	Torque
M	Mass properties
A	Vibration amplitude
ω_0	Angular frequency
$q(x)$	The distributed load exerted ($N.m^{-1}$)
X	Length

M_{xx}	Bending moment along the beam
N_{xx}	Shearing force along the beam
σ_{xx}	Stress along beam
W_0	Displacement of the beam in the z-direction
p	The axial force on the member
M_{ab}, M_{ba}	Bending moment in relevance to the primary and ending nodes a and b
Δ	The displacement of the end part of the member due to the end moment
$I(x)$	The function of the moment of inertia along the column
$E(x)$	The function of elastic modulus as a variable along the length of the member
α	Constant values for a reduction factor of the moment of inertia
ε	Elastic modulus along the length of the member
$A(x)$	The various logarithmic surface
$G(x)$	The logarithmic various heat generation
$K(T)$	Conductivity
$u(t)$	Vibrational displacement (m)
$\dot{u}(t)$	Vibrational velocity (m/s)
$\ddot{u}(t)$	Vibrational acceleration (m/s^2)
$\theta(t)$	Angular position as a function of time (rad)
A	The initial amplitude of vibration (m)
ω	Angular frequency (1/s)
$\ddot{\psi}(t)$	Angular acceleration (rad/s^2)

g	Gravity acceleration (m/s^2)
m	Mass of the cylinder (kg) or Mass (kg)
w	Weight of the cylinder (N)
I_c	Mass moment of inertia about the instantaneous centre ($kg.m^2$)
$\sum M_c$	Resultant moment about the instantaneous centre of rotation (N.m)
R	Radius of the circular path (m) or Radius of rotating circle (m)
r	The radius of the cylinder (m)
Ω	Angular velocity (rad/s)
w	Weight force (N)
N	Normal force (N)
α	Initial angular position (rad)
a	Acceleration of the mass (m/s^2)
$\sum M_O$	Resultant moment about the centre of the rotating circle (N.m)
H_O	Angular momentum about the centre of the rotating circle (N.m.s)
$x(t)$	Position of the particle as a function of time (m)

References

1. Mirgolbabaee, H., Tahernejad Ledari, S., Mohammad Zadeh, N., & Domiri Ganji, D. (2017). "Investigation of the nonlinear equation of the circular sector oscillator by Akbari-Ganji's method." *Journal of Taibah University for Science*, 11(6), 1110–1121.

2. Akbari, M. R., Nimafar, M., Ganji, D. D., & Chalmiani, H. K. (2017). "Investigation on non-linear vibration in arched beam for bridge construction via AGM." *Applied Mathematics and Computation*, 298, 95–110.

3. Mirgolbabaee, H., Ledari, S. T., & Ganji, D. D. (2016). "New approach method for solving Duffing-type nonlinear oscillator." *Alexandria Engineering Journal*, 55(2), 1695–1702.

4. Akbari, M. R., Nimafar, M., Ganji, D. D., & Akbarzade, M. M. (2014). "Scrutiny of non-linear differential equations Euler-Bernoulli beam with large rotational deviation by AGM." *Frontiers of Mechanical Engineering*, 9(4), 402–408.

5. Akbari, M. R., Ganji, D. D., Nimafar, M., & Ahmadi, A. R. (2014). "Significant progress in solution of nonlinear equations at displacement of structure and heat transfer extended surface by new AGM approach." *Frontiers of Mechanical Engineering*, 9(4), 390–401.

6. Akbari, M. R., Ganji, D. D., Rostami, A. K., & Nimafar, M. (2015). "Solving nonlinear differential equation governing on the rigid beams on viscoelastic foundation by AGM." *Journal of Marine Science and Application*, 14(1), 30–38.

CHAPTER 3

THE APPLICATION OF AGM IN HEAT TRANSFER AND FLUID DYNAMICS

3.1. Unsteady MHD Eyring-Powell Squeezing Flow

In this problem, the heat and mass transfer has been investigated in an unsteady two-dimensional squeezing flow of magnetohydromagnetic (MHD) radiative non-Newtonian Eyring-Powell fluid between two parallel infinite plates in the presence of heat generation/absorption, thermal radiation and Joule heating effect.

The study is based on the two-dimensional magnetohydrodynamic (MHD) squeezing flow of an incompressible Eyring- Powell fluid with velocity $u(x) = ax$ along the x-axis direction in the presence of a non-uniform magnetic field $B(x) = B_o(1 - \alpha t)^{-\frac{1}{2}}$ between two infinite parallel plates separated by a distance $a(x) = \pm l_o(1 - \alpha t)^{\frac{1}{2}}$, and investigating the characteristics of heat transfer and the flow of this Eyring- Powell fluid. Here, l is the initial distance between the plates at time t, and α is the characteristic parameter of the squeezing motion of the plate with the dimension of the inverse time. When α >0, both plates squeeze until they meet at $t = \frac{1}{\alpha}$, and when $\alpha < 0$ indicates that the plates are separated. Also, the effects of thermal radiation, heat generation/absorption, Lorentz force, and Joule heating are considered. The nonlinear differential equations that model this problem are divided into two categories of momentum equations and energy equations that are given below:

$$\frac{\partial u}{\partial x} + \frac{\partial v}{\partial y} = 0 \tag{3.1.1}$$

$$\frac{\partial u}{\partial t} + u\frac{\partial v}{\partial x} + v\frac{\partial u}{\partial y} = -\frac{1}{p}\frac{\partial p}{\partial x} + \left(v + \frac{1}{p\beta C}\right)\left(\frac{\partial^2 u}{\partial x^2} + \frac{\partial^2 u}{\partial y^2}\right) - \frac{1}{3p\beta c^3}\frac{\partial}{\partial x}\left[\left(2\left(\frac{\partial u}{\partial x}\right)^2 + 2\left(\frac{\partial v}{\partial y}\right)^2 + \left(\frac{\partial u}{\partial y} + \frac{\partial v}{\partial x}\right)^2\right)\frac{\partial u}{\partial x}\right] - \tag{3.1.2}$$

$$\frac{1}{6p\beta c^3}\frac{\partial}{\partial y}\left[\left(2\left(\frac{\partial u}{\partial x}\right)^2+2\left(\frac{\partial v}{\partial y}\right)^2+\left(\frac{\partial u}{\partial y}+\frac{\partial v}{\partial x}\right)^2\right)\left(\frac{\partial u}{\partial y}+\frac{\partial v}{\partial x}\right)\right]-\frac{\sigma B_0 2}{p(1-\alpha t)}u.$$

$$\frac{\partial v}{\partial t}+u\frac{\partial v}{\partial x}+v\frac{\partial v}{\partial y} = -\frac{1}{p}\frac{\partial p}{\partial y}+\left(v+\frac{1}{p\beta c}\right)\left(\frac{\partial^2 v}{\partial x^2}+\frac{\partial^2 v}{\partial y^2}\right) - \frac{1}{6p\beta c^3}\frac{\partial}{\partial x}\left[\left(2\left(\frac{\partial u}{\partial x}\right)^2+2\left(\frac{\partial v}{\partial y}\right)^2 + \left(\frac{\partial u}{\partial y}+\frac{\partial v}{\partial x}\right)^2\right)\left(\frac{\partial u}{\partial y}+\frac{\partial v}{\partial x}\right)\right] - \frac{1}{3p\beta c^3}\frac{\partial}{\partial y}\left[\left(2\left(\frac{\partial u}{\partial x}\right)^2+2\left(\frac{\partial v}{\partial y}\right)^2 + \left(\frac{\partial u}{\partial y}+\frac{\partial v}{\partial x}\right)^2\right)\frac{\partial v}{\partial y}\right] \quad (3.1.3)$$

$$\frac{\partial T}{\partial t}+u\frac{\partial T}{\partial x}+v\frac{\partial T}{\partial y} = \frac{k}{pC_P}\left(\frac{\partial^2 T}{\partial x^2}+\frac{\partial^2 T}{\partial y^2}\right)+\frac{16\sigma^* T_0^3}{3C_P k^*}\frac{\partial^2 T}{\partial y^2} + \frac{Q^*}{pC_P}(T-T_o)+\frac{\sigma B^2(x)}{pCp}u^2 \quad (3.1.4)$$

$$\frac{\partial C}{\partial t}+u\frac{\partial C}{\partial x}+v\frac{\partial C}{\partial y} = D_m\left(\frac{\partial^2 C}{\partial x^2}+\frac{\partial^2 C}{\partial y^2}\right)-\frac{k_1}{1-\alpha t}(C-C_0) \quad (3.1.5)$$

That u and v are the components of the velocity in the directions of x and y, respectively, ρ is fluid density, ν is the kinematic viscosity, β and c are characteristics of the Eyring-Powell fluid, σ is the electrical conductivity of the fluid, B_0 is the uniform magnetic field, k is the thermal conductivity of the fluid, C_p, is the specific heat capacity, σ* is the Stefan-Boltzman constant, T is the fluid temperature and T_0 is the referenced fluid temperature, k* is the mean absorption coefficient, Q* is the uniform volumetric heat generation/absorption coefficient, D_m, is the mass diffusion and C and C_0 are, respectively, the concentration and referenced concentration of the fluid.

Also, the appropriate boundary conditions for the above equations are given according to the assumptions and considered conditions in the problem as follows:

$$u = 0, v = v_w = \frac{da(t)}{dt}, T = T_0, C = C_0, at\ y = a(t) \tag{3.1.6}$$

$$v = \frac{\partial u}{\partial y} = \frac{dT}{dy} = \frac{dC}{dy} = 0, as\ y = 0 \tag{3.1.7}$$

Follow changes are applied to make velocity $f'(\eta)$, temperature θ (η), and concentration ϕ (η) functions dimensionless:

$$\eta = \frac{x}{l(1-\alpha t)^{\frac{1}{2}}}, u = \frac{\alpha x}{2(1-\alpha t)}\frac{df(\eta)}{d\eta}, v = -\frac{\alpha l}{2(1-\alpha t)^{\frac{1}{2}}} f(\eta), \theta = \frac{T - T_0}{T_1 - T_0}, \phi = \frac{C - C_0}{C_1 - C_0} \tag{3.1.8}$$

So, nonlinear differential equations govern the problem after making Eqs. (3.1.2)–(3.1.7) dimensionless, and changing PDEs to ODEs by Eq. (3.1.8) are given as follows:

$$(1+\Gamma)f^{iv} - s(\eta f''' + 3f'' + f'f'' - ff''') - \Gamma\delta(2f''(f''')^2 + (f'')^2 f^{iv}) - Ha^2 f'' = 0. \tag{3.1.9}$$

$$\left(1 + \frac{4}{3}R\right)\theta'' + PrS(f'\theta - \eta\theta' + Q\theta) + PrEcHa^2(f')^2 = 0 \tag{3.1.10}$$

$$\phi'' + ScS(f\phi' - \eta\phi') - Sc\gamma\phi = 0 \tag{3.1.11}$$

The dimensionless numbers and parameters are as follows:
$\Gamma = \frac{1}{\mu\beta C}$ and $\delta = \frac{\alpha^2 x^2}{8c^2 l^2 (1-\alpha t)^3}$, are the Eyring-Powell fluid parameters, $S = \frac{\alpha l}{2V}$ is the squeezing parameter; it is worth mentioning when the plates move apart ($S > 0$) and when the plates collapse together ($s < 0$), $Ha = \sqrt{\frac{\partial}{\mu}} B_0$ is the Hartman number (magnetic field parameter), $R = \frac{4\sigma^* T_0^3}{kk^*}$ is the

radiation parameter, $Q = \frac{2Q^{*}(1-\alpha t)}{\alpha p C p}$ is the heat generation $(Q > 0)$ or absorption $(Q < 0)$ parameter, $Pr = \frac{\mu Cp}{k}$, is the Prandtl number, $Ec = \frac{(u(x))^2}{CpT_0}$, is the Eckert number, $Sc = \frac{v}{D_m}$, is the Schmidt number and $\gamma = \frac{l^2 k_l}{v}$, is the chemical reaction parameter, wherein $(\gamma > 0)$ is the destructive chemical reaction and $(\gamma < 0)$ represents the generative chemical reaction. In the following, using Eq. (3.1.8) in Eqs. (3.1.6, 3.1.7), the boundary conditions of Eqs. (3.1.9–3.1.11) are as follows:

$$f(\eta) = 0, f''(\eta) = 0, \theta'(\eta) = 0, \phi'(\eta) = 0, at\ \eta = 0 \tag{3.1.12}$$

$$f'(\eta) = 0, f(\eta) = 1, \theta(\eta) = 1, \phi(\eta) = 1, at\ \eta = 1 \tag{3.1.13}$$

Also, the skin fraction coefficient (Cf), local Nusselt number (Nu_x) and local Sherwood number (Sh) are defined as follows:

$$cf = \frac{\tau_w}{pv_w^2} \tag{3.1.14}$$

$$u = \frac{lq_w}{k(T_1 - T_0)} \tag{3.1.15}$$

$$Sh = \frac{lj_w}{D_m(C_1 - C_0)} \tag{3.1.16}$$

That τ_w the wall shear stress, q_w the wall heat flux and j_w, the mass flux are respectively equal to:

$$\tau_w = \left\{\left(\mu + \frac{1}{\beta c}\right)\frac{\partial u}{\partial y} - \frac{1}{6pc^3}\left(\frac{\partial u}{\partial y}\right)^2\right\}_{y=a(t)} \tag{3.1.17}$$

$$q_w = \left\{-k\frac{\partial T}{\partial y} - \frac{16\sigma^* T_0^3}{3k^*}\frac{\partial T}{\partial y}\right\}_{y=a(t)} \tag{3.1.18}$$

$$j_w = \left\{D_m \frac{\partial C}{\partial y}\right\}_{y=a(t)} \tag{3.1.19}$$

Finally, by replacing Eqs. (3.1.17–3.1.19) in Eqs. (3.1.14–3.1.16), the surface drag force, heat transfer rate, and mass transfer rate are as follows:

$$Re_x \frac{l^2}{x^2} Cf = (1+\Gamma) f''(1) - \frac{\Gamma\delta}{3}\left(f''(1)\right)^3 \tag{3.1.20}$$

$$(1-\alpha t)^{\frac{1}{2}} Nu = -\left(1+\frac{4}{3}R\right)\theta'(1) \tag{3.1.21}$$

$$(1-\alpha t)^{\frac{1}{2}} Sh = -\phi'(1) \tag{3.1.22}$$

That $Re_x = \frac{\alpha l x (1-\alpha t)^{\frac{1}{2}}}{2v}$, is the local Reynolds number.

To solve the nonlinear ordinary differential Eqs. (3.1.9)–(3.1.11) concerning the boundary conditions (Eqs. (3.1.12, 3.1.13)), here the applied method is AGM. Solving nonlinear equations with AGM is as follows: suppose the nonlinear differential equation (P) is as follows:

$$p_k : f(u.u'.u''.\ldots.u^m) = 0. u = u(x) \tag{3.1.23}$$

The boundary conditions are:

$$\begin{cases} u(0) = u_0. u'(0) = u_1. \ldots. u^{(m-1)}(0) = u_{m-1} \\ u(L) = u_{L0}. u'(L) = u_{L1}. \ldots. u^{(m-1)}(L) = u_{L(m-1)} \end{cases} \tag{3.1.24}$$

Suppose the answer to the equation is as follows:

$$u(x) = \lim_{n\to\infty} \sum a_i x^i = \lim_{n\to\infty}(a_0 + a_1 x^1 + a_2 x^2 + \cdots + a_n x^n). \tag{3.1.25}$$

Now we consider the equation for two following conditions:

If x=0, the boundary conditions are:

$$\begin{cases} u(0) = a_0 = u_0 \\ u'(0) = a_1 = u_1 \\ u''(0) = a_2 = u_2 \\ \quad \ldots \end{cases} \tag{3.1.26}$$

Also, when x=L, the boundary conditions are:

$$\begin{cases} u(0) = a_0 + a_1 L + a_2 L^2 + \cdots + a_n L^n = u_{L0} \\ u'(0) = a_1 + 2a_2 L + 3a_3 L^2 + \cdots + n a_n L^{n-1} = u_{L1} \\ u''(0) = 2a_2 + 6a_3 L + 12 a_4 L^2 + \cdots + n(n-1) a_n L^{n-2} = \\ \cdots \end{cases} \quad (3.1.27)$$

After replacing the boundary conditions Eq. (3.1.27) in the Eq.(3.1.23) we will have:

$$\begin{cases} p_0 : f\big(u(0).u'(0).u''(0).\ldots.u^m(0)\big). \\ p_1 : f\big(u(L).u'(L).u''(L).\ldots.u^m(L)\big). \\ \cdots \end{cases} \quad (3.1.28)$$

We form the number of the required equations for solving the preferred nonlinear equation by using the boundary conditions according to the selection of the sentencing rate in the answer and the number of fixed indexes obtained for an answer. If we differentiate m times from Eq. (3.1.23):

$$\begin{cases} p_k' : f'\big(u'.u''.u'''.\ldots.u^{(m+1)}\big). \\ p_k'' : f''\big(u''.u'''.u^{(4)}.\ldots.u^{(m+2)}\big). \\ \cdots \end{cases} \quad (3.1.29)$$

And after replacing the boundary equation into Eq. (3.1.29), it is observed that:

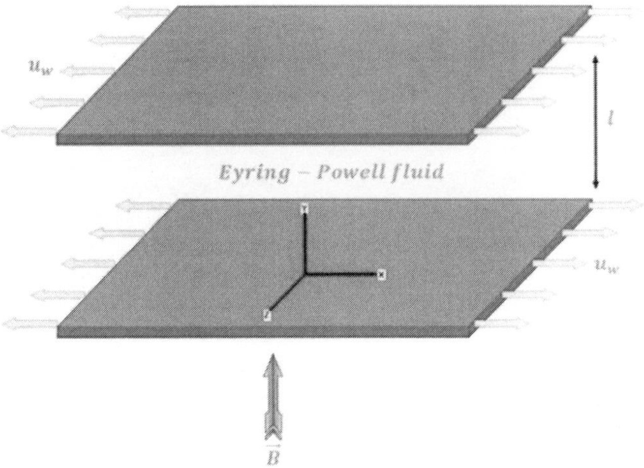

Fig. 3.1.1. Geometry of the problem.

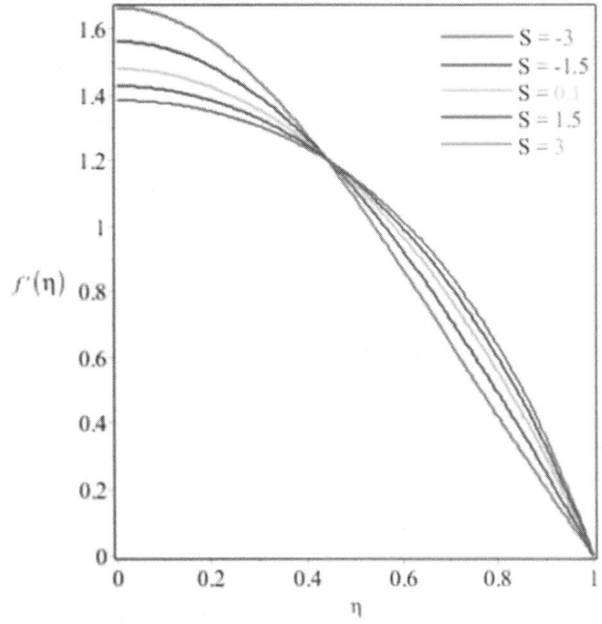

Fig. 3.1.2. Influence of S on velocity profile.

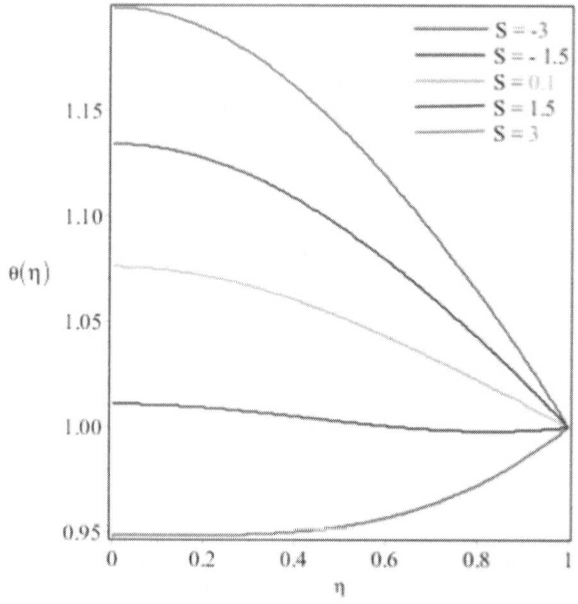

Fig. 3.1.3. Influence of S on temperature profile.

$$p_k' : \begin{cases} f'\left(u'(0).u''(0).u'''(0).\dots.u^{(m+1)}(0)\right). \\ f''\left(u''(L).u'''(L).u^{(4)}(L).\dots.u^{(m+2)}(L)\right). \\ \dots \end{cases} \tag{3.1.30}$$

$$p_k'' : \begin{cases} f''\left(u''(0).u'''(0).u^{(4)}(0).\dots.u^{(m+2)}(0)\right). \\ f''\left(u''(L).u'''(L).u^{(4)}(L).\dots.u^{(m+2)}(L)\right). \\ \dots \end{cases} \tag{3.1.31}$$

Finally, the equation's answer is calculated by obtaining the fixed indexes, which are gained by replacing the boundary conditions in the equation in the total form of the answer in the sum of the sentences to n times. After the solution, the Eqs. (3.1.9)–(3.1.11). With AGM, the nonlinear equations are as follows:

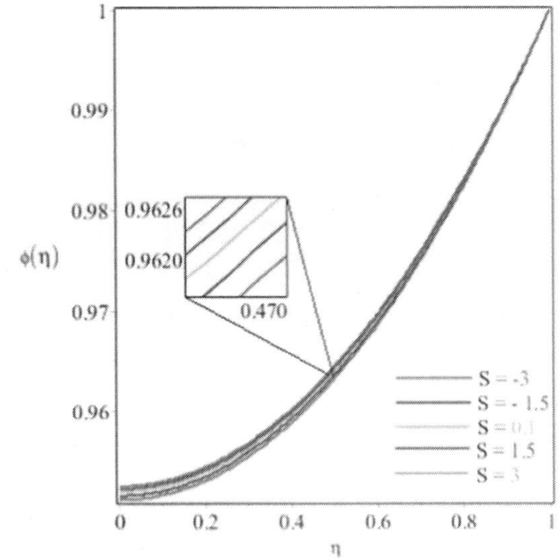

Fig. 3.1.4. Influence of S on the concentration profile.

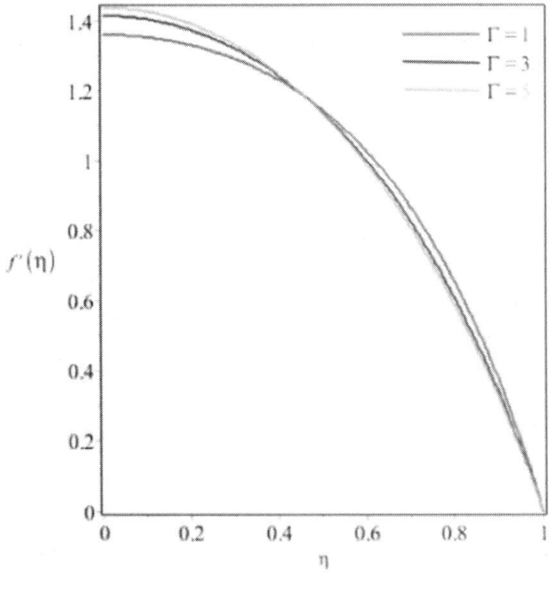

Fig. 3.1.5. Influence of Γ on the velocity profile.

$$
\begin{cases}
F\colon (1+L)\dfrac{d^4}{d\eta^4}f(\eta) - \\
S\left(\eta\dfrac{d^3}{d\eta^3}f(\eta) + 3\dfrac{d^2}{d\eta^2}f(\eta) + \dfrac{d}{d\eta}f(\eta)\dfrac{d^2}{d\eta^2}f(\eta) - f\right. \\
-L\delta\left((2\dfrac{d^2}{d\eta^2}f(\eta)(\dfrac{d^3}{d\eta^3}f(\eta))^2\right) \\
+\left(\left(\dfrac{d^2}{d\eta^2}f(\eta)\right)^2\left(\dfrac{d^4}{d\eta^4}f(\eta)\right)\right) - \left((Ha)^2\dfrac{d^2}{d\eta^2}f\right. \\
\Theta := \left(1+\dfrac{4}{3}R\right)\dfrac{d^2}{d\eta^2}\theta(\eta) \\
+PrS\left(f(\eta)\dfrac{d}{d\eta}\theta(\eta) - \eta\dfrac{d}{d\eta}\theta(\eta) + Q\theta(\eta)\right) + PrEcHa \\
\Phi := \dfrac{d^2}{d\eta^2}\emptyset(\eta) + ScS\left(f(\eta)\dfrac{d}{d\eta}\emptyset(\eta) - \eta\dfrac{d}{d\eta}\emptyset(\eta)\right) -
\end{cases}
\tag{3.1.32}
$$

The supposed answer for each of the above equations is in the form of multi sentences:

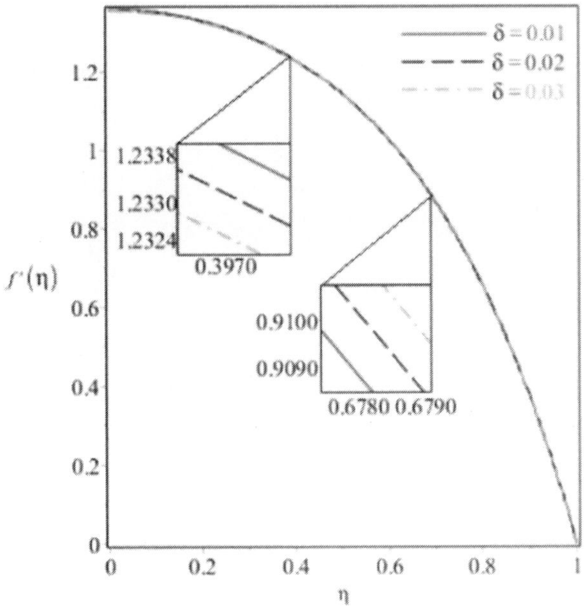

Fig. 3.1.6. Influence of δ on the velocity profile.

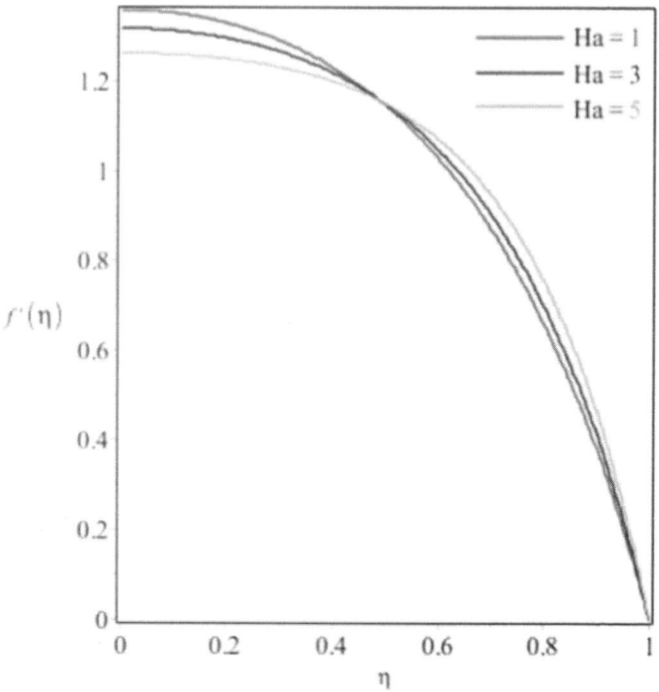

Fig. 3.1.7. Influence of Ha on the velocity profile.

$$\begin{cases} f(\eta) = a_0 + a_1\eta + a_2\eta^2 + a_3\eta^3 + a_4\eta^4 + a_5\eta^5 + a_6\eta^6 + a \\ \theta(\eta) = b_0 + b_1\eta + b_2\eta^2 + b_3\eta^3 + b_4\eta^4 + b_5\eta^5 \\ \phi(\eta) = c_0 + c_1\eta + c_2\eta^2 + c_3\eta^3 + c_4\eta^4 + c_5\eta^5 \end{cases} \quad (3.1.33)$$

By replacing the above answers (3.1.33) in Eq. (3.1.32) and according to the boundary conditions of this equation as follows, we will have:

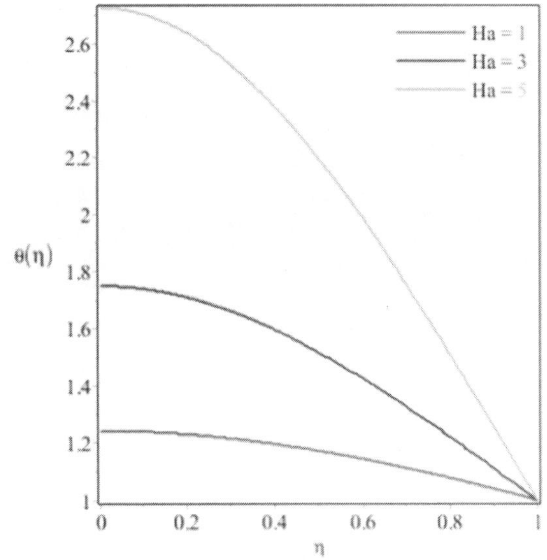

Fig. 3.1.8. Influence of Ha on the temperature profile.

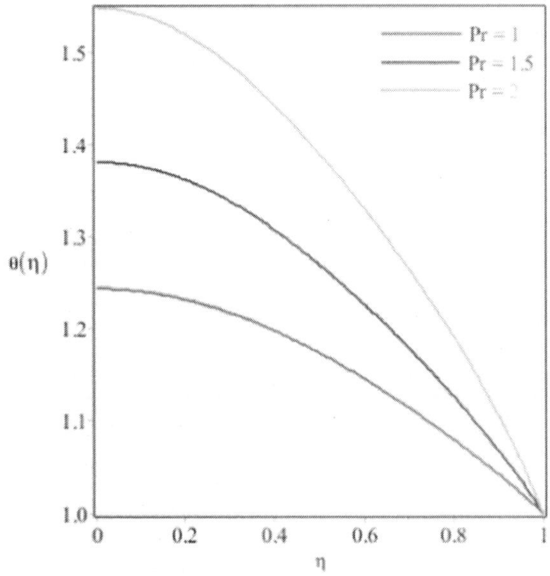

Fig. 3.1.9. Influence of Pr on the temperature profile.

$$\begin{cases} f(0) = 0 \rightarrow a_0 = 0, \\ f''(0) = 0 \rightarrow 2a_2 = 0 \\ f(1) = 1 \rightarrow a_0 + a_1 + a_2 + a_3 + a_4 + a_5 + a_6 + a_7 + a_8 \\ f'(1) = 0 \rightarrow a_1 + 2a_2 + 3a_3 + 4a_4 + 5a_5 + 6a_6 + 7a_7 + 8a \\ \theta'(0) = 0 \rightarrow b_1 = 0, \\ \theta(1) = 1 \rightarrow b_0 + b_1 + b_2 + b_3 + b_4 + b_5 + b_6 = \\ \phi'(0) = 0 \rightarrow c_1 = 0, \\ \phi(1) = 1 \rightarrow c_0 + c_1 + c_2 + c_3 + c_4 + c_5 + c_6 = \end{cases} \quad (3.1.34)$$

The boundary conditions which have been used in the answer in the main differential equations to obtain the fixed index are as follows:

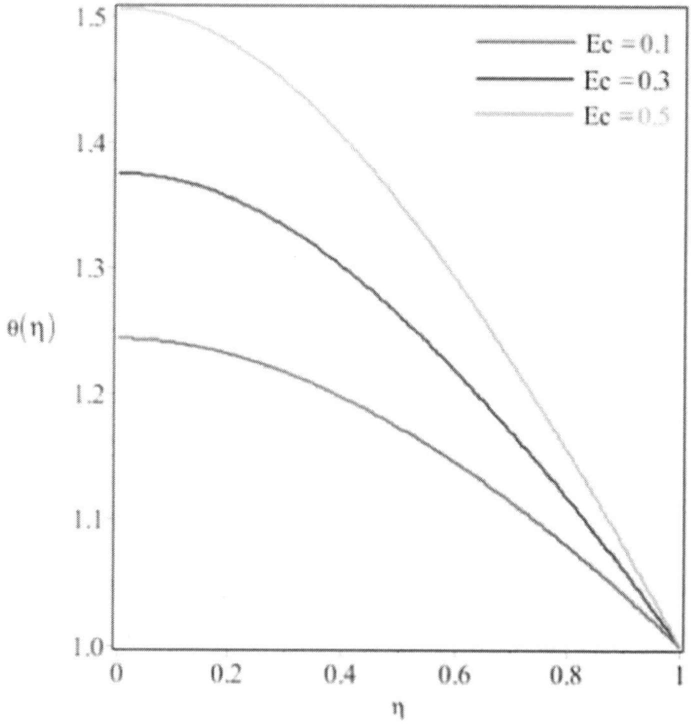

Fig. 3.1.10. Influence of Ec on the temperature profile.

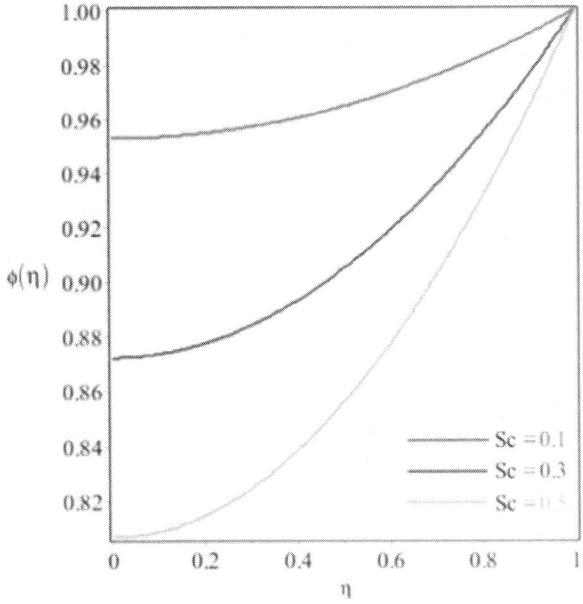

Fig. 3.1.11. Influence of Sc on the concentration profile.

$$\begin{cases} F(0), F'(0), F''(0), F'''(0), F^{(4)}(0), F^{(5)}(0), \\ \theta(0), \theta'(0), \theta''(0), \theta'''(0), \theta^{(4)}(0), \\ \phi(0), \phi'(0), \phi''(0), \phi'''(0), \phi^{(4)}(0). \end{cases} \tag{3.1.35}$$

After solving the Eqs. (3.1.34, 3.1.35), the values obtained were $a_i, b_i, c_i,$. We replaced these fixed indexes in the total form of the answer of the nonlinear differential equations (Fig. 3.1.1).

After solving nonlinear differential Eqs. (3.1.9)–(3.1.11) with the boundary condition of Eqs. (3.1.12, 3.1.13) using AGM, numerical results are presented in several figures and tables, in which the effect of change in known dimensionless numbers and various parameters on velocity $f'(\eta)$, temperature $\theta(\eta)$, and concentration $\phi(\eta)$ profiles can be concluded. Also, following this section, the behaviour of skin friction coefficient and local Nusselt and Sherwood numbers are investigated by the changes in the values of dimensionless numbers and the desired parameters.

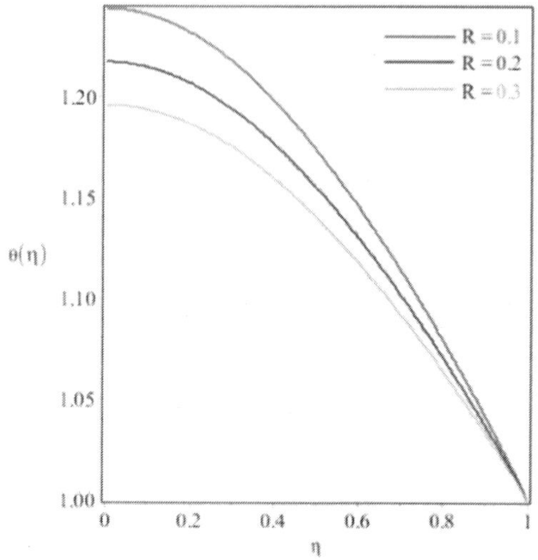

Fig. 3.1.12. Influence of R on the temperature profile.

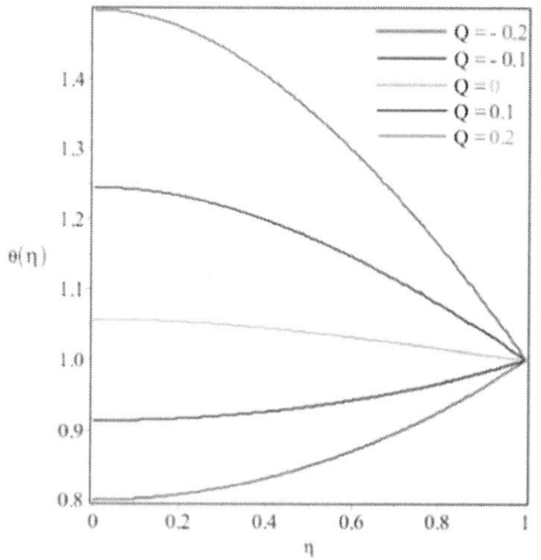

Fig. 3.1.13. Influence of Q on the temperature profile.

At the beginning of this section, in Figs. 3.1.2–3.1.4, the behaviour of velocity, temperature, and concentration profiles are shown in terms of the changes in squeezing parameter (S), respectively. In Fig. 3.1.2, observations indicate dual behaviour of the velocity function $f'(\eta)$ in terms of changes in S. It can be seen that initially, the velocity profile decreases with increasing S. Squeezing the fluid more and more causes a reduction of fluid kinematic viscosity; therefore, fluid velocity decreases. But then, with increasing η, the velocity profile shows a completely different behaviour than before and $f'(\eta)$ increases with increasing S. In Fig. 3.1.3, which shows the effect of changes in S on the temperature function $\theta(\eta)$, a type of convergence is observed in η =1. It means that during the motion of channel walls, with increasing the fluid compression, which reduces fluid molecules motion, fluid temperature distribution reduces. Also, from Fig. 3.1.4, which shows the effect of changes in the value of squeezing parameters on concentration profiles, it can be concluded that since the fluid is dense outside the flow channel, so when plates move apart and space, an increase in S leads to reduction of fluid concentration. However, in contrast, when two plates get closer together, "l decreases"—a completely inverse result occurs, that is, with increasing S, fluid concentration increases showing the accuracy of the relation of $S = \dfrac{\alpha . l}{2\nu}$.

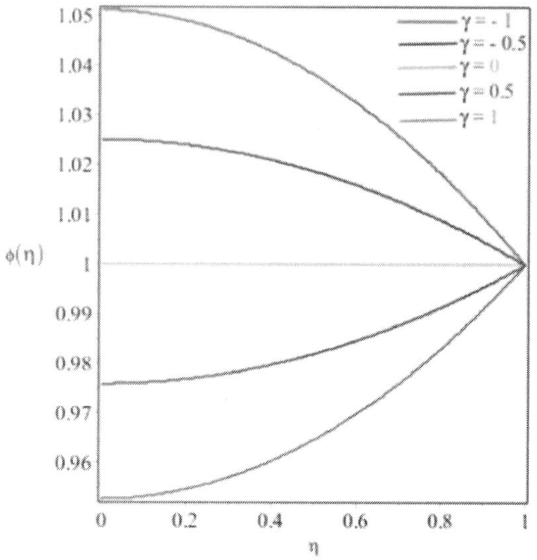

Fig. 3.1.14. Influence of γ on the concentration profile.

Table 3.1.1. Skin friction coefficient $(\mathrm{Re}_x \frac{l^2}{x^2} Cf)$, influence of S, τ, δ and Ha when $\Pr = \gamma = 1$, $Es = Sc = R = Q = 0.1$.

S	Γ	δ	Ha	$\mathrm{Re}_x \frac{l^2}{x^2} Cf$
− 3				− 3.520215
− 1.5				− 5.000191
0.1				− 6.208625
1.5				− 7.080987
3				− 7.889254
	1			− 8.374338
	3			− 14.489402
	5			− 20.475782
		0.01		− 8.374338
		0.03		− 8.267387
		0.05		− 8.153736
			1	− 8.374338
			3	− 9.397386
			5	− 11.130746

Figs. 3.1.5 and 3.1.6, which show the effect of changes in the fluid parameters on velocity profile, have been discussed. Fig. 3.1.2 illustrates that with the increase in the value of Γ, the velocity profile rises, but in the range of $0,5 \le \eta \le 1$, the increase in τ shows a reduction in velocity profile. In Fig. 3.1.3, which shows the effect of changes δ on the velocity profile, the obtained result contrasts the above results. That is, there is an inverse relation between δ and $f'(\eta)$ in the range of $0 \le \eta < 0.5$, but in the range of, $0.5 \le \eta \le 1$ they have a direct relation. The effect of changes in Hartman number (Ha) on fluid velocity and temperature profiles are shown in Figs. 3.1.7 and 3.1.8, respectively. First, in Fig. 3.1.7, the velocity function $f'(\eta)$ has dual behaviour regarding changes in the incremental value of Ha. In the range of $0 \le \eta < 0.5$ increasing Ha, $f'(\eta)$ function reduces. However, this is obvious because Ha plays the role of the magnetic field parameter here. Hence, an increase in Ha causes

a resistant drag force called Lorentz force in fluid flow field that since this force is in the opposite direction of the fluid motion, thus, reduces the velocity in the desired range with its resistance, but in the range of $0.5 \leq \eta \leq 1$ completely different behaviour of $f'(\eta)$ function with changes in Ha are seen. It means that the function also increases with increasing Ha in this range. Also, in Fig. 3.1.8, the effect of changes in incremental values of the Hartman number on temperature function $\theta(\eta)$ is shown; observations indicate an increase in the fluid temperature due to an increase in Ha. The behaviour of the fluid temperature profile in terms of changes in the incremental value of Prandtl number (Pr) is shown in Fig. 3.1.9. It can be seen that fluid temperature distribution in terms of changes of Pr has shown positive and incremental behaviour. In fact, according to the relation of $\Pr = \frac{\mu.C_p}{k}$, it can be described that, as Pr increases, the heat capacity of the fluid (C_p) increases, and it also increases the fluid temperature. The effect of increasing Eckert number (Ec) on fluid temperature distribution is shown in Fig. 3.1.10. Observations indicate direct relation between Ec and temperature function $\theta(\eta)$. It is clear that by increasing Ec, the viscous dissipation and squeezing reduces, so the collision between the fluid molecules increases, and fluid temperature rises. Fig. 3.1.11 shows the effect of changes in incremental values of Schmidt number (Sc) on fluid concentration profile. It is observed that there is an inverse relation between Sc and concentration function $(\phi(\eta))$. That is, with increasing Sc, $\phi(\eta)$ the function is also reduced. In fact, regarding the definition, $Sc = \frac{\nu}{D_m}$ it reduces the mean absorption coefficient (k^*). Therefore, increasing Sc leads to a reduction in the mass diffusion (D_m), thus reducing the Eyring-Powell fluid concentration. Fig. 3.1.12 shows the effect of changes in incremental values of the radiation parameter (R) on fluid temperature distribution. Observations indicate a decrease in fluid temperature due to an increase in R. Since, according to the relation

$R = \frac{4\sigma T_0^{\,3}}{kk^*}$, an increase in the thermal radiation results in Fig. 3.1.12 are completely reasonable and acceptable.

Table 3.1.2. Local Nusselt number $\left((1-\alpha t)^{\frac{1}{2}}.Nu\right)$, Influence of S. Ha, Pr, Ec, R, and Q when $\tau = \gamma = 1$, $Sc = 0.1$, $\delta = 0.01$.

S	Ha	Pr	Ec	R	Q	$(1-\alpha t)^{\frac{1}{2}}. Nu$
− 3						− 0.180357
− 1.5						− 0.024435
0.1						0.128587
1.5						0.254679
3						0.383983
	1					0.467646
	3					1.284601
	5					2.948141
		1				0.467646
		1.5				0.665190
		2				0.838304
			0.1			0.467646
			0.3			0.668621
			0.5			0.869595
				0.1		0.467646
				0.2		0.472686
				0.3		0.476747
					− 0.2	− 0.490503
					− 0.1	− 0.224875
					0	0.088161
					0.1	0.467646
					0.2	0.944712

Table 3.1.3. Local Sherwood number $\left((1-\alpha t)^{\frac{1}{2}}.Sh\right)$**, influence of S, Sc and** γ **when** $\tau = Ha = \Pr = 1$, $Ec = R = Q = 0.1$, $\delta = 0.01$.

S	Sc	γ	$(1-\alpha t)^{\frac{1}{2}}.Sh$
− 3			− 0.099060
− 1.5			− 0.097801
0.1			− 0.096734
1.5			− 0.095964
3			− 0.095205
	0.1		− 0.094756
	0.3		− 0.257493
	0.5		− 0.392604
		− 1	0.101130
		− 0.5	0.049720
		0	0
		0.5	− 0.048129
		1	− 0.094756

Table 3.1.4. Comparison of $f(\eta)$ between Runge-Kutta Fehlberg and AGM solution.

η	$f_{RKF}(\eta)$	$f_{AGM}(\eta)$	$\lvert Error \rvert$
0	0.00000000	0.00000000	0.00000000
0.1	0.13570349	0.13586036	1.568743×10^{-4}
0.2	0.26993918	0.27024429	3.051154×10^{-4}
0.3	0.40111409	0.40154944	4.353539×10^{-4}
0.4	0.52737667	0.52791339	5.367204×10^{-4}
0.5	0.64646613	0.64706222	5.960935×10^{-4}
0.6	0.75553451	0.75613223	5.977254×10^{-4}
0.7	0.85092861	0.85145329	5.246875×10^{-4}
0.8	0.92791395	0.92828058	3.666305×10^{-4}
0.9	0.98031233	0.98045878	1.464529×10^{-4}
1	1.00000000	1.00000000	0.00000000

It is clear that increasing Sc leads to a reduction in the mass diffusion (D_m), thus reducing the Eyring- Powell fluid concentration. Fig. 3.1.12 shows the effect of changes in incremental values of the radiation parameter (R) on fluid temperature distribution. Observations indicate a decrease in fluid temperature due to an increase in R. According to the relation $R = \frac{4\sigma T_0^{\ 3}}{kk^*}$, an increase in thermal radiation reduces the mean absorption coefficient (k*). Therefore, the results obtained in Fig. 3.1.12 are reasonable and acceptable.

Table 3.1.5. Comparison of $\theta(\eta)$ between Runge-Kutta Fehlberg and AGM solution.

η	$\theta_{RFK}(\eta)$	$\theta_{AGM}(\eta)$	$\lvert Error \rvert$
0	1.24057491	1.24436958	$3.79467073 \times 10^{-3}$
0.1	1.23757897	1.24136511	$3.78613404 \times 10^{-3}$
0.2	1.22869345	1.23245465	$3.76119513 \times 10^{-3}$
0.3	1.21421461	1.21793614	$3.72153240 \times 10^{-3}$
0.4	1.19460186	1.19826972	$3.66785455 \times 10^{-3}$
0.5	1.17042909	1.17402303	$3.59393479 \times 10^{-3}$
0.6	1.14232106	1.14579471	$3.47364692 \times 10^{-3}$
0.7	1.11087683	1.11411598	$3.23915155 \times 10^{-3}$
0.8	1.07657820	1.07933042	$2.75221559 \times 10^{-3}$
0.9	1.03967242	1.04145177	$1.77934971 \times 10^{-3}$
1	1.00000000	0.9999999991	$9.00000000 \times 10^{-10}$

Table 3.1.6. Comparison of $\phi(\eta)$ between Runge-Kutta Fehlberg and the AGM solution.

η	$\phi_{RFK}(\eta)$	$\phi_{AGM}(\eta)$	$\lvert Error \rvert$
0	0.95272406	0.95279202	$6.79564601 \times 10^{-5}$
0.1	0.95320035	0.95326834	$6.79898635 \times 10^{-5}$
0.2	0.95462834	0.95469642	$6.80841870 \times 10^{-5}$
0.3	0.95700552	0.95707374	$6.82185399 \times 10^{-5}$
0.4	0.96032813	0.96039646	$6.83380048 \times 10^{-5}$
0.5	0.96459176	0.96466003	$6.82776020 \times 10^{-5}$
0.6	0.96979240	0.96985996	$6.75606404 \times 10^{-5}$
0.7	0.97592789	0.97599285	$6.49624536 \times 10^{-5}$
0.8	0.98300000	0.98305768	$5.76754208 \times 10^{-5}$
0.9	0.99101744	0.99105727	$3.98302281 \times 10^{-5}$
1	1.00000000	1.00000000	0.00000000

The behaviour of the temperature function with changes in the values of the heat generation/absorption parameter (Q) is shown in Fig. 3.1.13. It can be seen that results obtained from the graph can be divided into two categories. Increasing the heat generation ($Q > 0$) due to an increase in an exothermic chemical reaction will increase the $\theta(\eta)$, and in contrast with increasing heat absorption ($Q < 0$), $\theta(\eta)$ shows a reduction in behaviour. Fig. 3.1.14 shows the effect of changes in chemical reaction parameters on concentration function $\phi(\eta)$. Observations show that by increasing the distractive chemical reaction ($\gamma > 0$), fluid concentration is decreased, and by increasing the generative chemical reaction ($\gamma < 0$), fluid concentration is increased. In general, it can be said that increasing the chemical reaction parameter causes reduced concentration, and decreasing it will also increase the concentration. In the following section, the discussion of results obtained from the effect of changes in the values of various parameters and different dimensionless numbers on skin friction coefficient, local Nusselt, and Sherwood numbers, as presented in Tables 3.1.1–3.1.3, are done.

First, Table 3.1.1, which describes the effect of fluid parameters, squeezing parameters, and Hartman number on the skin friction coefficient, has been investigated. It is observed that the skin friction coefficient increases by increasing Γ, but it is decreased by increasing δ. Also, the skin friction coefficient indicates positive and incremental behaviour by changes in S and Ha. Table 3.1.2 shows the effect of changes in the squeezing parameter, the Hartman number, the Prandtl number, the Eckert number, the thermal radiation parameter, and the heat generation/absorption parameter on the local Nusselt number. Results show positive and incremental behaviour of heat transfer rate in terms of changes in incremental values of S, Ha, Pr, Ec, R, and Q. Table 3.1.3 presents the behaviour of the local Sherwood number regarding changes in the squeezing parameter, the Schmidt number, and the chemical reaction. The results show a decrease in mass transfer rate due to the increase in S and an increase in mass transfer rate due to the increase in Sc and γ.

At the end of this section, in Tables 3.1.4–3.1.6, a comparison has been made for the $f(\eta)$, $\theta(\eta)$, $\phi(\eta)$ functions between Runge-Kutta Fehlberg and AGM solution, respectively. Results obtained show that AGM error is very little compared to RKF. Hence, AGM is a high-accuracy method in nonlinear differential equations solution, especially in the field of heat and mass transfer.

This study has been done on the heat and mass transfer in an unsteady two-dimensional squeezing flow of magneto hydromagnetic (MHD) radiative non-Newtonian Eyring-Powell fluid between two parallel infinite plates in the presence of heat generation/absorption, thermal radiation, and Joule heating effect. After the transformation of PDEs into ODEs and solving simplified governing equations using AGM, the results are as follows:

1) The velocity profile first is increased with increasing Γ and then is decreased, but with increasing δ, S, and Ha first is decreased and then is increased.

2) The temperature profile decreases with increasing S and R and decreases with decreasing Ha, Pr, and Ec. Also, the temperature profile increases with increasing $Q > 0$ and decreases with increasing $Q < 0$.

3) The concentration profile decreases with increasing S when the plates are spaced apart, but the concentration profile increases with increasing S when the plates get closer. Also, the concentration profile decreases with increasing γ and Sc.

4) The skin friction coefficient has a direct relation with Γ, S, and Ha and inverse relation with δ.

5) The local Nusselt numbers relate to S, Ha, Pr, Ec, R, and Q.

6) The local Sherwood number has an inverse relation with S and a direct relation with Sc and γ.

7) The comparison between results obtained from solving nonlinear differential equations governing the problem of the two methods demonstrates the high accuracy of AGM in nonlinear differential equation solutions, especially in the field of heat and mass transfer.

3.2. Condensation Film on an Inclined Rotating Disk

The steady three-dimensional problem of condensation film on an inclined rotating disk is illustrated in this problem. A similar transform reduces the governing nonlinear partial differential equations to the nonlinear ordinary differential equations system. The analytical solution applied to solve this system is AGM. Fig. 3.2.1 shows a disk rotating in its plane with angular velocity Ω. The angle between the horizontal axis and the disk is β. With W velocity, a fluid film of thickness t is formed by spraying. We assume the disk radius is larger than the film thickness, so the end effects can be ignored. Vapour shear effects at the vapour interface and on the fluids are usually also unimportant. The gravitational acceleration, g, acts in a downward direction.

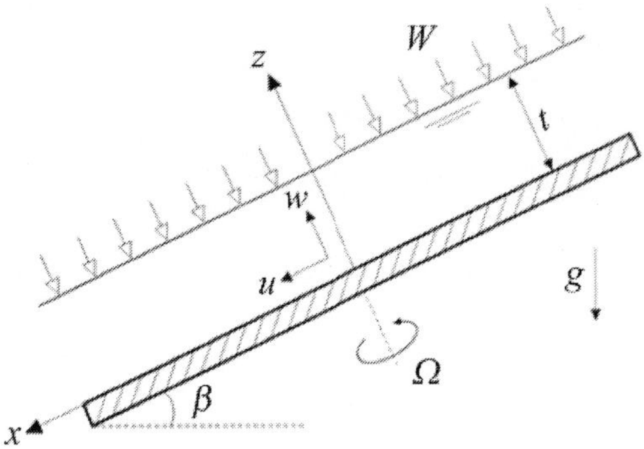

Fig. 3.2.1. Schematic diagram of the problem.

The temperature on the disk is T_w, and the temperature on the film surface is T_0. Besides, the ambient pressure on the film surface is constant at P_0, and we can safely say the pressure is a function of z only. With neglecting viscous dissipation, the continuity, momentum, and energy equations for the steady-state are given in the following:

$$u_x + u_y + u_z = 0, \tag{3.2.1}$$

$$uu_x + vu_y + wu_z = \vartheta(u_{xx} + v_{yy} + u_{zz}) + \bar{g}\sin\beta, \tag{3.2.2}$$

$$uv_x + vv_y + wv_z = \vartheta(v_{xx} + v_{yy} + v_{zz}), \tag{3.2.3}$$

$$uw_x + vw_y + ww_z = \vartheta(w_{xx} + w_{yy} + w_{zz}) - \bar{g}\cos\beta - \frac{p_z}{\rho}, \tag{3.2.4}$$

$$uT_x + vT_y + wT_z = \alpha(T_{xx} + T_{yy} + T_{zz}). \tag{3.2.5}$$

In the above equations, u, v, and w indicate the velocity components in the x, y, and z directions, T denotes the temperature, and ρ, ϑ, and α denote the density, kinematic viscosity, and thermal diffusivity of the fluid, respectively. Assuming zero slip on the disk and zero shear stress on the film surface, the boundary conditions are:

$$\begin{aligned} &u=-\Omega y,\ v=\Omega x,\ w=0,\ T=T_w \ at\ z=0 \\ &u_z=0,\ v_z=0,\ w=-W,\ T=T_0,\ P=P_0\ at\ z=1 \end{aligned} \tag{3.2.6}$$

For this problem, Wang introduced the following transform:

$$\begin{aligned} &u=-\Omega y g(\eta)+\Omega x f'(\eta)+\overline{g}k(\eta)\sin\frac{\beta}{\Omega}, \\ &v=\Omega x g(\eta)+\Omega y f'(\eta)+\overline{g}s(\eta)\sin\frac{\beta}{\Omega}, \\ &w=-2\sqrt{\Omega\vartheta}f(\eta), \\ &T=(T_0-T_w)\theta(\eta)+T_w, \end{aligned} \tag{3.2.7}$$

Where

$$\eta=z\sqrt{\frac{\Omega}{\vartheta}} \tag{3.2.8}$$

Continuity Eq. (3.2.1) is satisfied automatically. Eqs. (3.2.2) and (3.2.3) can be written as follows:

$$f'''-(f')^2+g^2+2ff''=0, \tag{3.2.9}$$

$$g''-2gf'+2fg'=0, \tag{3.2.10}$$

$$k''-kf'+sg+2fk'+1=0, \tag{3.2.11}$$

$$s''-kg-sf'+2fs'=0, \tag{3.2.12}$$

If the temperature is a function of the distance z only, Eq. (3.2.5) becomes:

$$\theta''+2\Pr f\,\theta'=0, \tag{3.2.13}$$

Where $Pr = \frac{\vartheta}{\alpha}$, is the Prandtl number. The boundary conditions for Eqs. (3.2.9)-(3.2.13) are:

$$\begin{aligned} &f(0)=0, \quad f'(0)=0, \quad f'''(\delta)=0 \\ &g(0)=1, \quad g'(\delta)=0 \\ &k(0)=0, \quad k'(\delta)=0 \\ &s(0)=0, \quad s'(\delta)=0 \\ &\theta(0)=0, \quad \theta(\delta)=1 \end{aligned} \tag{3.2.14}$$

And δ is the constant normalised thickness as:

$$\delta = t\sqrt{\frac{\Omega}{\vartheta}}, \tag{3.2.15}$$

Which is known through the condensation or spraying velocity by:

$$f(\delta) = \frac{W}{2\sqrt{\Omega\vartheta}} = \alpha. \tag{3.2.16}$$

After the flow field is found, the pressure can be obtained by integrating Eq. (3.2.4)

$$\begin{aligned} &\mathrm{p(z)}=\mathrm{P}_0+\rho\{\vartheta[\mathrm{w_z(z)}-\mathrm{w_z(t)}] \\ &-\frac{\left[w^2(z)-w^2(t)\right]}{2}-\bar{g}(\mathrm{z-t})\cos\beta\} \end{aligned} \tag{3.2.17}$$

The exact solution to Eqs. (3.2.13) and (3.2.14) can be expressed in double integrals, but it is easier to integrate numerically for a similarity temperature profile. For a given δ, the function $f(\eta)$ was determined previously.

For $\Pr = 0,$ using $\theta(\delta)=1$ the exact solution is:

$$\theta'(0) = \frac{1}{\delta}. \tag{3.2.18}$$

Eq. (3.2.18) serves as an asymptotic limit for small δ. The decrease of $\theta'(0)$ for increasing δ is not monotonic, as seen from the waviness of the curves for large Pr.

$$Nu = \frac{(\frac{\partial T}{\partial z})_w}{T_0 - T_w} = \delta\theta'(0). \tag{3.2.19}$$

In the condensation of saturated vapour, the latent heat is a factor. Consider a film of radius r. The latent heat input is equal to the heat transferred to the disk plus the heat transported by the radial flow:

$$HW\pi r^2 = k\frac{\partial T}{\partial z}(0)\pi r \int \rho C_p [T_0 - T(z)]V(z)2\pi r dz \tag{3.2.20}$$

Here, H is the latent heat per mass, C_p is the specific heat, and $V = \Omega r f'$, is radial velocity. Using, $T = (T_0 - T_w)\theta(\eta) + T_w$, and Eq. (3.2.13), Eq. (3.2.20) is simplified to:

$$\frac{H_v}{k(T_0 - T_w)} = \frac{\theta'(0) - \theta'(\delta)/2}{\alpha} = j \tag{3.2.21}$$

To show the main idea of AGM, the general form of the differential equation that governs the problem is considered as follows:

$$P: f(u, u', u''', \ldots, u^{(m)}) = 0; u = u(x) \tag{3.2.22}$$

With boundary conditions:

$$\begin{Bmatrix} u(x) = u_0, u'(x) = u_1, \ldots, u^{(m-1)}(x) = u_{m-1} \\ at\ x = 0 \\ u(x) = u_{L0}, u'(x) = u_{L1}, \ldots, u^{(m-1)}(x) = u_{Lm-1} \\ at\ x = L \end{Bmatrix} \tag{3.2.23}$$

To solve this equation, we consider the answer as a polynomial of degree n with unknown constant coefficients:

$$u(x) = a_0 + a_1x^1 + a_2x^2 + \ldots + a_nx^n = \sum_{i}^{n} = 0^{a_ix^i} \tag{3.2.24}$$

The greater the number of sentences in the above series, the more the final answer is more precise.

To find the coefficients of a polynomial of degree n, we face n+1 unknown coefficients, where their determination requires the same number of equations. The boundary conditions of the problem achieve a number of these equations:

$$\begin{Bmatrix} u(0) = a_0 = u_0 \\ u'(0) = a_1 = u_1 \\ u'''(0) = a_2 = u_2 \\ \cdot \quad \cdot \quad \cdot \\ \cdot \quad \cdot \quad \cdot \\ \cdot \quad \cdot \quad \cdot \end{Bmatrix} \tag{3.2.25}$$

$$\begin{Bmatrix} u(L) = a_0 + a_1L + a_2L^2 + \ldots + a_nL^n = uL_0 \\ u'(L) = a_1 + 2a_2L + 3a_3L^2 + \ldots + na_nL^{n-1} = u_{L1} \\ u''(L) = 2a_2 + 6a_3L + 12a_4L^2 + \ldots + n(n-1)a_nL^{n-2} = u^{L2} \\ \cdot \quad \cdot \quad \cdot \quad \cdot \quad \cdot \quad \cdot \\ \cdot \quad \cdot \quad \cdot \quad \cdot \quad \cdot \quad \cdot \\ \cdot \quad \cdot \quad \cdot \quad \cdot \quad \cdot \quad \cdot \end{Bmatrix} \tag{3.2.26}$$

Further, to achieve the remaining unknown coefficients and create a system with n+1 equations and n+1 unknowns, we will first place series (3.2.24) in the equation Eq. (3.2.22). Then we will place the boundary points in the equation and its derivatives to create new equations.

$$P:\begin{Bmatrix} f\,(u(0),u'(0),\ldots,u^{(m-1)}(0)) \\ f\,(u(L),u'(L),\ldots,u^{(m-1)}(L)) \end{Bmatrix}$$
$$P':\begin{Bmatrix} f\,(u'(0),u''(0),\ldots,u^{(m-1)}(0)) \\ f\,(u'(L),u''(L),\ldots,u^{(m-1)}(L)) \end{Bmatrix} \tag{3.2.27}$$
$$P'':\begin{Bmatrix} f\,(u''(0),u'''(0),\ldots,u^{(m-1)}(0)) \\ f\,(u''(L),u'''(L),\ldots,u^{(m-1)}(L)) \end{Bmatrix}$$

Finally, series constant coefficients and, thus, the solution to the problem will be determined by solving the equations.

The final solution is assumed to be in the form of polynomials with constant coefficients. As shown in Fig. 3.2.2, choosing a polynomial of the 12th degree yields an accurate and appropriate result for the current study.

$$\begin{aligned}
f(x) &= \sum_{i}^{12}{}_{1}(a_i x^{i}) = a_{12}x^{12} + a_{11}x^{11} + \ldots + a_2 x^{2} + a_1 x^{1} + a_0,\\
g(x) &= \sum_{i}^{12}{}_{1}(b_i\, x^{i}) = b_{12}x^{12} + b_{11}x^{11} + \ldots + b_2 x^{2} + b_1 x^{1} + b_0,\\
k(x) &= \sum_{i}^{12}{}_{1}(c_i\, x^{i}) = c_{12}x^{12} + c_{11}x^{11} + \ldots + c_2 x^{2} + c_1 x^{1} + c_0,\\
s(x) &= \sum_{i}^{12}{}_{1}(d_i\, x^{i}) = d_{12}x^{12} + d_{11}x^{11} + \ldots + d\, x^{2} + d_1 x^{1} + d_0,\\
\theta(x) &= \sum_{i}^{12}{}_{1}(e_i\, x^{i}) = e_{12}x^{12} + e_{11}x^{11} + \ldots + e_2 x^{2} + e_1 x^{1} + e_0
\end{aligned} \tag{3.2.28}$$

Applying the boundary conditions of the problem to the above series results in 11 equations for determining the unknown coefficients:

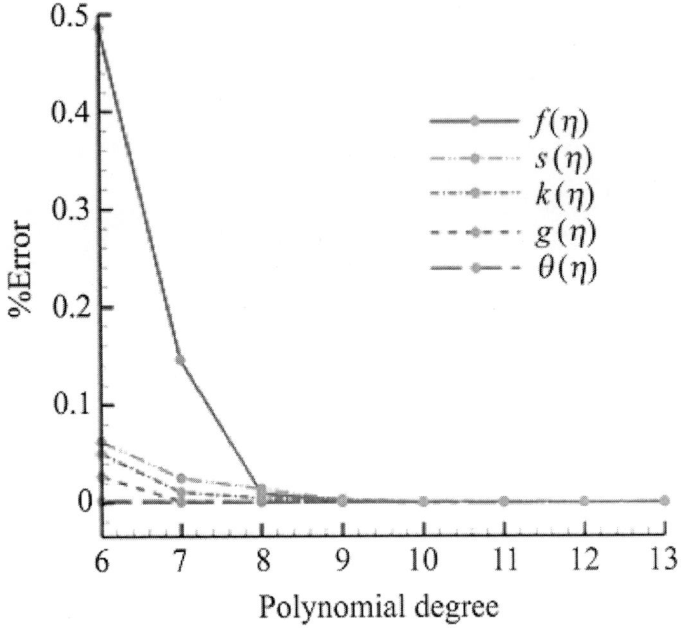

Fig. 3.2.2. %Error of $f, g, k, s \; and \; \theta$ **at different polynomial degrees when** $pr = 0.7 \; and \; \delta = 0.5.$

$$f(0) = 0 \rightarrow a_0 = 0 \tag{3.2.29}$$

$$f'(0) = 0 \rightarrow a_1 = 0 \tag{3.2.30}$$

$$\begin{aligned} f''(\delta) = 0 \rightarrow 132\delta^{10}a_{12} + 110\delta^{9}a_{11} + 90\delta^{8}a_{10} \\ +72\delta a_9 + 56\delta^6 a_8 + 42\delta^5 a_7 + \\ 30\delta^4 a_6 + 20\delta^3 a_5 + 12\delta^2 a_4 + 6\delta a_3 + 2a_2 = 0 \end{aligned} \tag{3.2.31}$$

$$g(0) = 1 \rightarrow b_0 = 1 \tag{3.2.32}$$

$$g'(\delta)=0 \rightarrow 12\delta^{11}b_{12}+11\delta^{10}b_{11}+10\delta^{9}b_{10}$$
$$+9\delta^{8}b_{9}+8\delta^{7}b_{8}+7\delta^{6}b_{7}+ \tag{3.2.33}$$
$$6\delta^{5}b_{6}+5\delta^{4}b_{5}+4\delta^{3}b_{4}+3\delta^{2}b_{3}+2\delta b_{2}+b_{1}=0$$

$$k(0)=0 \rightarrow c_{0}=1 \tag{3.2.34}$$

$$k'(\delta)=0 \rightarrow 12\delta^{11}c_{12}+11\delta^{10}c_{11}+10\delta^{9}c_{10}$$
$$+9\delta^{8}c_{9}+7\delta^{7}c_{8}+6\delta^{6}c_{7}+ \tag{3.2.35}$$
$$5\delta^{5}c_{6}+4\delta^{4}c_{5}+3\delta^{3}c_{4}+3\delta^{2}c_{3}+2\delta c_{2}+c_{1}=0$$

$$s(0)=0 \rightarrow d_{0}=0 \tag{3.2.36}$$

$$s'(\delta)=0 \rightarrow 12\delta^{11}d_{12}+11\delta^{10}d_{11}+10\delta^{9}d_{10}$$
$$+9\delta^{8}d_{9}+8\delta^{7}d_{8}+7\delta^{6}d_{7}+ \tag{3.2.37}$$
$$6\delta^{5}d_{6}+5\delta^{4}d_{5}+4\delta^{3}d_{4}+3\delta^{2}d_{3}+2\delta d_{2}+d_{1}=0$$

$$\theta(0)=0 \rightarrow e_{0}=0 \tag{3.2.38}$$

$$\theta'(\delta)=1 \rightarrow \delta^{12}e_{12}+\delta^{11}e_{11}+\delta^{10}e_{10}$$
$$+\delta^{9}e_{9}+\delta^{8}e_{8}+\delta^{7}e^{7}+ \tag{3.2.39}$$
$$\delta^{6}e_{6}+\delta^{5}e_{5}+\delta^{4}e_{4}+\delta^{3}e_{3}+\delta^{2}e_{2}+\delta e_{1}+e_{0}=1$$

New equations are required to determine the remaining unknowns. To this end, by substituting the assumed solutions and then the boundary points into the main equation, we obtain the following:

$$P(\eta = 0) = 0 \rightarrow$$

$$\begin{aligned}
&4a_0a_2 - a_1^{\ 2} - b_0^{\ 2} + 6a_3 = 0 \\
&2a_0b_1 - 6a_1b_0 + 2b_2 = 0 \\
&2a_0c_1 - a_1c_0 + d_0b_0 + 2c_2 + 1 = 0 \\
&2a_0d_1 - a_1d_0 - b_0c_0 + 2b_2 = 0 \\
&1.4a_0e_1 + 2e_2 = 0
\end{aligned} \tag{3.2.40}$$

$$P'(\eta = 0) = 0 \rightarrow$$

$$\begin{aligned}
&12a_0a_3 - 2b_0b_1 + 24a_4 = 0 \\
&4a_0b_2 - 4a_2b_0 + 6b_3 = 0 \\
&4a_0c_2 - a_1c_1 - 2a_2c_0 + b_0d_1 + d_0b_1 + 6c_3 = 0 \\
&4a_0d_2 + a_1d_1 - 2a_2d_0 - b_0c_1 - c_0b_1 - 6d_3 = 0 \\
&2.8a_0e_2 + 1.4a_0e_1 + 6e_3 = 0
\end{aligned} \tag{3.2.41}$$

Following the same procedure, by applying the boundary conditions to the main equation and its higher order derivatives, a system of equations consisting of 60 equations and 60 unknowns was obtained, through the solution of which one may determine all the constant coefficients of the series, which are the solutions to the problem. For the values Pr=0.7 and δ=0.5, the yielded solution is:

$$
\begin{aligned}
f(\eta) &= 0.000001011729441\eta^{12} - 0.000002929460859\eta^{11} \\
&+0.00001325221023\eta^{10} - 0.00001464931110\eta^{9} \\
&-0.00003091887095\eta^{8} + 0.0004570750926\eta^{7} \\
&-0.001340379877\eta^{6} - 0.0001031285153\eta^{5} \\
&+0.006555167023\eta^{4} - 0.16666666\eta^{3} \\
&+0.2412683778\eta^{2} \\
g(\eta) &= -0.000012208780069\eta^{12} + 0.00004488203203\eta^{11} \\
&-0.00009081928830\eta^{10} + 0.0001078128991\eta^{9} \\
&-0.0004796687285\eta^{8} + 0.001606089647\eta^{7} \\
&-0.002724634915\eta^{6} + 0.005244133619\eta^{5} \\
&-0.08649644236\eta^{4} + 0.1608455852\eta^{3} \\
&-0.07866200428\eta + 1 \\
k(\eta) &= -0.00001734806557\eta^{12} + 0.00001701458056\eta^{11} \\
&-0.00007251186687\eta^{10} + 0.0003269115881\eta^{9} \\
&-0.000412056255\eta^{8} + 0.0001444379910\eta^{7} \\
&-0.0008490558309\eta^{6} + 0.003828189247\eta^{5} \\
&-0.0002628319267\eta^{4} + 0.006682563687\eta^{3} \\
&-0.5\eta^{2} + 0.494113778\eta \\
s(\eta) &= +0.00003302754465\eta^{12} - 0.00001781743806\eta^{11} \\
&+0.00003508011197\eta^{10} - 0.000005449217763\eta^{9} \\
&+0.00004272225942\eta^{8} - 0.0003123466090\eta^{7} \\
&-0.0000438053211\eta^{6} + 0.002634806476\eta^{5} \\
&-0.04490566501\eta^{4} + 0.08235229645\eta^{3} - 0.04009538212\eta \\
\theta(\eta) &= -0.0000002351113387\eta^{12} + 0.000002015240046\eta^{11} \\
&-0.00001583402893\eta^{10} + 0.0001131002067\eta^{9} \\
&-0.0007270727571\eta^{8} + 0.004154701469\eta^{7} \\
&-0.02077350734\eta^{6} + 0.08902931719\eta^{5} \\
&-0.3179618471\eta^{4} + 0.9084624203\eta^{3} \\
&-1.9466705186\eta^{2} + 2.781007409\eta
\end{aligned}
\qquad (3.2.42)
$$

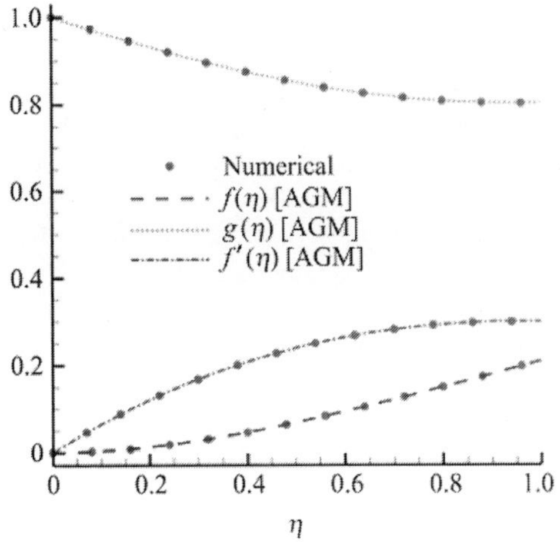

Fig. 3.2.3. The normalised radial velocity profiles for the rotating flow obtained by AGM in comparison with the numerical solution when $\Pr = 0.7 \; and \; \delta = 0.5.$

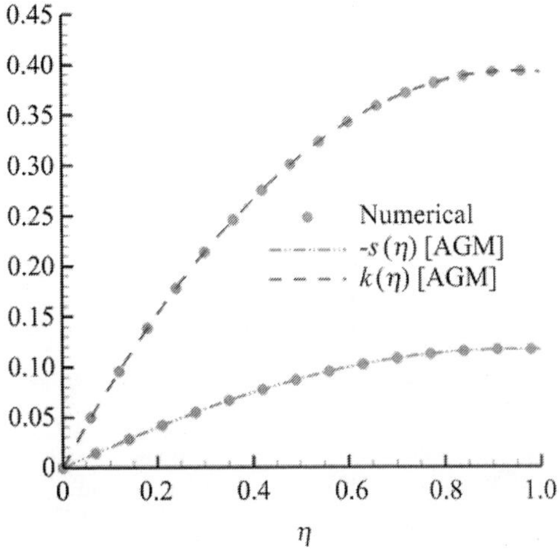

Fig. 3.2.4. The normalised velocity profiles for the draining flow (k(η)) and lateral flow (s(η)) obtained by AGM in comparison with the numerical solution when $\Pr = 0.7 \; and \; \delta = 0.5.$

The objective of the present study is to apply the new efficient AGM to obtain an explicit analytic solution for the problem of condensation film on an inclined rotating disk. Fig. 3.2.2 shows the average % error for different functions compared to numerical solutions (4th order Runge-Kutta). As can be seen, by selecting a polynomial of order 12, the error of this method is minimised, and the solution converges. Figs. 3.2.3 and 3.2.4 and Tables 3.2.1 and 3.2.2 ensure that the results obtained from analytic and numerical solutions are in excellent agreement. In Tables 3.2.1 and 3.2.2 error is introduced as follows:

$$\%Error = \left| \frac{f(\eta)_{Num} - f(\eta)_{AGM}}{f(\eta)_{Num}} \right| \times 100 \tag{3.2.43}$$

The flow rate, due to the tilting of the disk, includes two components. $\int_0^\delta k(\eta)d\eta$, is the draining flow rate per width in the x direction, normalised by $\bar{g}\sin\beta\sqrt{\vartheta/\Omega^3}$. Furthermore, lateral flow in y direction is represented by $\int_0^\delta s(\eta)d\eta$.

Table 3.2.1. Comparison between numerical results and AGM solution for f, f′, and g when Pr=0.7 and δ=0.5.

η	$f(\eta)$			$f'(\eta)$			$g(\eta)$		
	AGM	NUM	%Error	AGM	NUM	%Error	AGM	NUM	%Error
0	0	0	0	0	0	0	1	1	0
0.05	0.000582	0.000582	8.39E-04	0.022880	0.02288	5.03E-04	0.996086	0.996086	1.06E-06
0.1	0.002247	0.002247	1.43E-04	0.043279	0.04328	5.13E-04	0.992286	0.992286	5.70E-07
0.15	0.004869	0.004869	7.13E-05	0.061218	0.061218	3.08E-04	0.988700	0.988700	4.44E-07
0.2	0.008328	0.008328	7.77E-05	0.076714	0.076714	7.70E-05	0.985417	0.985417	7.10E-07
0.25	0.012500	0.012500	7.01E-05	0.089785	0.089785	2.11E-04	0.982514	0.982514	7.27E-07
0.3	0.017266	0.017266	6.84E-05	0.100448	0.100448	4.28E-04	0.980055	0.980055	8.35E-07
0.35	0.022505	0.022505	7.20E-05	0.108718	0.108718	1.48E-04	0.978090	0.9780900	1.00E-06
0.4	0.028099	0.028098	6.32E-05	0.114601	0.11461	3.36E-05	0.976660	0.9766608	7.52E-07
0.45	0.033927	0.033927	4.77E-05	0.118137	0.118137	7.30E-05	0.975792	0.9757920	1.68E-07
0.5	0.039873	0.039873	6.99E-05	0.119310	0.11931	2.03E-04	0.975501	0.9755007	1.09E-06

Table 3.2.2. Comparison between numerical results and AGM solution for k, s, and θ when Pr=0.7 and δ=0.5.

η	$k(\eta)$			$-s(\eta)$			$\theta(\eta)$		
	AGM	NUM	%Error	AGM	NUM	%Error	AGM	NUM	%Error
0	0	0	0	0	0	0	0	0	0
0.05	0.023457	0.023457	6.61E-06	0.001995	0.001995	1.99E-04	0.100280	0.100279	3.40E-05
0.1	0.044418	0.044418	8.13E-06	0.003932	0.003932	8.32E-05	0.200554	0.200554	5.20E-06
0.15	0.062890	0.062890	9.26E-06	0.005759	0.005760	6.74E-05	0.300813	0.300813	1.80E-07
0.2	0.078877	0.078877	1.09E-05	0.007431	0.007431	8.21E-05	0.401036	0.401036	6.47E-07
0.25	0.092385	0.092385	1.28E-05	0.008910	0.008910	8.61E-05	0.501201	0.501201	5.23E-07
0.3	0.103421	0.103421	1.48E-05	0.010163	0.010163	9.52E-05	0.601278	0.601278	6.53E-07
0.35	0.111991	0.111991	1.75E-05	0.011163	0.011163	1.09E-04	0.701234	0.701234	1.80E-07
0.4	0.118102	0.118102	1.98E-05	0.011891	0.011891	1.09E-04	0.801032	0.801032	1.61E-06
0.45	0.121762	0.121762	2.09E-05	0.012333	0.012333	9.93E-05	0.900634	0.900634	4.16E-06
0.5	0.122980	0.122980	2.57E-05	0.012481	0.012481	1.36E-04	1	1	2.52E-08

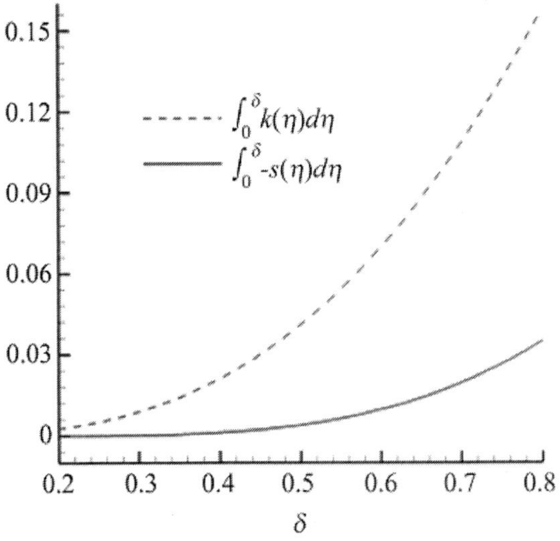

Fig. 3.2.5. The normalised draining flow rate in the x direction and induced lateral flow rate in the y direction versus normalised film thickness (δ)

As Fig. 3.2.5 shows, both flow rates increase by increasing thickness; however, changes seem more sensible in $\int_0^\delta k(\eta)$.

The shear stress applied on the disk in the x and y directions is $k'(0)$ and $s'(0)$, respectively. It can be seen from Fig. 3.2.6 that the $k'(0)$ increases linearly as thickness increases, and the increase is also larger than $s'(0)$. The normalised temperature profile for different Prandtl numbers is illustrated in Fig. 3.2.7. The value of the Prandtl number ranges from about zero for liquid metals to seven for water and 100 or higher for some specific liquid metals. $\theta'(0)$, represents the local heat on the disk normalised by $k(T_0 - T_w)\sqrt{\Omega/\vartheta}$, in which k represents thermal conductivity coefficient. This parameter is illustrated in Fig. 3.2.8 against different Prandtl numbers. Fig. 3.2.9 shows that as long as a constant value is reached, the reduction in latent heat continues as the thickness increases; however, it increases with the Prandtl number. Although the effect of the Nusselt number is not sensible in small Prandtl numbers, it increases with the increase in thickness (Fig. 3.2.10); ultimately, in thicknesses of about zero, it reaches a maximum of 1.

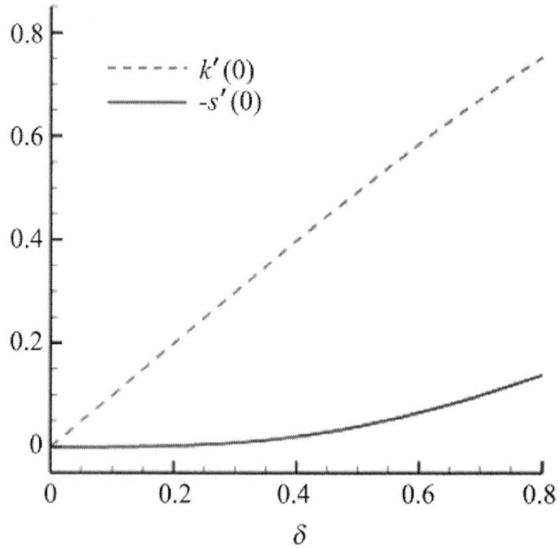

Fig. 3.2.6. The normalised shear stress on the disk in x and y directions versus normalised film thickness (δ).

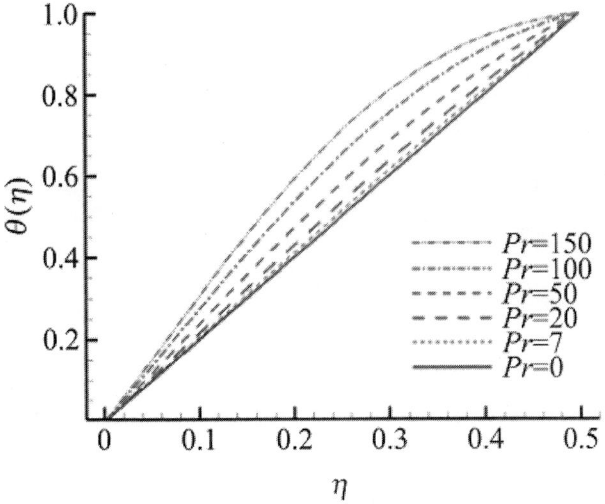

Fig. 3.2.7. The normalised temperature profiles $(\theta(\eta))$ obtained by AGM for different Pr numbers when δ= 0.5.

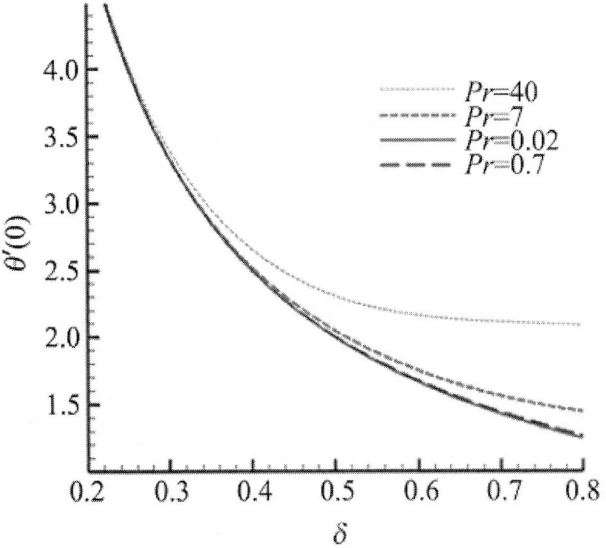

Fig. 3.2.8. The temperature gradient on the disk $(\theta'(0))$ obtained by AGM for different Pr numbers versus normalised film thickness (δ).

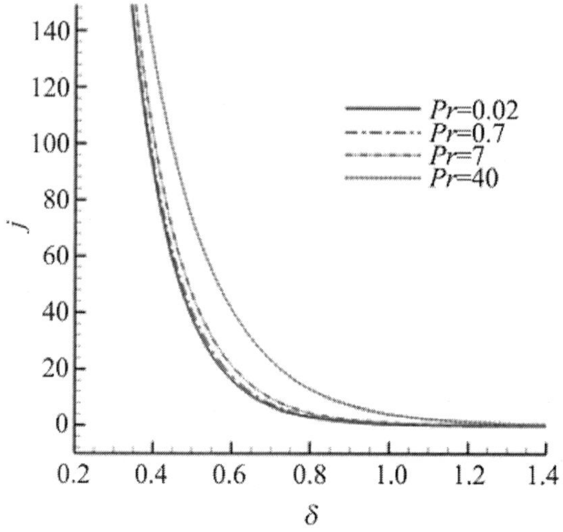

Fig. 3.2.9. Latent heat (j(α)) in different Pr numbers versus normalised film thickness (δ).

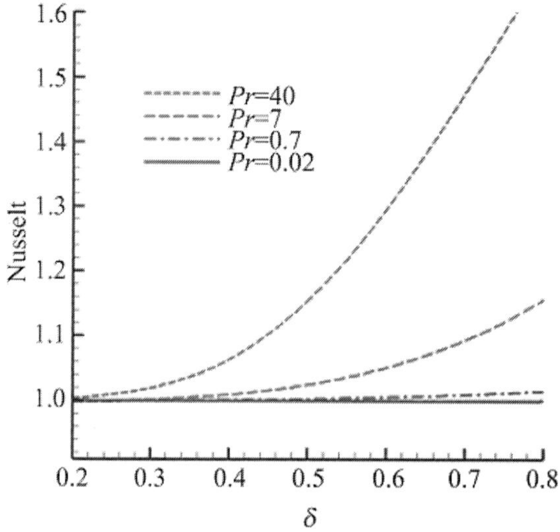

Fig. 3.2.10. The Nusselt number in different Pr numbers versus normalised film thickness (δ).

The present study used AGM to find an analytic solution for a system of nonlinear ordinary differential equations obtained by applying similarity transformation on the steady three-dimensional problem of condensation film on an inclined rotating disk. Additionally, the solution was validated by numerical results obtained from the 4th-order Runge-Kutta. Some conclusions were summarized as follows:

1. Draining and lateral flow rates increase as normalised thickness increases.

2. Increasing normalised thickness increases normalised shear stress on the disk in both the x and y directions.

3. Decrease in normalised thickness decreases the Nusselt number until the constant value of 1 is reached. However, for small Prandtl numbers, these changes are not as noticeable as large ones.

4. Increasing normalised thickness leads to a decrease in local heat transfer on the disk and the latent heat, but in constant thicknesses, increasing the Prandtl number increases both effects.

5. AGM is a powerful, efficient, fast, and accurate method of solving various nonlinear problems and agrees well with numerical results.

3.3. Nanofluid Heat Transfer between Two Pipes

The nanofluid flow between two circular cylinders is studied in the magnetic field. The Koo-Kleinstreuer-Li (KKL) model is applied to nanofluids. The thermal radiation effect has been considered in the energy equation. AGM is selected for solving ODEs. Semi-analytical procedures are examined for various active parameters: aspect ratio, Hartmann number, Eckert number, and Reynolds number.

Laminar 2D flow in cylindrical coordinates is studied (Fig. 3.3.1).

The governing formulation is as follows:

$$\upsilon_{nf}\left(\frac{\partial^2 v}{\partial r^2}+\frac{1}{r}\frac{\partial v}{\partial r}-\frac{v}{r^2}\right)-\frac{\sigma_{nf}VB_0^2}{p_{nf}}=v\frac{\partial v}{\partial r} \tag{3.3.1}$$

$$\frac{k_{nf}}{r}\frac{\partial}{\partial r}\left(r\frac{\partial T}{\partial r}\right)+\mu_{nf}\left(\frac{\partial v}{\partial r}-\frac{v}{r}\right)^2-\frac{\partial q_r}{\partial r}=(\rho C_p)_{nf}\,v\frac{\partial T}{\partial r},$$

$$q_r=\frac{4\sigma_e}{3\beta_R}\frac{\partial T^4}{\partial y},T^4\cong 4T_c^3T-3T_c^4 \quad (3.3.2)$$

$$r=r_1: v(r)=\Omega_1 r_1, T=T_1$$

$$r=r_2: v(r)=0_1, T=T_2 \quad (3.3.3)$$

$(\sigma)_{nf}, (\rho C_p)_{nf}, (\rho\beta)_{nf}$, and $(\rho)_{nf}$ can be introduced as follows:

$$\frac{\sigma_{nf}}{\sigma_f}=1+\frac{3(\sigma_p/\sigma_f-1)\phi}{(\sigma_p/\sigma_f+2))-(\sigma_p/\sigma_f-1))\phi}$$

$$(\rho\beta)_{nf}=(1-\phi)(\rho\beta)_f+(\rho\beta)_p\phi \quad (3.3.4)$$

$$(\rho C_p)_{nf}=\varphi(\rho C_p)_f(1-\phi),\rho_{nf}=\phi\rho_p+\rho_f(1-\varphi)$$

(k_{nf}) and (μ_{nf}) are obtained according to the KKL model:

$$k_{nf}=\frac{3(k_p/k_f-1)\phi}{-(k_p/k_f-1)\phi+(k_p/k_f+2)}+1$$

$$+5\phi\times10^4 c_{p,f}\,g'(d_p,T,\phi)\rho_f\sqrt{\frac{\kappa_b T}{\rho_p d_p}} \quad (3.3.5)$$

$$g'(d_p,T,\phi)=(a_1+a_2Ln(d_p)+a_5Ln(d_p)^2+a_3Ln(\phi)$$
$$+a_4Ln(d_p)Ln(\phi)Ln(T)+(a_6+a_7Ln(d_p)$$
$$+a_{10}Ln(d_p)^2+a_8Ln(\phi)+a_9Ln(d_p)Ln(\phi))$$
$$R_f=d_p(1/k_{p,eff}-1/k_p)\ ,\ R_f=4\times10^{-8}km^2/w$$

$$\mu_{nf}=\frac{\mu_f}{(1-\phi)^{2.5}}+\frac{k_{Brownian}}{k_f}\times\frac{\mu_f}{\Pr}$$

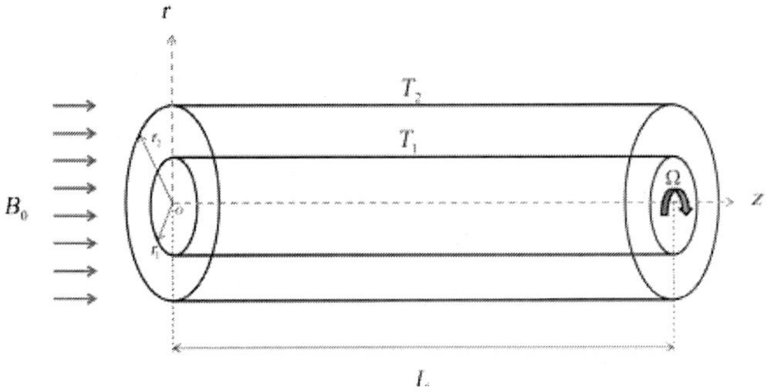

Fig. 3.3.1. Geometry of the problem.

Table 3.3.1. Thermophysical properties of water and nanoparticles.

	ρ (kg/m^3)	C_p (J/kg k)	k (W/m.k)	d_p (nm)	σ (Ωm)$^{-1}$
Pure water	997.1	4179	0.613	-	0.05
CuO	6500	540	18	29	10^{-10}

Table 3.3.2. The coefficient values of CuO-Water nanofluids.

Coefficient values	CuO-Water
a_1	-26.593310846
a_2	-0.40381833
a_3	-33.3516805
a_4	-1.915825591
a_5	6.421858E-02
a_6	48.40336955
a_7	-9.787756683
a_8	190.245610009
a_9	10.9285386565
a_{10}	-0.72009983664

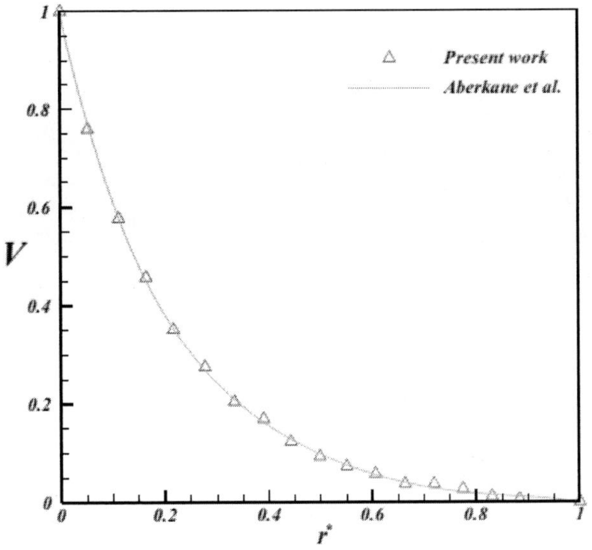

Fig. 3.3.2. Comparison of Aberkane et al. and present velocity profile results, for η=0.5, Ha=4.

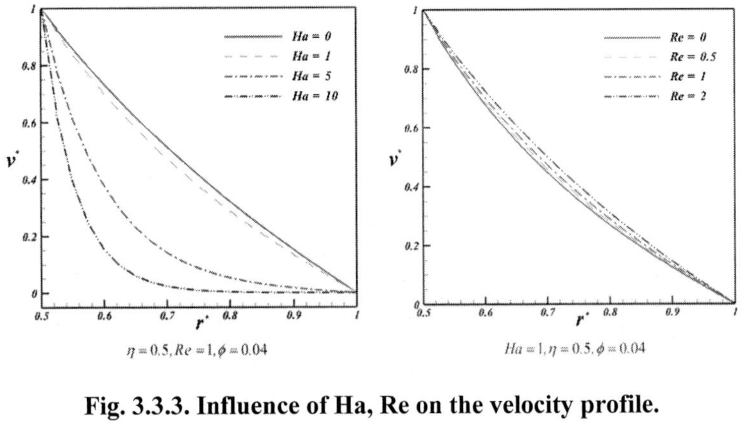

Fig. 3.3.3. Influence of Ha, Re on the velocity profile.

All needed coefficients and properties are illustrated in Tables 3.3.1 and 3.3.2.

The dimensionless forms of the above equations are as follows:

$$\frac{\partial^2 v^*}{\partial r^{*2}}+\frac{1}{r^*}\frac{\partial v^*}{\partial r^*}-\left(\frac{Ha2}{(1-\eta)^2}\frac{A^5}{A^2}+\frac{1}{r^{*2}}\right)v^*-\mathrm{Re}\frac{A_1}{A_2}v^*\frac{\partial v^*}{\partial r^*}=0 \quad (3.3.6)$$

$$\frac{1}{r^*}\frac{\partial}{\partial r^*}\left(r^*\frac{\partial 0}{\partial r^*}\right)+Ec\,\mathsf{Pr}\frac{A_2}{A_4}\left(\frac{\partial v^*}{\partial r^*}-\frac{v^*}{r^*}\right)^2+\frac{4}{3A_4}Rd\frac{\partial 20}{\partial r^{*2}}$$
$$-\mathsf{PrRe}\frac{A_3}{A_4}v^*\frac{\partial 0}{\partial r^*}=0 \quad (3.3.7)$$

$$r*=\eta : v^*(r^*)=1{,}0=1$$
$$r*=1 : v^*(r^*)=0{,}0=0 \quad (3.3.8)$$

Where

$$r*=\frac{r}{r_2},v*=\frac{v}{\Omega_1 r_1},\eta=\frac{F_1}{F_2}Ha=B_0 d\sqrt{\frac{\sigma_f}{\mu_f}}0=\frac{T-T_2}{T_1-T_2}$$

$$\mathrm{Re}=\frac{p_f\Omega r_1 r_2}{\mu_f},pr=\frac{\mu_f(pCp)}{p_f k_f},Ec=\frac{p_r(\Omega_1 r_1)^2}{(pCp)\Delta T}$$

$$Rd=4aeT_e^3/(B_8 k_f),A_1=\frac{p_{nf}}{p_f},A2=\frac{\mu_{nf}}{\mu_f} \quad (3.3.9)$$

$$A3=\frac{pCp)_{nf}}{(pCp)_f},A4=\frac{k_{nf}}{k_f},A5=\frac{\sigma_{nf}}{\sigma_f}$$

Nu over the hot cylinder is as follows:

$$Nu=a4\frac{\partial_0}{\partial_4}\Big|_{r^*}=\eta \quad (3.3.10)$$

Basic Idea of AGM:

$$Pk=f(u,u',u'',\ldots u^{(m)})=0;\ u=u(x) \quad (3.3.11)$$

$$\begin{cases} u(x)=u0, u'(x)=u1,\ldots,u^{(m-1)}(x)=u_{m-1} at \quad x=0 \\ u(x)=u_{L0}, u'(x)=u_{L1},\ldots,u^{(m-1)}(x)=u_{Lm-1} at \quad x=L \end{cases} \tag{3.3.12}$$

We assume that the solution to this equation is as follows:

$$u(x)=\sum_{i=0}^{n} a_i x^i = a_0 + a_1 x^1 + a_2 x^2 + \cdots + a_n x^n \tag{3.3.13}$$

The larger n makes more considerable accuracy of the solution. The residual can be obtained by inserting Eq. (3.3.13) into (3.3.11). According to boundary conditions and values of residual at boundaries, the constant parameters in Eq. (3.3.12) can be obtained.

First, we introduce the residuals:

$$\begin{aligned} G &= \frac{1}{r^*}\frac{\partial}{\partial r*}-\left(r^*\frac{\partial 0}{\partial r^*}\right)+Rv\ \Pr\frac{A_{21}}{A_4}\left(\frac{\partial v^*}{\partial r^*}-\frac{v^*}{r^*}\right)^2 \\ &+\frac{4}{3A_4}Rd\partial\frac{\partial^2 0}{\partial r^{*2}}-\Pr\mathrm{Re}\frac{A_3}{A_4}v*\frac{\partial 0}{\partial r^*}=0 \\ F &= \frac{\partial^2 v^*}{\partial r^{*2}}+\frac{1}{r^*}\frac{\partial v^*}{\partial r}-\left(\frac{Ha^2}{(1-\eta)^2}\frac{A_5}{A_2}+\frac{1}{r^{*2}}\right)v^* \\ &-\mathrm{Re}\frac{A_1}{A_2}v^*\frac{\partial v^*}{\partial r^*}=0 \end{aligned} \tag{3.3.14}$$

We assume that the solutions to these equations are as follows:

$$v^*=\sum_{i=0}^{7} a_i (r^*)^i,\ \theta=\sum_{i=0}^{7} b_i (r^*)^i \tag{3.3.15}$$

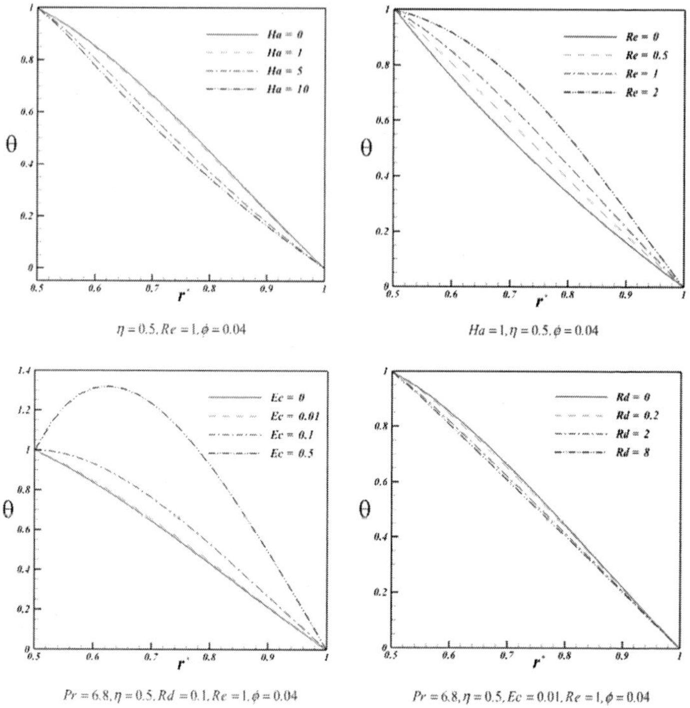

Fig. 3.3.4. Influence of Ha, Re, Ec, Rd on the temperature profile.

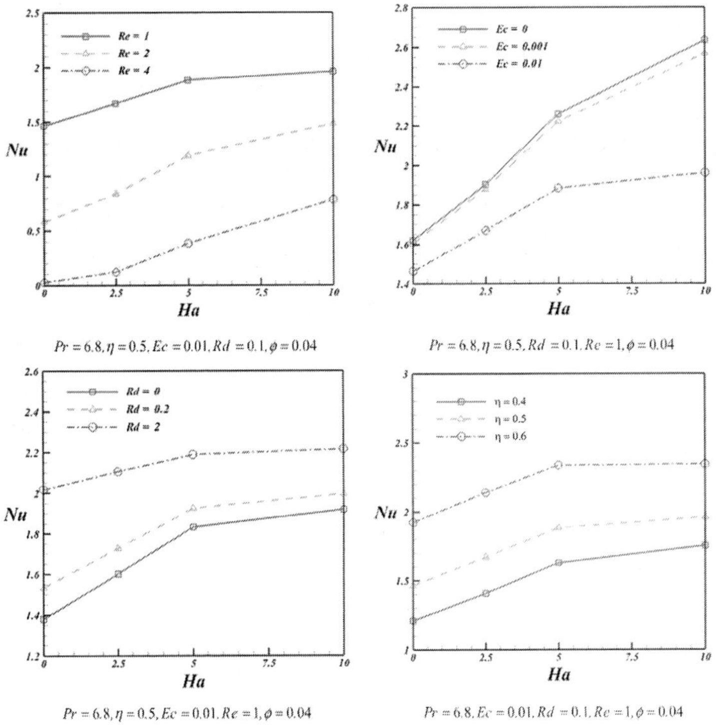

Fig. 3.3.5. Influence of Re, Ha, Ec, Rd, η on Nusselt number.

According to the below equations, all constant parameters can be obtained.

$$
\begin{aligned}
&v^{*} = v^{*}(B.C), \theta = \theta(B.C), \\
&F(B.C) = 0, F^{'}(B.C) = 0 \qquad (3.3.16) \\
&G(B.C) = 0, G^{'}(B.C) = 0
\end{aligned}
$$

CuO-water nanofluid flow in an annulus is studied analytically using AGM. Horizontal magnetic field and thermal radiation impacts are taken into account. Influences of effective parameters are depicted as graphs. Fig. 3.3.2 proved the accuracy of AGM Fig. 3.3.3. Depicts the influence of Ha, Re on the velocity profile as inertial forces dominate viscous forces velocity augments. As the electromagnetic forces dominate the viscous force, the velocity decreases. So velocity enhances with the rise of Rebut it reduces with the rise of Ha. The impacts of Ha, Re, Rd, and Ec on the temperature profile are illustrated in Fig 3.3.4. Lorentz forces reduce the velocity and generate secondary flow; the temperature reduces with the rise of Ha. The temperature gradient near the inner pipe reduces with the rise of Ha, Rd, but it enhances with the rise of Re. Higher Ec provides higher viscous dissipation, so temperature augments with increased Ec. Fig. 3.3.5 shows the impacts of Re, Ec. Rd, n, and Ha on Nu. Nusselt numbers augment with the rise of Ha and Rd, but they decrease with the rise of Re and Ec. Distance between the two pipes reduces with augments on n, so the heat transfer rate increases with the increase in aspect ratio.

The effect of radiation heat transfer on nanofluid heat transfer between two pipes is examined in a magnetic field. The KKL model is selected for nanofluid. Governing formulate is solved using AGM. Roles of aspect ratio with the Reynolds number, the Eckert number, and the Hartmann number are illustrated as graphs. Results revealed that temperature increases with the rise of Eckert and Reynolds numbers but reduces with the augment of the Hartmann number and radiation parameter. The Nusselt number augments with the rise of Lorentz forces but decreases with the rise of viscous dissipation.

3.4. Non-Newtonian Fluid Flow for Specific Turbine Cooling Application

This problem has been considered a non-Newtonian fluid flow in an axisymmetric channel with a porous wall for specific turbine cooling applications.

The simultaneous development of flow and heat transfer for a non-Newtonian viscoelastic fluid flow on the turbine disc for cooling purposes has been considered in this study. Fig. 3.4.1 represents a schematic of the proposed problem in which the r-axis is parallel to the disk's surface, and the z-axis is normal. The porous disc of the channel is $z = +L$. As depicted, although the wall that coincides with the r-axis is heated due to the external gas flow, non-Newtonian fluid is injected uniformly from the other perforated wall to cool the heated wall.

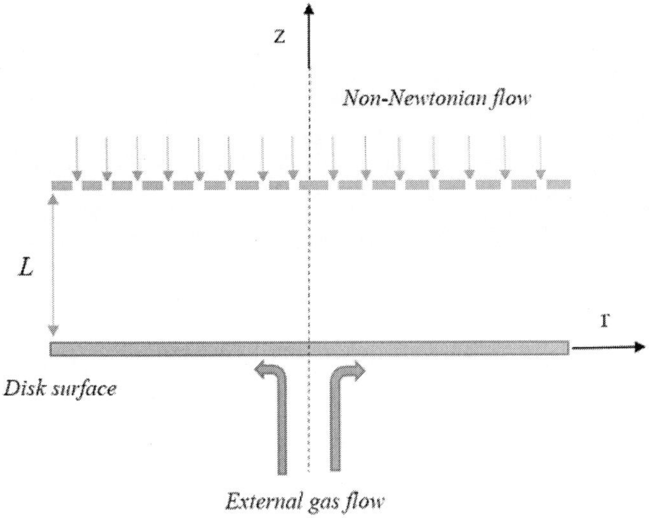

Fig. 3.4.1. Schematic diagram of the considered physical system

The following equations could be written in cylindrical coordinates for a steady, axisymmetric, non-Newtonian fluid flow. The continuity equation:

$$\frac{\partial(ru_r)}{\partial r}+\frac{\partial(ru_z)}{\partial z}=0 \tag{3.4.1}$$

And the momentum equations:

$$\begin{aligned} &u_r\frac{\partial(u_r)}{\partial r}+u_z\frac{\partial(u_r)}{\partial z}= \\ &-\frac{1}{\rho}\frac{\partial P}{\partial r}+\frac{1}{P}\left[\frac{\partial\tau_{rr}}{\partial r}+\frac{1}{r}(\tau_{rr}-\tau_{\theta\theta})+\frac{\partial\tau_{rz}}{\partial z}\right]\end{aligned} \tag{3.4.2}$$

$$\begin{aligned} &u_r\frac{\partial(u_z)}{\partial r}+u_z\frac{\partial(u_z)}{\partial z}= \\ &-\frac{1}{\rho}\frac{\partial P}{\partial z}+\frac{1}{P}\left[\frac{\partial\tau_{zr}}{\partial r}+\frac{1}{r}\tau_{rz}+\frac{\partial\tau_{zz}}{\partial z}\right]\end{aligned} \tag{3.4.3}$$

Here ρ, and P, are density and pressure, and also it is noteworthy to mention that τ_{rr}, τ_{rz}, τ_{zr}, τ_{zz} are the components of the stress matrix and u_r, u_z are the velocity components in the r and z directions. The analytical model under consideration leads to the following boundary conditions:

$$z = 0 \rightarrow u_r = u_z = 0 \tag{3.4.4}$$

$$z = L \rightarrow u_r = 0, u_z = -v \tag{3.4.5}$$

In which V is the injection velocity. For a particular class of viscoelastic and viscoelastic fluids, researchers represented that if the stress components τ_{ij} at a point $x_k\,(k = 1,2,3)$ and time t are assumed to be polynomials in the velocity gradient $\frac{\delta \upsilon_m}{\delta x_n}(m,n = 1,2,3)$ and the acceleration gradients $\frac{\delta a_m}{\delta x_n}(m,n = 1,2,3)$, and if in addition, the medium is assumed to be isentropic the stress matrix can be demonstrated in the following form:

$$\left\|\tau_{ij}\right\| = \phi_0 I + \phi_2 A + \phi_2 B + \phi_3 A^2 + \ldots \tag{3.4.6}$$

Here I is the unit matrices, A and B are symmetric kinematic matrixes defined by:

$$A = \left\|\frac{\delta \upsilon_i}{\delta x_j} + \frac{\delta \upsilon_j}{\delta x_i}\right\|, B = \left\|\frac{\delta a_i}{\delta x_j} + \frac{\delta a_j}{\delta x_i} + 2\frac{\delta \upsilon_m}{\delta x_i}\frac{\delta \upsilon_m}{\delta x_j}\right\| \tag{3.4.7}$$

Moreover, $\phi_k\,(k = 0,1,2,3)$ are polynomials in the invariants of A, B, A^2. This study is restricted to second-order fluids, in which $\phi_k\,(k = 0,1,2,3)$ are constant and $\phi_k\,(k = 4,5,\ldots)$, are zero. So that the stress components are as follows:

$$\tau_{rr} = \phi_1 A_{rr} + \phi_2 A_{rr}^2 + \phi_3 B_{rr} \tag{3.4.8}$$

$$\tau_{zz} = \phi_1 A_{zz} + \phi_2 A_{zz}^2 + \phi_3 B_{zz} \tag{3.4.9}$$

$$\tau_{\theta\theta} = \phi_1 A_{\theta\theta} + \phi_2 A_{\theta\theta}^2 + \phi_3 B_{\theta\theta} \tag{3.4.10}$$

$$\tau_{rz} = \phi_1 A_{rz} + \phi_2 A_{rz}^2 + \phi_3 B_{rz} \tag{3.4.11}$$

For solving the aforementioned problem depicted in Fig. 3.4.1 in the case of axially symmetric flow, it would be much more convenient to define a stream function so that the continuity equation is satisfied:

$$\psi = Vr^2 f(\eta) \tag{3.4.12}$$

Where $\eta = \frac{z}{L}$ and the velocity components can be derived as:

$$u_r = \frac{Vr}{L} f'(\eta) \tag{3.4.13}$$

$$u_z = -2Vf(\eta) \tag{3.4.14}$$

Using Eqs. (3.4.12)–(3.4.14) the equations of motion reduce to:

$$\begin{aligned} f'^2 - 2ff'' &= -\frac{L^2}{\rho V^2 r}\frac{\partial P}{\partial r} + \frac{\phi_1}{\rho V L} f''' \\ &+ \frac{\phi_2}{\rho L^2}(f''^2 - 2ff''') + \frac{\phi_3}{\rho L^2}(f''^2 - 2ff^{iv}) \end{aligned} \tag{3.4.15}$$

$$\begin{aligned} 4ff' &= -\frac{L^2}{\rho V^2}\frac{\partial P}{\partial z} - 2\frac{\phi_1}{\rho V L} f'' + 2\frac{\phi_2}{\rho L^2}(14ff'' + \frac{r^2}{L} ff''') \\ &+ 4\frac{\phi_3}{\rho L^2}(11ff'' + 2ff''' + \frac{r^2}{L} ff''') \end{aligned} \tag{3.4.16}$$

The pressure term can be eliminated by differentiating Eq. (3.4.15) with respect to z and Eq. (3.4.16) with respect to r and subtracting the resulting equations. Therefore, the following equations would be obtained:

$$-2ff''' = \frac{f^{iv}}{\text{Re}} - K_1(4f'f''' + 2ff^{iv}) \\ -K_2(4f'f''' + 2ff^{iv} + 2ff^{v}) \tag{3.4.17}$$

Where $K_1 = \frac{\phi_2}{\rho L^2}, K_2 = \frac{\phi_3}{\rho L^2}$ is the injection Reynolds number. When $K_2 = 0$ the equation turned to:

$$f^{iv} + 2\text{Re}ff''' - K_1\text{Re}(4f'f''' + 2ff^{iv}) = 0 \tag{3.4.18}$$

The boundary conditions are:

$$f(0) = 0, f'(0) = 0, f(1) = 1, f'(1) = 0 \tag{3.4.19}$$

The energy equation for the present problem with viscous dissipation in non-dimensional form is given by:

$$\rho C\left(u_r\frac{\partial T}{\partial r} + u_z\frac{\partial T}{\partial z}\right) = \bar{k}\nabla^2 T + \varphi \tag{3.4.20}$$

$$\varphi = \tau_{rr}\frac{\partial u_r}{\partial r} + \tau_{\theta\theta}\frac{u_r}{r} + \tau_{zz}\frac{\partial u_z}{\partial z} + \tau_{rz}\left(\frac{\partial u_r}{\partial z} + \frac{\partial u_z}{\partial r}\right) \tag{3.4.21}$$

Here u_r, u_z are the velocity components in the r and z directions, and V is the injection velocity; ρ, P, T, C, K are the density, pressure, temperature, specific heat, and heat conduction coefficient of fluid, respectively. φ, is the dissipation function. Letting the blade-wall $z = 0$ temperature distribution be $T_\omega = T_0 + \sum_{n=0}^{\infty} C_n (\frac{r}{L})^n$ and assuming the fluid temperature to have the form of equation (3.4.22):

$$T = T_0 + \sum_{n=0}^{\infty} C_n \left(\frac{r}{L}\right)^n q_n(\eta) \tag{3.4.22}$$

Where T_0 is the temperature of the incoming coolant $z = L$ and neglecting the dissipation effect, the following equations and boundary conditions are obtained:

$$q_n'' - \Pr.\mathrm{Re}(f' q_n - 2fq_n') = 0, (n = 0, 2, 3, 4, ...) \tag{3.4.23}$$

$$q_n(0) = 1, q_n(1) = 0 \tag{3.4.24}$$

According to the coupled system of nonlinear differential equations and by considering the method's basic idea, we rewrite Eqs. (3.4.18) and (3.4.23) in the following order:

$$F(\eta) = f^{iv} + 2\mathrm{Re} ff''' - K_1 \mathrm{Re}(4f'f''' + 2ff^{iv}) = 0 \tag{3.4.25}$$

$$\Theta(\eta) = q_n'' - \Pr.\mathrm{Re}(f' q_n - 2fq_n') = 0, (n = 0, 2, 3, 4, ...) \tag{3.4.26}$$

Due to the basic idea of AGM, we have utilised a proper trial function as the solution of the considered differential equation, which is a finite series of polynomials with constant coefficients, as follows:

$$f(\eta) = \sum_{i=0}^{9} a_i \eta^i = (a_0 + a_1\eta^1 + a_2\eta^2 + a_3\eta^3 + a_4\eta^4 + a_5\eta^5 + a_6\eta^6 + a_7\eta^7 + a_8\eta^8 + a_9\eta^9) \tag{3.4.27}$$

$$\theta(\eta) = \sum_{i=0}^{12} b_i \eta^i = b_0 + b_1\eta^1 + b_2\eta^2 + b_3\eta^3 + b_4\eta^4 + b_5\eta^5 + b_6\eta^6 + b_7\eta^7 + b_8\eta^8 + b_9\eta^9 + b_{10}\eta^{10} + b_{11}\eta^{11} + b_{12}\eta^{12} \tag{3.4.28}$$

In AGM, the boundary conditions are applied to calculate constant coefficients of Eqs. (3.4.27) and (3.4.28) according to the following approaches:

(a) Applying the boundary conditions on Eqs. (3.4.27) and (3.4.28) are expressed as follows:

$$u = u(B.C) \tag{3.4.29}$$

According to the above explanations, the boundary conditions are applied on Eqs where B.C. is the abbreviation of boundary condition. (3.4.27) and (3.4.28) in the following form:

$$f(0) = 0 \to a_0 = 0 \tag{3.4.30}$$

$$\begin{aligned} &f(1) = 0 \to \\ &a_9 + a_8 + a_7 + a_6 + a_5 + a_4 + a_3 + a_2 + a_1 + a_0 = 0 \end{aligned} \tag{3.4.31}$$

$$f'(0) = 0 \to a_1 = 0 \tag{3.4.32}$$

$$\begin{aligned} &f'(1) = 0 \to \\ &9a_9 + 8a_8 + 7a_7 + 6a_6 + 5a_5 + 4a_4 \\ &+3a_3 + 2a_2 + a_1 = 0 \end{aligned} \tag{3.4.33}$$

$$\theta(0) = 1 \to b_0 = 1 \tag{3.4.34}$$

$$\begin{aligned} &\theta(1) = 0 \to \\ &b_{12} + b_{11} + b_{10} + b_9 + b_8 + b_7 + b_6 + b_5 + b_4 \\ &+b_3 + b_2 + b_1 + b_0 = 0 \end{aligned} \tag{3.4.35}$$

(b) Applying the boundary conditions on the main differential equations, in this case, the study is Eqs. (3.4.25) and (3.4.26), and also on their derivatives is done after substituting Eqs. (3.4.27) and (3.4.28) into the main differential equations as follows:

$$F(f(\eta)) \to F(f(B.C)) = 0, F'(f(B.C)) = 0, \ldots \tag{3.4.36}$$

$$\Theta(\theta(\eta)) \to \Theta(\theta(B.C)) = 0, \Theta'(\theta(B.C)) = 0, \ldots \tag{3.4.37}$$

The boundary conditions are applied to the above differential equation. In fact, due to the excellence of AGM compared to other methods, we have to reach a set of polynomials in the processes of solution that are the same number as the overall constant coefficients in our trial functions. Therefore, we could obtain these only by simple and mundane calculations. In the proposed problem, we have assumed two trial functions containing 23 constant coefficients and six equations according to Eqs. (3.4.30)–(3.4.35). We must create 17 additional equations from Eqs. (3.4.36) and (3.4.37) to achieve a set of polynomials containing 23 equations and 23 constants.

According to the above explanations, we have created additional equations Eqs. (3.4.36) and (3.4.37) in the following order:

Six equations have been created by calculating equations obtained from $F(0)=0$, $F(1)=0$, $F'(0)=0$, $F'(1)=0$, $F''(0)=0$, $F''(1)=0$

11 equations have been created by calculating equations obtained from $\Theta(0)=0$, $\Theta(1)=0$, $\Theta'(0)=0$, $\Theta'(1)=0$, $\Theta''(0)=0$, $\Theta''(1)=0$, $\Theta'''(0)=0$, $\Theta'''(1)=0$, $\Theta''''(0)=0$, $\Theta''''(1)=0$, $\Theta'''''(0)=0$

The equations in the (I) and (II) subsections are too large to be displayed graphically. So by utilising the above procedures, we have obtained a set of polynomials containing 23 equations and 23 constants which, by solving them, we would be able to obtain Eqs. (3.4.27) and (3.4.28). For instance, when $\mathrm{Re}=0.5,\ k=0.1,\ \mathrm{Pr}=2.5,\ n=0$ substitution is obtained.

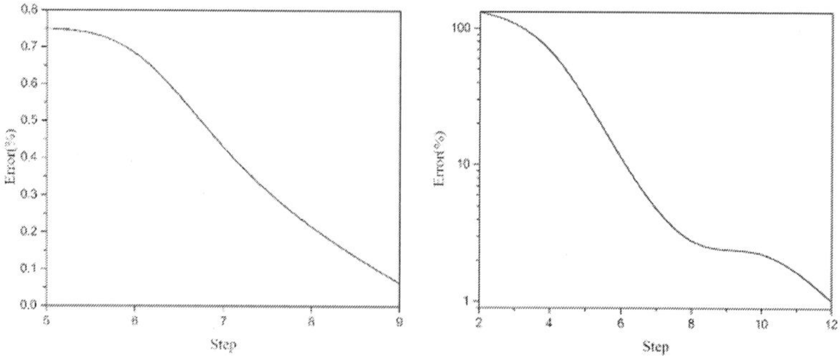

(a) (b)

Fig. 3.4.2. (a) Convergence figure for $f(\eta)$ in varies terms of AGM. (b) Convergence figure for $q_n(\eta)$ in various terms of AGM

Constant coefficients from mentioned procedures Eqs. (3.4.27) and (3.4.28) could easily be yielded as follows:

$$f(\eta) = -0.03009057971\eta^7 + 0.09356884058\eta^6 \\ +0.002083333333\eta^4 - 2.227989130\eta^3 + 3.162427536\eta^2 \quad (3.4.38)$$

$$\theta(\eta) = -0.06730148307\eta^{12} + 0.4123487273\eta^{11} \\ -0.9074026402\eta^{10} + 0.6454071134\eta^9 + 04475609223\eta^8 \\ -0.6806457857\eta^7 + 0.0002382034452\eta^6 \\ -0.3821145744\eta^5 + 0.09039613610\eta^4 + 1 - 1.372051844\eta \quad (3.4.39)$$

As can be detected, AGM has been properly and efficiently applied through the considered problem. So in this section, we have presented some detailed investigations of the problem using different concepts to support our work.

Fig. 3.4.2 is divided into two parts, both of which represent the difference between the solution obtained by AGM and the numerical method (Runge-Kutte 4th) in which we have introduced error percentages as follows:

$$\%Error = \left| \frac{u(\eta)_{NM} - u(\eta)_{AGM}}{u(\eta)_{NM}} \right| \quad (3.4.40)$$

Where $u(\eta)_{NM}$ is the value obtained by the numerical method (Runge-Kutte 4th) and $u(\eta)_{AGM}$ is the value obtained by AGM. It is worth mentioning that both Fig. 3.4.2a and Fig. 3.4.2b depict the convergence issue. It could be deduced that while we utilised more terms for our proposed AGM solution, the obtained error when $\eta = 0.5$ would decrease.

Moreover, a comparison between AGM and numerical results has been done for different values of active parameters in Fig. 3.4.3. The effects of these parameters on the velocity $(f(\eta))$ and the temperature $(q_n(\eta))$ are shown. The results represent that AGM has enough accuracy and efficiency, so it would be applicable for solving nonlinear equations of a coupled system.

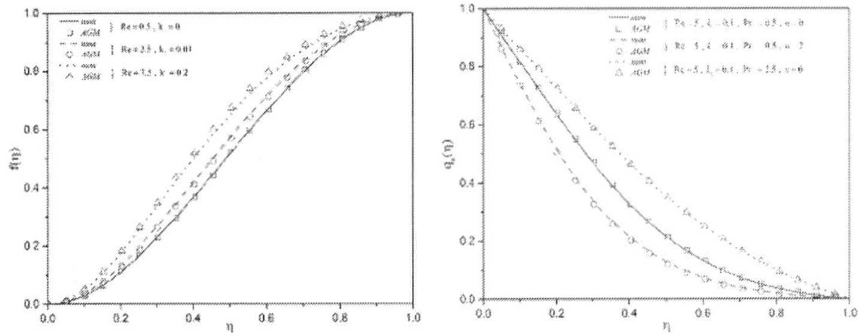

(a) (b)

Fig. 3.4.3. (a, b) Comparison between numerical results and AGM solution for $f(\eta)$ and $q_n(\eta)$

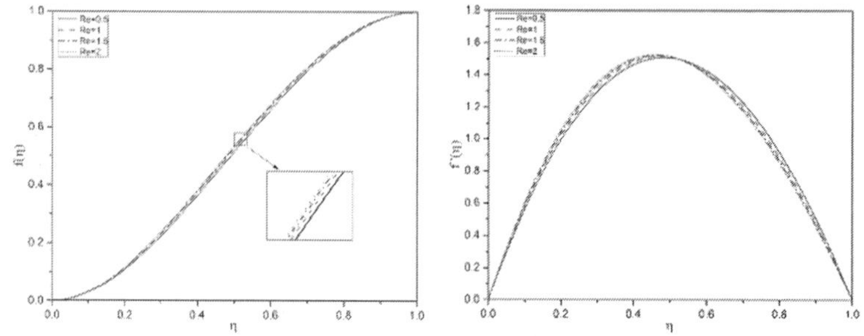

(a) (b)

Fig. 3.4.4. Velocity component profiles a $f(\eta)$ and b $f'(\eta)$ for variable Re at $k_1 = 0.01$

Due to the importance of the Reynolds number, a dimensionless number that illustrates the correlation of inertial forces to viscous forces, Fig. 3.4.4a–b and Fig. 3.4.5 have been considered. Figs. 3.4.4a–b represent the effect of the Reynolds number on velocity profiles ($f(\eta)$, $f'(\eta)$) where they directly correlate. Additionally, the effects of the Reynolds number and injection Reynolds number on skin friction $(f''(\eta))$ have been shown in Fig. 3.4.5, in which it is crystal clear that by increasing both Re and k_1 parameters, the skin friction would increase substantially.

Afterwards, the effect of the power-law index (n) and the Reynolds number on the temperature profile are represented. Fig. 3.4.6 illustrates that the thermal boundary layer would decrease by increasing the Reynolds number and power-law index (n).

Finally, in Fig. 3.4.7, the effects of considered parameters on the Nusselt number, which is equal to the dimensionless temperature gradient at the surface and provides a measure of the convection heat transfer occurring at the surface, have been discussed. Therefore, the Nusselt number on the horizontal axis is plotted against the Reynolds number on the vertical axis. Then, the various profiles are depicted due to the Prandtl number and power-law index. This figure demonstrates that when the Prandtl number or power-law index amounts increase in the constant value of the Reynolds number, the Nusselt number will climb dramatically.

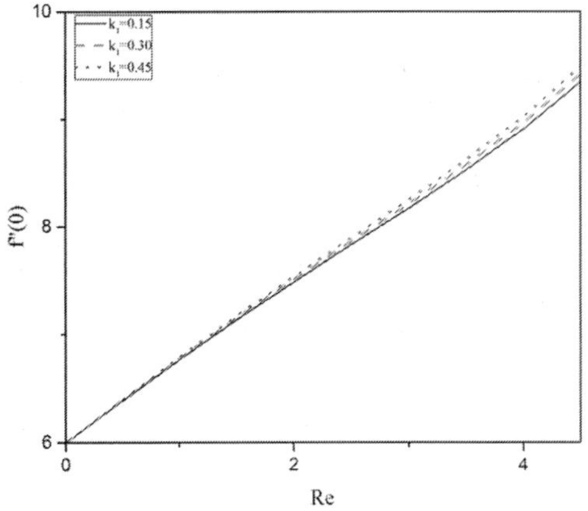

Fig. 3.4.5. Skin friction under the effect of k_1 and Re

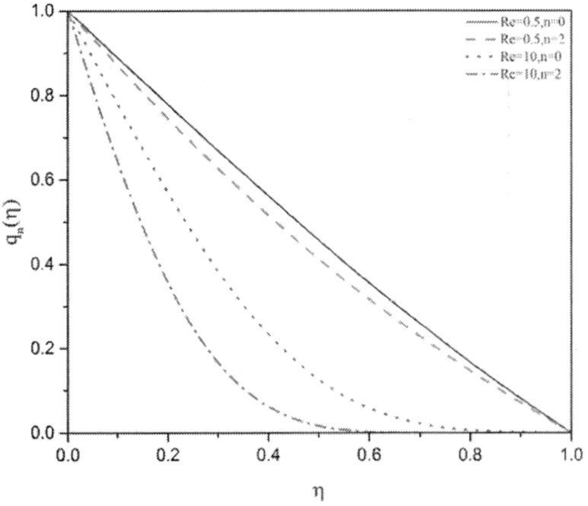

Fig. 3.4.6. Temperature profile (q_n) at $K_1 = 0.01$

Because the Nusselt number and the thermal boundary thickness layer are negatively correlated, it is obvious that by rising the Nusselt number, with respect to its relationship with other factors, we would obtain better cooling efficiency in industrial applications. One could also refer to previous articles in this field and obtain more details and results through other analytical and semi-analytical methods.

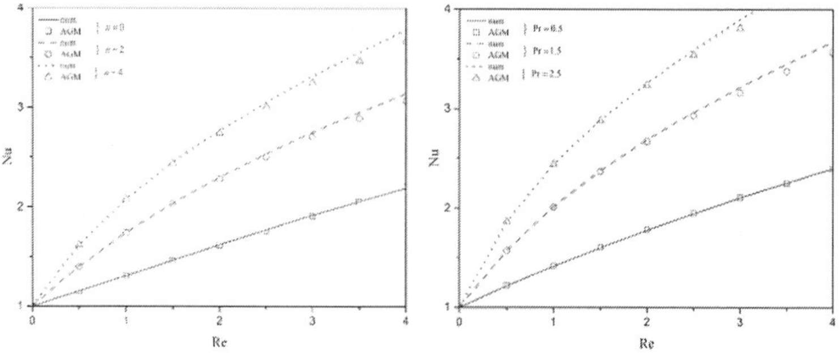

(a) (b)

Fig. 3.4.7. Nusselt number for (a) variable n at $K_1 = 0.01$, $\Pr = 1$ **(b) variable Pr at** $K_1 = 0.1$, $n = 2$

This study uses AGM to solve nonlinear differential equations of momentum and heat transfer of a non-Newtonian fluid flow for specific turbine cooling applications. Different parameter values have been compared among AGM and the numerical methods (Runge-Kutte 4th). The obtained data revealed that AGM solutions are in excellent agreement with numerical solutions. Moreover, convergence figures represent that we could obtain more accurate solutions by applying more terms of AGM. Therefore, it would be obvious that AGM is a convenient and qualified method for solving nonlinear problems due to its accuracy and efficiency.

Furthermore, according to the results, it is crystal clear that the Nusselt number is an increasing function of each Reynolds number, Prandtl number, and power-law index. Therefore, when these properties increase during the pertinent procedures, the Nusselt number would rise substantially. Finally, we could achieve more appropriate cooling characteristics in industrial applications due to the reverse relation between the Nusselt number and the thermal boundary layer thickness.

3.5. Micropolar Fluid Flow and Heat Transfer in a Permeable Channel

In this problem, micropolar fluid flow and heat transfer in a permeable channel have been investigated.

We consider the steady laminar flow of a micropolar fluid along a two-dimensional channel with parallel porous walls through which fluid is uniformly injected or removed with speed v_0 which is represented in Fig. 3.5.1. The geometry of the problem has defined clearly in Fig. 3.5.1. By utilizing Cartesian coordinates, the governing equations for flow are:

$$\frac{\partial u}{\partial x}+\frac{\partial v}{\partial y}=0 \tag{3.5.1}$$

$$\rho\left(u\frac{\partial u}{\partial x}+v\frac{\partial u}{\partial y}\right)=-\frac{\partial P}{\partial x}+(\mu+\kappa)\left(\frac{\partial^2 u}{\partial x^2}+\frac{\partial^2 u}{\partial y^2}\right)+\kappa\frac{\partial N}{\partial y} \tag{3.5.2}$$

$$\rho\left(u\frac{\partial u}{\partial x}+v\frac{\partial v}{\partial y}\right)=-\frac{\partial P}{\partial x}+(\mu+\kappa)\left(\frac{\partial^2 v}{\partial x^2}+\frac{\partial^2 v}{\partial y^2}\right)-\kappa\frac{\partial N}{\partial x} \tag{3.5.3}$$

$$\rho\left(u\frac{\partial N}{\partial x}+v\frac{\partial N}{\partial y}\right)=-\frac{\kappa}{j}\left(2N+\frac{\partial u}{\partial y}-\frac{\partial v}{\partial x}\right)+\left(\frac{\upsilon_s}{j}\right)\left(\frac{\partial^2 N}{\partial x^2}+\frac{\partial^2 N}{\partial y^2}\right) \tag{3.5.4}$$

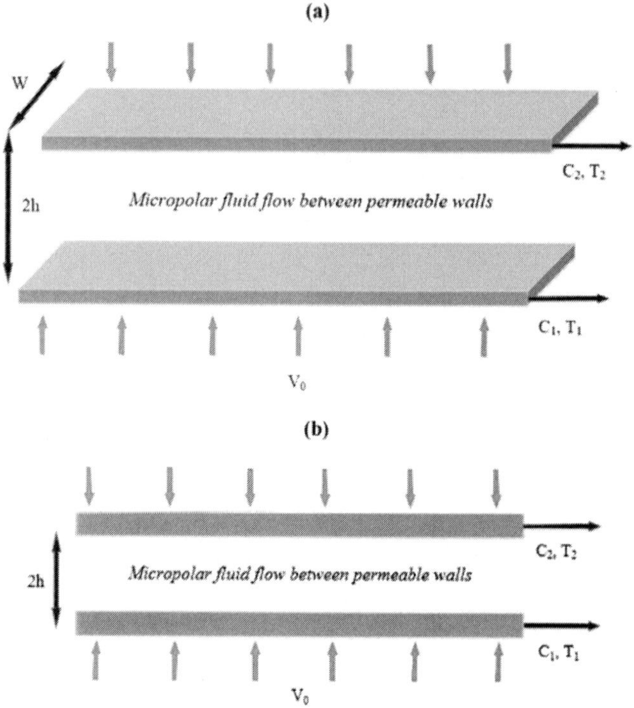

Fig. 3.5.1. (a) The geometry of the problem (b) x-y view.

$$\rho\left(u\frac{\partial T}{\partial x}+v\frac{\partial T}{\partial y}\right)=\frac{k_1}{c_p}\frac{\partial^2 T}{\partial y^2} \tag{3.5.5}$$

$$\rho\left(u\frac{\partial C}{\partial x}+v\frac{\partial C}{\partial y}\right)=D^*\frac{\partial^2 C}{\partial y^2} \tag{3.5.6}$$

Where $\upsilon_s=\left(\mu+\frac{k}{2}\right)$ Also, because we have defined the constants in Eqs. (3.5.1)–(3.5.7) in the nomenclature section, so we have refused to announce these again for celerity and brevity. The appropriate boundary conditions are:

$$y=-h \Rightarrow v=u=0, n=-s\frac{\partial u}{\partial y}$$
$$y=+h \Rightarrow v=0, u=\frac{v\circ x}{h}, n=\frac{v\circ x}{h^2} \tag{3.5.7}$$

Where s is a boundary parameter, declare the degree to which the microelements can rotate near channel walls.

The case s= 0 represents non-rotatable concentrated microelements close to the wall. Also, s =0.5 represents weak concentrations and the vanishing of the antisymmetric part of the stress tensor, and s= 1 represents turbulent flow. We introduce the following dimensionless variables:

$$\eta=\frac{y}{h}, \psi=-v\circ xf(\eta), N=\frac{v\circ x}{h^2}g(\eta),$$
$$\theta(\eta)=\frac{T-T_2}{T_1-T_2}, \phi(\eta)=\frac{C-C_2}{C_1-C_2} \tag{3.5.8}$$

Where $T_2=T_1-Ax, C_2=C_1-Bx$ that A and B as constants. The stream function is defined as its original form as follows:

$$u=\frac{\partial\psi}{\partial y}, v=\frac{\partial\psi}{\partial x} \tag{3.5.9}$$

Eqs. (3.5.1)–(3.5.7) will reduce to the following coupled system of nonlinear differential equations:

$$(1+N_1)f^{IV}-N_1 g-\operatorname{Re}\left(ff'''-ff''\right)=0 \tag{3.5.10}$$

$$N_2 g''-N_1\left(f''-2g\right)-N_3\operatorname{Re}\left(fg'-f'g\right)=0 \tag{3.5.11}$$

$$\theta''+Pe_h\left(f'\theta-f\theta'\right)=0 \tag{3.5.12}$$

$$\phi''+Pe_m\left(f'\phi-f\phi'\right)=0 \tag{3.5.13}$$

The boundary conditions are listed as follows:

$$\begin{aligned}&\eta=-1 \Rightarrow f=f^{'}=g=0, \theta=\phi=1\\&\eta=+1 \Rightarrow f=\theta=\phi=0, f^{'}=-1, g=1\end{aligned} \tag{3.5.14}$$

The parameters of primary interest are the buoyancy ratio N, the Peclet numbers for the diffusion of heat Pe_h and mass Pe_m respectively, the Reynolds number Re where for suction Re> 0 and injection Re <0 also Grashof number Gr given by:

$$\begin{aligned}&N_1=\frac{\kappa}{\mu}, N_2=\frac{\upsilon_s}{\mu h^2}, N_3=\frac{j}{h^2}, \mathrm{Re}=\frac{v_0}{\upsilon}h\\&\mathrm{Pr}=\frac{\upsilon\rho c_p}{k_1}, Sc=\frac{\upsilon}{D^*}, Gr=\frac{g\beta_T A h^4}{\upsilon^2}\\&Pe_h=\mathrm{Pr}\,\mathrm{Re}, Pe_m=Sc\,\mathrm{Re}\end{aligned} \tag{3.5.15}$$

Where Pr is the Prandtl number, Sc is the generalised Schmidt number, N_1 is the coupling parameter, and N_2 is the spin gradient viscosity parameter. In technological processes, Nusselt and Sherwood's numbers are considered widely, which are defined as follows:

$$Nu_X=\frac{q^{''}_{y=-h^x}}{(T_1-T_2)k_1}=-\theta^{'}(-1) \tag{3.5.16}$$

$$Sh_X=\frac{m^{''}_{y=-h^x}}{(C_1-C_2)D^*}=-\phi(-1) \tag{3.5.17}$$

where q'' and m'', are local heat flux and mass flux, respectively.

Physics of the problems in every field of engineering sciences leads to a set of linear or nonlinear differential equations as its governing equations. According to the physics of these problems and their obtained mathematical formulation, sufficient boundary or initial conditions should be applied to achieve solutions for considered problems. Since applying analytical methods for obtaining the solution to linear and nonlinear differential equations are not an exception, we could recognize the importance of these boundary and initial conditions in determining the accuracy and efficiency of analytical methods in achieving acceptable solution due to the physic of problems. The entire process has been declared clearly to comprehend the given method in this problem. By the boundary conditions, the general manner of a differential equation is as follows:

$$p_k : f\left(u, u', u'', \ldots, u^{(m)}\right) = 0; \; u = u\left(x\right) \tag{3.5.18}$$

The nonlinear differential equation of p, which is a function of u, the parameter u, which is a function of x, and their derivatives are considered as follows:

Boundary conditions:

$$\begin{cases} u(x) = u_0, u'(x) = u_1, \ldots, u^{(m-1)}(x) = u_{m-1} \quad at \quad x = 0 \\ u(x) = u_{L_0}, u'(x) = u_{L_1}, \ldots, u^{(m-1)}(x) = u_{L_{m-1}} \quad at \quad x = L \end{cases} \tag{3.5.19}$$

To solve the first differential equation with respect to the boundary conditions in x = L in Eq. (3.5.19), the series of letters in the nth order with constant coefficients,

which we assume as the solution of the first differential equation is considered as follows:

$$u\left(x\right) = \sum_{i=0}^{n} a_i x^i = a_0 + a_1 x^1 + a_2 x^2 + \ldots + a_n x^n \tag{3.5.20}$$

The more choice of series sentences from Eq. (3.5.20) causes a more precise solution for Eq. (3.5.18). For obtaining the solution of differential Eq. (3.5.18) regarding the series from degree (n), there are (n + 1) unknown coefficients that need (n + 1) equations to be specified. The boundary conditions of Eq. (3.5.19) are used to solve a set of equations consisting of (n + 1).

Applying the boundary conditions:

(a) The application of the boundary conditions for the answer of differential Eq. (3.5.20) is in the form of:

When x = 0:

$$\begin{cases} u(0) = a_0 = u_0 \\ u'(0) = a_1 = u_1 \\ u''(0) = a_2 = u_2 \\ \vdots \end{cases} \tag{3.5.21}$$

And when x= L:

$$\begin{cases} u(L) = a_0 + a_1 L + a_2 L^2 + \ldots + a_n L^n = u_{L_0} \\ u'(L) = a_1 + 2a_2 L + 3a_3 L^2 + \ldots + n a_n L^{n-1} = u_{L_1} \\ u''(L) = 2a_2 + 6a_3 L + 12a_4 L^2 + \ldots + n(n-1) a_n L^{n-2} = u_{L_{m-1}} \\ \vdots \end{cases} \tag{3.5.22}$$

(b) After substituting Eq. (3.5.22) into Eq. (3.5.18), the application of the boundary conditions on differential Eq. (3.5.18) is done according to the following procedure:

$$\begin{aligned} &p_0 : f\left(u(0), u'(0), u''(0), \ldots, u^{(m)}(0)\right) \\ &p_1 : f\left(u(L), u'(L), u''(L), \ldots, u^{(m)}(L)\right) \\ &\vdots \quad \vdots \quad \vdots \quad \vdots \quad \vdots \end{aligned} \tag{3.5.23}$$

Concerning the choice of n; (n< m) sentences from Eq. (3.5.20) and to make a set of equations which consist of (n + 1) equations and (n + 1); unknowns, we are confronted with several additional unknowns which are indeed the same coefficients of Eq. (3.5.20). Therefore, to remove this problem, we should derive m times from Eq. (3.5.18) according to the additional unknowns in the aforementioned sets of differential equations and then apply the boundary conditions to them.

$$\begin{aligned} &p_k' : f\left(u', u'', u''', \ldots, u^{(m+1)}\right) \\ &p_k'' : f\left(u'', u''', u^{(IV)}, \ldots, u^{(m+2)}\right) \\ &\vdots \quad \vdots \quad \vdots \quad \vdots \quad \vdots \end{aligned} \tag{3.5.24}$$

(c) The application of the boundary conditions on the derivatives of the differential equation P_k in Eq. (3.5.24) is done in the form of:

$$p_k^{'}:\begin{cases} f\left(u^{'}(0),u^{''}(0),u^{'''}(0),\ldots,u^{(m+1)}(0)\right) \\ f\left(u^{'}(L),u^{''}(L),u^{'''}(L),\ldots,u^{(m+1)}(L)\right) \end{cases} \tag{3.5.25}$$

$$p_k^{''}:\begin{cases} f\left(u^{''}(0),u^{'''}(0),\ldots,u^{(m+2)}(0)\right) \\ f\left(u^{''}(L),u^{'''}(L),\ldots,u^{(m+2)}(L)\right) \end{cases} \tag{3.5.26}$$

(n + 1) equations can be made from Eq. (3.5.21) to Eq. (3.5.26) so that (n + 1) unknown coefficients of Eq. (3.5.20) such as a_0, a_1, a_2, …, a_n. will be computed. The solution of the nonlinear differential Eq. (3.5.18) will be gained by determining Eq. (3.5.20) coefficients. To comprehend the following explanation, we have presented the relevant process step by step as follows:

According to the coupled system of nonlinear differential equations and by considering the method's basic idea, we rewrite Eqs. (3.5.10)–(3.5.13) in the following order:

$$F(\eta)=(1+N_1)f^{IV}-N_1 g-\mathrm{Re}\left(ff^{'''}-f^{'}f^{''}\right)=0 \tag{3.5.27}$$

$$G(\eta)=N_2 g^{''}-N_1\left(f^{''}-2g\right)-N_3\,\mathrm{Re}\left(fg^{'}-f^{'}g\right)=0 \tag{3.5.28}$$

$$\Theta(\eta)=\theta^{''}+P_{e_h}\left(f^{'}\theta-f\theta^{'}\right)=0 \tag{3.5.29}$$

$$\Phi(\eta)=\phi^{''}+P_{e_m}\left(f^{'}\phi-f\phi^{'}\right)=0 \tag{3.5.30}$$

Due to the basic idea of AGM, we have utilised a proper trial function as the solution of the considered differential equation, which is a finite series of polynomials with constant coefficients, as follows:

$$f(\eta)=\sum_{i=0}^{9}a_i\eta^i$$
$$=a_0+a_1\eta^1+a_2\eta^2+a_3\eta^3+a_4\eta^4+a_5\eta^5+a_6\eta^6$$
$$+a_7\eta^7+a_8\eta^8+a_9\eta^9 \tag{3.5.31}$$

$$g(\eta)=\sum_{i=0}^{9}b_i\eta^i$$
$$=b_0++b_1\eta^1+b_2\eta^2+b_3\eta^3+b_4\eta^4+b_5\eta^5+b_6\eta^6$$
$$+b_7\eta^7+b_8\eta^8+b_9\eta^9 \tag{3.5.32}$$

$$\theta(\eta)=\sum_{i=0}^{7}c_i\eta^i$$
$$=c_0++c_1\eta^1+c_2\eta^2+c_3\eta^3+c_4\eta^4+c_5\eta^5+c_6\eta^6+c_7\eta^7 \tag{3.5.33}$$

$$\phi(\eta)=\sum_{i=0}^{7}d_i\eta^i$$
$$=d_0++d_1\eta^1+d_2\eta^2+d_3\eta^3+d_4\eta^4+d_5\eta^5+d_6\eta^6+d_7\eta^7 \tag{3.5.34}$$

In AGM, the boundary conditions are applied in order to compute the constant coefficients of Eqs. (3.5.31)–(3.5.34) according to the following approaches:

(a) The boundary conditions on Eqs. (3.5.31)–(3.5.34) are expressed as follows:

$$u=u(B.C) \tag{3.5.35}$$

Where B.C is the abbreviation of boundary condition, according to the above explanations, the boundary conditions are applied on Eqs. (3.5.31)–(3.5.34); in the following form:

$$f(-1)=0\rightarrow -a_9+a_8-a_7+a_6-a_5+a_4-a_3$$
$$+a_2-a_1+a_0=0 \tag{3.5.36}$$

$$f(+1)=0 \rightarrow a_9+a_8+a_7+a_6+a_5+a_4+a_3 \\ +a_2+a_1+a_0=0 \tag{3.5.37}$$

$$f'(-1)=0 \rightarrow 9a_9-8a_8+7a_7-6a_6 \\ +5a_5-4a_4+3a_3-2a_2+a_1=0 \tag{3.5.38}$$

$$f'(+1)=-1 \rightarrow 9a_9+8a_8+7a_7+6a_6 \\ +5a_5+4a_4+3a_3+2a_2+a_1=-1 \tag{3.5.39}$$

$$g(-1)=0 \rightarrow -b_9+b_8-b_7+b_6-b_5+b_4-b_3 \\ +b_2-b_1+b_0=0 \tag{3.5.40}$$

$$g(+1)=1 \rightarrow b_9+b_8+b_7+b_6+b_5+b_4+b_3 \\ +b_2+b_1+b_0=1 \tag{3.5.41}$$

$$\theta(-1)=1 \rightarrow -c_7+c_6-c_5+c_4-c_3+c_2-c_1+c_0=1 \tag{3.5.42}$$

$$\theta(+1)=0 \rightarrow c_7+c_6+c_5+c_4+c_3+c_2+c_1+c_0=0 \tag{3.5.43}$$

$$\theta(-1)=1 \rightarrow -d_7+d_6-d_5+d_4-d_3+d_2-d_1+d_0=1 \tag{3.5.44}$$

$$\phi(+1)=0 \rightarrow d_7+d_6+d_5+d_4+d_3+d_2+d_1+d_0=0 \tag{3.5.45}$$

(b) Applying the boundary conditions on the main differential equations, in this case, the study on Eqs. (3.5.27)–(3.5.30), and also on their derivatives is done after substituting Eqs. (3.5.31)–(3.5.34) into the main differential equations as follows:

$$F(f(\eta)) \rightarrow F(f(B.C))=0, F'(f(B.C))=0,\ldots \tag{3.5.46}$$

$$G(g(\eta)) \rightarrow G(g(B.C))=0, G'(g(B.C))=0,\ldots \tag{3.5.47}$$

$$\Theta(\theta(\eta)) \to \Theta(\theta(B.C)) = 0, \Theta'(\theta(B.C)) = 0, \ldots \quad (3.5.48)$$

$$\Phi(\phi(\eta)) \to \Phi(\phi(B.C)) = 0, \Phi'(\phi(B.C)) = 0, \ldots \quad (3.5.49)$$

The boundary conditions on the achieved differential equation are applied based on the above equations. In fact, due to the excellence of AGM compared to other methods, we have to reach a set of polynomials in the solution processes according to the overall number of used constant coefficients in trial functions, which we would normally obtain by simple calculations only. In the proposed problem, we have engaged with four trial functions containing 36 constant coefficients and ten equations according to Eqs. (3.5.36)–(3.5.45); we must create 26 additional equations from Eqs. (3.5.46)–(3.5.49) to achieve a set of polynomials containing 36 equations and 36 constants. According to the above explanations, we have created additional equations (3.5.46)–(3.5.49) in the following order:

I. Six equations have been created by calculating equations from:

$$F(-1) = 0, F(+1) = 0, F'(-1) = 0, F'(+1) = 0, F''(-1) = 0, F''(+1) = 0$$

II. Eight equations have been created by calculating equations from:

$$G(-1) = 0, G(+1) = 0, G'(-1) = 0, G'(+1) = 0, G''(-1) = 0, G''(+1) = 0$$
$$, G'''(-1) = 0, G'''(+1) = 0$$

III. Six equations have been created by calculating equations from:

$$\Theta(-1) = 0, \Theta(+1) = 0, \Theta'(-1) = 0, \Theta'(+1) = 0, \Theta''(-1) = 0, \Theta''(+1) = 0$$

IV. Six equations have been created by calculating equations from:

$$\Phi(-1) = 0, \Phi(+1) = 0, \Phi'(-1) = 0, \Phi'(+1) = 0, \Phi''(-1) = 0, \Phi''(+1) = 0$$

The equations in the (I)-(IV) subsections are too large to be displayed graphically. So by utilising the above procedures, we have obtained a set of polynomials containing 36 equations and 36 constants which, by solving them, we can obtain Eqs. (3.5.31)–(3.5.34). For instance, when $\mathrm{Re} = 0.1, N_1 = 0.1, N_2 = 0.1, N_3 = 0.1, P_{e_h} = 0.1, P_{e_m} = 0.1,$

substituting constant coefficients from procedures, Eqs. (3.5.31)–(3.5.34) could easily be yielded as follows:

$$f(\eta) = 0.0000011226\eta^9 + 0.00007769\eta^8 - 0.0016237\eta^7 \\ -0.004568\eta^6 - 0.0321635\eta^5 - 0.060965\eta^4 - 0.306566\eta^3 \\ -0.341988\eta^2 + 0.1596354\eta + 0.0924003 \tag{3.5.50}$$

$$g(\eta) = 0.00001106\eta^9 + 0.0000777\eta^8 \\ -0.0016237\eta^7 - 0.004568\eta^6 - 0.0321635\eta^5 - 0.060966\eta^4 \\ -0.306566\eta^3 - 0.341988\eta^2 + 0.159635\eta + 0.0924004 \tag{3.5.51}$$

$$\theta(\eta) = -0.000001704\eta^7 - 0.00001289\eta^6 \\ -0.0012362\eta^5 + 0.060966\eta^4 + 0.00413266\eta^3 \\ -0.0126272\eta^2 - 0.50289\eta + 0.510503 \tag{3.5.52}$$

$$\phi(\eta) = -0.000001704\eta^7 - 0.00001289\eta^6 \\ -0.0012362\eta^5 + 0.0021367\eta^4 + 0.00413266\eta^3 \\ -0.0126272\eta^2 - 0.50289\eta + 0.510503 \tag{3.5.53}$$

In this study, AGM has been utilised to solve the nonlinear differential equation of heat and mass transfer of the steady laminar flow of a micropolar fluid along a two-dimensional channel with porous walls. The geometry of the problem is shown in Fig. 3.5.1, and the processes of obtaining an analytical solution for the problem have been explained clearly in previous sections. The trial functions have been chosen logically in which the boundary conditions (due to the basic idea of AGM) can be satisfied. Also, a symmetric condition could apply at both boundary points: *start point* =-1 and *endpoint*= +1 in this case. We have shown AGM efficiency and accuracy through the proper figures and tables.

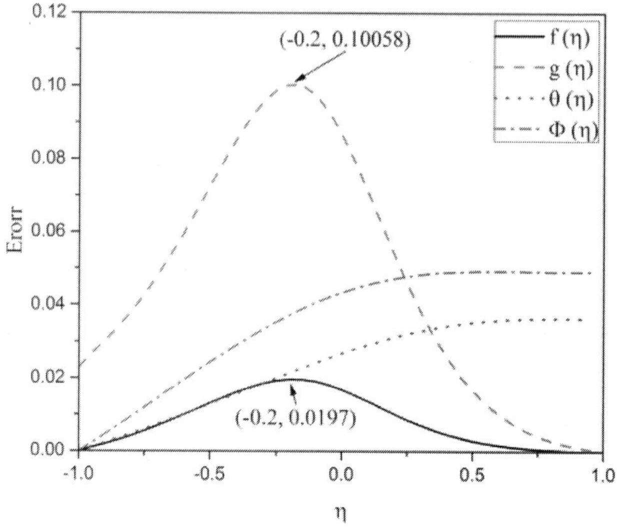

Fig. 3.5.2. Obtained error for $f(\eta), g(\eta), \theta(\eta), \phi(\eta)$ **when** $\mathrm{Re}=1, N_1=N_2=N_3=0.1, pe_h=0.2, pe_m=0.5$

Fig. 3.5.2 shows the difference between the solution obtained by AGM and the numerical method (Runge-Kutte 4th) in which we have introduced an error percentage as follows:

$$\%Error = \left|\frac{u(\eta)_{NM} - u(\eta)_{AGM}}{u(\eta)_{NM}}\right| \times 100 \tag{3.5.54}$$

Where $u(\eta)_{NM}$ is the value obtained by the numerical method (Runge-Kutte 4th) and $u(\eta)_{AGM}$ is the value obtained by AGM. Eq. (3.5.54) has been applied through the functions of Eqs. (3.5.31)–(3.5.34), so the parameter u is defined as a symbol of data in this case. In Fig. 3.5.3(a)–(d), the convergence issue has been considered, which shows that by increasing steps in our assumed trial functions, we will obtain more accurate solutions. In these figures, we have obtained our results due to critical points in Fig. 3.5.2. which are shown as follows:

A comparison between AGM and numerical results for different values of active parameters is shown in Figs. 3.5.4–3.5.6 and Table 3.5.1. Compared with numerical results, the results obtained represented that AGM has enough accuracy and efficiency to solve nonlinear equations of a coupled system. Afterwards, the effect of different parameters, such as the Reynolds number, the micro rotation/angular velocity, and the Peclet number on the flow, heat transfer, and concentration characteristics are discussed. Fig. 3.5.7 shows a set of figures in each of these effects of one parameter. Generally, the micro rotation profile (g) values decrease with the increase of Re, N_1, and N_3. However, it increases when N_2 increases. It is noteworthy to mention that when $N_1 > 1$ and $N_2 < 1$, the behaviour of the angular velocity is oscillatory and irregular. Since Nusselt and Sherwood's numbers have great usage in technological processes, we have shown changes in these dimensionless numbers in Figs. 3.5.8 and 3.5.9. The effects of the Peclet number on the fluid temperature and concentration profiles are shown in Figs. 3.5.8(a) and 3.5.9(a). As shown in Fig. 3.5.8, (a) the fluid temperature increases with the increase of Peclet number and also due to Fig. 3.5.9(a), the concentration profile and the Peclet number increase. On the other hand, according to Fig. 3.5.8(b), an increase in the Peclet number and the Reynolds number leads to an increase in the Nusselt number. Also, according to Fig. 3.5.9(b), an increase in the Peclet and Reynolds numbers leads to an increase in the Sherwood number.

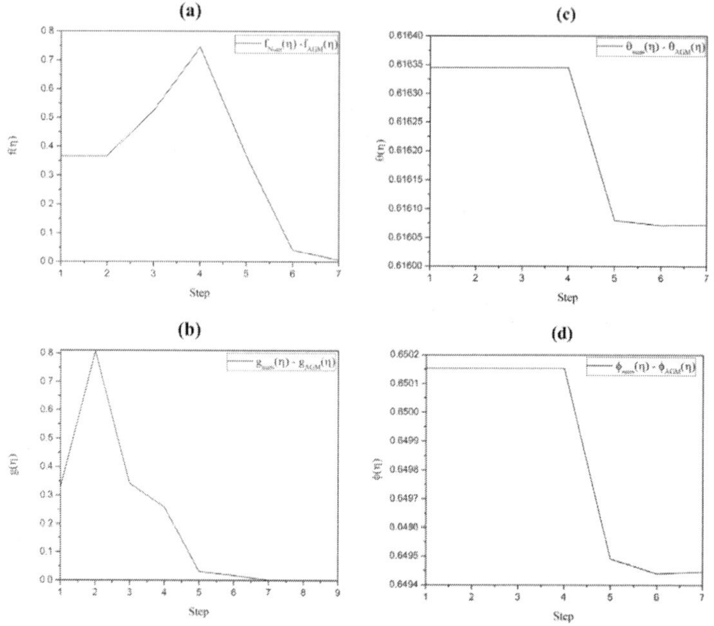

Fig. 3.5.3. Obtained error at different time steps for (a) f(η), (b) g(η), (c) θ(η), (d) Φ(η).

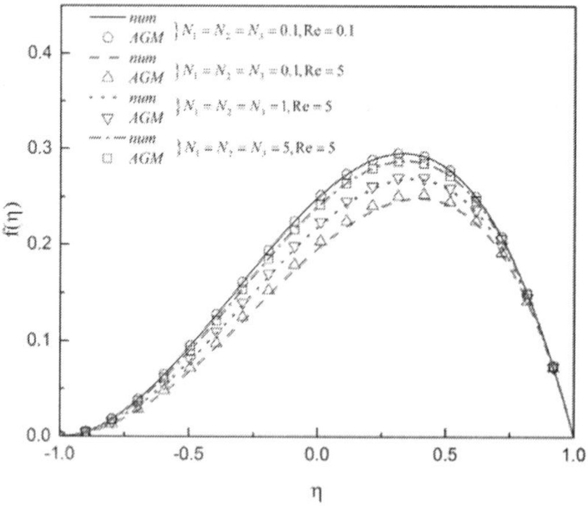

Fig. 3.5.4. Comparison between numerical and AGM solution results for f(η).

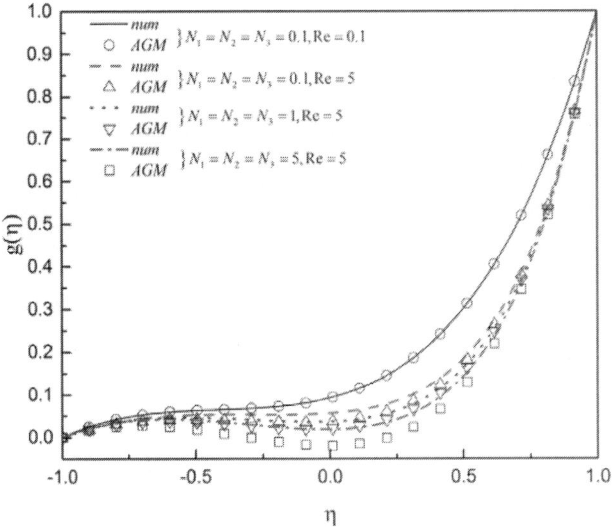

Fig. 3.5.5. Comparison between numerical and AGM solution results for g(η).

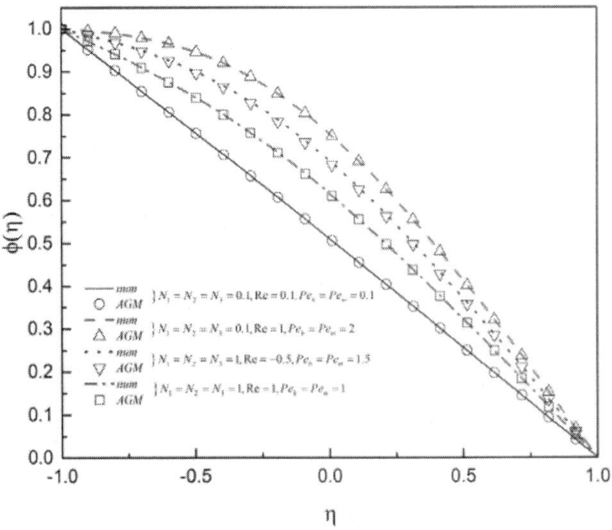

Fig. 3.5.6. Comparison between numerical and AGM solution results for θ(η).

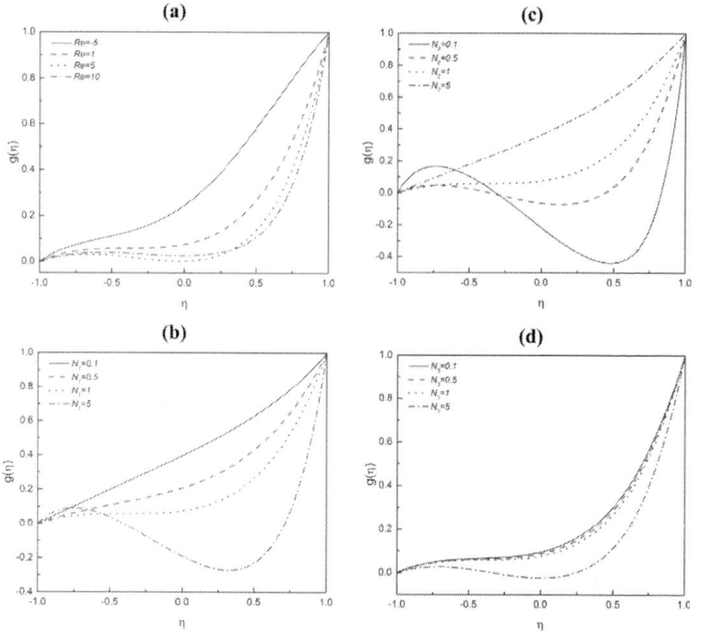

Fig. 3.5.7. The effects of Re, N_1, N_2, N_3 on micro rotation profile (g) when (a) $N_1=N_2=N_3=1$ (b) Re= $N_2=N_3=1$; (c) N_1=Re =$N_3=1$ (d) $N_1=N_2$=Re=1.

Table 3.5.1. Comparison between the numerical results and AGM solution for $\Phi(\eta)$ at various Re, Pe_m when $Pe_h=0.2$, $N_1=N_2=N_3=0.1$.

η	Re = 1, Pe_m = 0.5			Re = 0.5, Pe_m = 0.2		
	Num	AGM	Error	Num	AGM	Error
−1	1	1	0	1	1	0
−0.8	0.9192939269	0.9193811169	0.0000948446	0.9077116680	0.907720689	0.000009938
−0.6	0.8356460368	0.8358088531	0.0001948388	0.8142102267	0.814227024	0.000020630
−0.4	0.7471790040	0.7473962040	0.0002906933	0.7187421955	0.718764563	0.000031120
−0.2	0.6530205940	0.6532642196	0.0003730750	0.6209800569	0.621005159	0.000040423
0	0.5531286373	0.5533690723	0.0004346819	0.5209385166	0.520963368	0.000047706
0.2	0.4481088763	0.4483206457	0.0004725848	0.4188891701	0.418911173	0.000052526
0.4	0.3390200357	0.3391859739	0.0004894642	0.3152725867	0.315289927	0.000055001
0.6	0.2271589176	0.2272708614	0.0004927993	0.2106066932	0.210618442	0.000055789
0.8	0.1138170680	0.1138730124	0.0004915290	0.1053902317	0.105396114	0.000055816
1	0	0	0	0	0	0

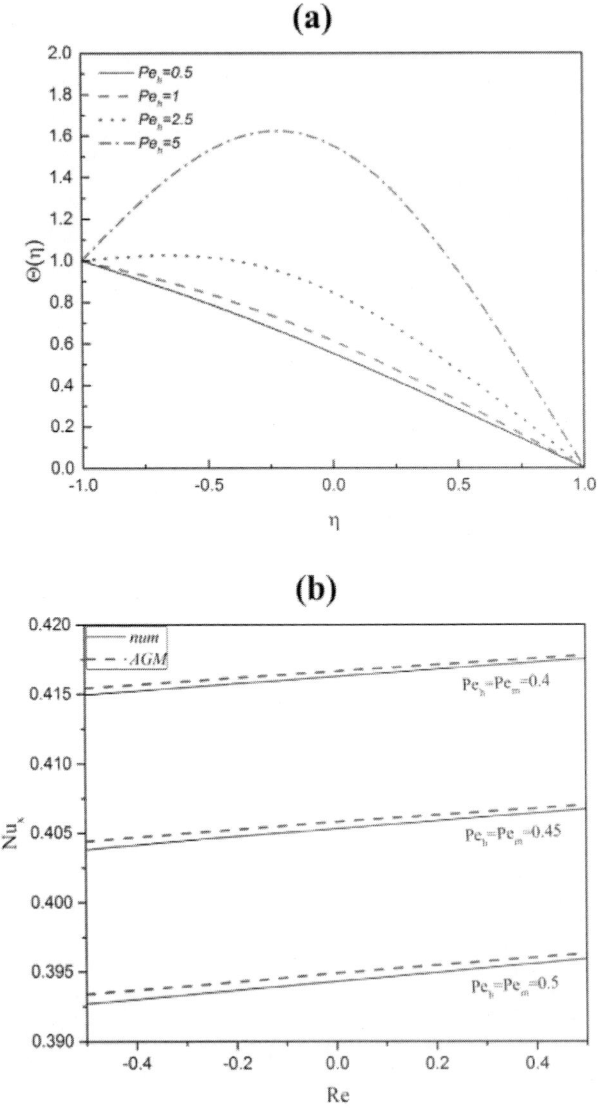

Fig. 3.5.8. (a) The effects of Pe_h on temperature profile at Re $=N_1=N_2=N_3=Pe_m=1$, (b) effects of Re and $Pe_h= Pe_m$ on Nusselt number when $N_1= N_2=N_3=1$.

(a)

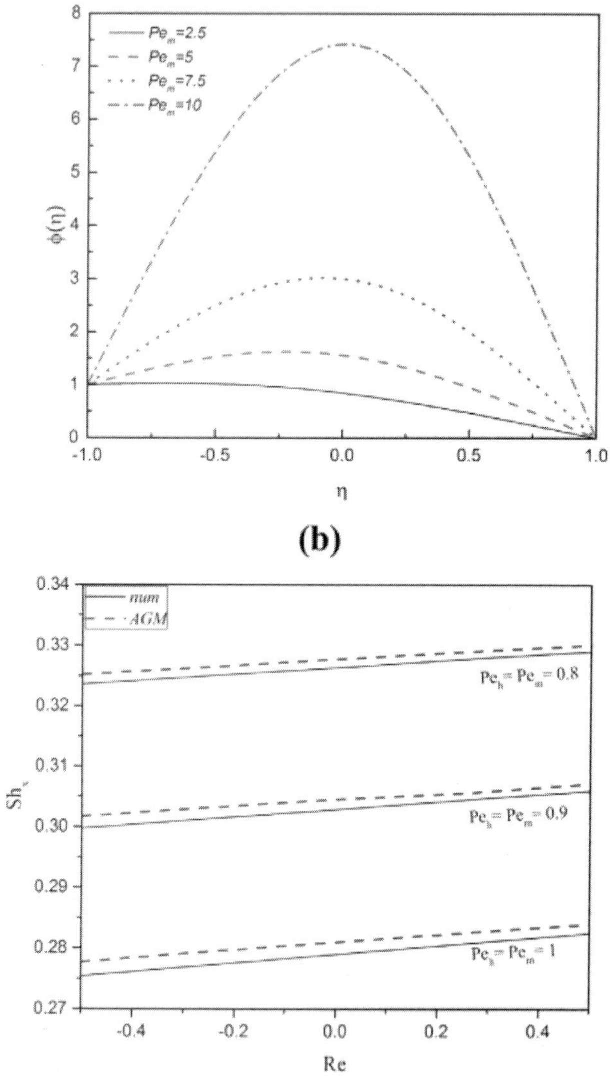

(b)

Fig. 3.5.9. (a) The effects of Pe_m on concentration profile at Re= N_1=N_2=N_3=Pe_h=1, (b) Re and Pe_h=Pe_m on Sherwood number when N_1= N_2= N_3=1.

This study uses AGM to solve the nonlinear differential equation of heat and mass transfer of a steady laminar flow of a micropolar fluid along a two-dimensional channel with porous walls. Different parameter values have been compared among AGM and numerical methods (Runge-Kutte 4th). Data from the error figure show that solutions obtained with AGM have minor differences from exact solutions. Also, the convergence figure shows that we would obtain more accurate solutions by applying more AGM terms. Furthermore, according to the results, the Reynolds number is directly related to the Nusselt and Sherwood numbers. However, the Peclet number has an inverse relationship. Finally, it will be obvious that AGM is a convenient analytical method, and due to its accuracy, efficiency, and convergence, it could be applied to solving nonlinear problems.

List of symbols

B	Non-uniform magnetic field
B_0	uniform magnetic field
C	fluid concentration
C_0	initial fluid concentration
D_m	mass diffusion
Q	heat generation/absorption parameter
Q^*	uniform volumetric heat generation/absorption coefficient
T	fluid temperature
T_0	initial fluid temperature
R	radiation parameter
S	squeezing parameter
Ha	Hartman number
Pr	Prandtl number
Ec	Eckert number

Sc	Schmidt number
Cf	Skin friction coefficient
Nu_x	Nusselt number
Sh	Sherwood number
Re_x	Reynolds number
C_p	specific heat capacity
c	characteristic of the fluid
j_w	the mass flux
k	thermal conductivity
k^*	mean absorption coefficient
l	the initial distance between the plates
q_w	the wall heat flux
u, v	velocity components
x, y	space coordinates
α	characteristic parameter
β	characteristic of the fluid
γ	chemical reaction parameter
μ	dynamic viscosity
ν	kinematic viscosity
σ	electrical conductivity of the fluid
σ^*	Stefan-Boltzmann constant
ρ	fluid density
Γ, δ	Eyring- Powell fluid parameters

θ	dimensionless temperature
$\emptyset$	dimensionless concentration
τ_w	the wall shear stress
η	dimensionless variable
AGM	Akbari-Ganji's Method
ρ	fluid density
ρ_s	particle density
P_0	ambient pressure on the film surface
T	temperature
T_0	the temperature on the film surface
T_w	the temperature on the disk
$\bar{g}$	gravitational acceleration
t	the thickness of the fluid film
W	velocity of spraying
u, v, w	velocity components along the x, y, and z axes, respectively
Ω	angular velocity
α	thermal diffusivity of the fluid
β	the angle between the horizontal axis and the disk
η	dimensionless variable
δ	normalised thickness
ϑ	kinematic viscosity
Ec	Eckret number

v^*	Dimensionless velocity
C_p	Specific heat capacity
B	Constant applied magnetic field
P_r	Prandtl number
Ha	Hartmann number
Nu	Nusslet number
Re	Reynolds number
r^*	Dimensionless radius
Rd	Radiation parameter
k	Thermal conductivity
σ	Electrical conductivity
Ω	Constant rotation velocity
η	Aspect ratio
ϕ	Nanoparticle volume fraction
ρ	Fluid density
a	Thermal diffusivity
θ	Dimensionless temperature
μ	Dynamic viscosity
A, B	Symmetric kinematic matrices
C	Species heat
C_n	Blade-wall temperature coefficients
$\frac{\partial v_m}{\partial x_n}$	Velocity gradients

$\frac{\partial a_m}{\partial x_n}$	Acceleration gradients
x_k	General coordinates
$f(\eta)$	Velocity function
F	Transformation of f
$\bar{k}$	Fluid thermal conductivity
N	Power-law index in temperature distribution
K_r	Rotation parameter
AGM	Akbari-Ganji's method (AGM)
p	Pressure
T	Fluid temperature
$q_n(\eta)$	Temperature function
τ_{ij}	Stress tensor component
V	Injection velocity
u_r, u_z	Velocity components in r and z directions, respectively
η	Dimensionless coordinates in the z-direction
r, θ, z	Cylindrical-coordinate symbols
ϕ_k	Viscosity coefficients
φ	Dissipation function
C	species concentration
D^*	thermal conductivity and molecular diffusivity
f	dimensionless stream function
g	dimensionless micro rotation

h	width of channel
j	micro-inertia density
N	micro rotation/angular velocity
$N_{1,2,3}$	dimensionless parameter
Sh	Sherwood number
Sc	Schmidt number
Pe	Peclet number
q	mass transfer parameter
s	micro rotation boundary condition
(u, v)	Cartesian velocity components
(x, y)	Cartesian coordinate components
η	similarity variable
θ	dimensionless temperature
κ	coupling coefficient
ρ_s	micro rotation/spin-gradient viscosity
ψ	stream function

References

1. Ghadikolaei, S. S., Hosseinzadeh, K., & Ganji, D. D. (2017). "Analysis of unsteady MHD Eyring-Powell squeezing flow in stretching channel with considering thermal radiation and Joule heating effect using AGM." *Case studies in thermal engineering*, 10, 579-594.

2. Berkan, S., Hoseini, S. R., & Ganji, D. D. (2017). "Analytical investigation of steady three-dimensional problem of condensation film on inclined rotating disk by Akbari-Ganji's method." *Propulsion and Power Research*, 6(4), 277-284.

3. Sheikholeslami, M., Nimafar, M., & Ganji, D. D. (2017). "Nanofluid heat transfer between two pipes considering Brownian motion using AGM." *Alexandria Engineering Journal*, 56(2), 277-283.

4. Mirgolbabaee, H., Ledari, S. T., Sheikholeslami, M., & Ganji, D. D. (2017). "Semi-analytical investigation of momentum and heat transfer of a non-Newtonian fluid flow for specific turbine cooling application using AGM." *International Journal of Applied and Computational Mathematics*, 3(1), 1463-1475.

5. Mirgolbabaee, H., Ledari, S. T., & Ganji, D. D. (2017). "Semi-analytical investigation on micropolar fluid flow and heat transfer in a permeable channel using AGM." *Journal of the Association of Arab Universities for Basic and Applied Sciences*, 24, 213-222.

6. Jalili, B., Emad, M., Jalili, P., Ganji, D. D., Saleem, S., & Tag-eldin, E. M. (2023). Thermal analysis of boundary layer nanofluid flow over the movable plate with internal heat generation, radiation, and viscous dissipation. *Case Studies in Thermal Engineering*, 103203.

7. Jalili, P., Narimisa, H., Jalili, B., Shateri, A., & Ganji, D. D. (2023). A novel analytical approach to micro-polar nanofluid thermal analysis in the presence of thermophoresis, Brownian motion and Hall currents. *Soft Computing*, *27*(2), 677-689.

8. Jalili, B., Ganji, A. D., Jalili, P., Nourazar, S. S., & Ganji, D. D. (2022). Thermal analysis of Williamson fluid flow with Lorentz force on the stretching plate. *Case Studies in Thermal Engineering*, *39*, 102374.

CHAPTER 4

APPLICATION OF AGM IN NANOFLUID FLOW

4.1. Magneto Hydrodynamic Nanofluid Flow between Two Parallel

In this problem, the heat and mass transfer process of steady nanofluid flow between two parallel plates is investigated in a uniform magnetic field. The impact of thermophoresis and Brownian motion has been considered in the nanofluid model.

Consider the steady flow of a nanofluid between two horizontal parallel plates when plates and fluid rotate together around the y-axis, which is normal to the plates with an angular velocity (see Fig. 4.1.1). The lower plate is located at y=0. The upper one is located at y=h. Two opposite and equal forces stretch the plate at y=0. Therefore the position of the point (0, 0, 0) remains unchanged. A uniform magnetic flux with density B_0 is applied along the y-axis. Under these assumptions, the governing equations are as follows:

$$\frac{\partial_u}{\partial_x}+\frac{\partial_v}{\partial_y}=0 \tag{4.1.1}$$

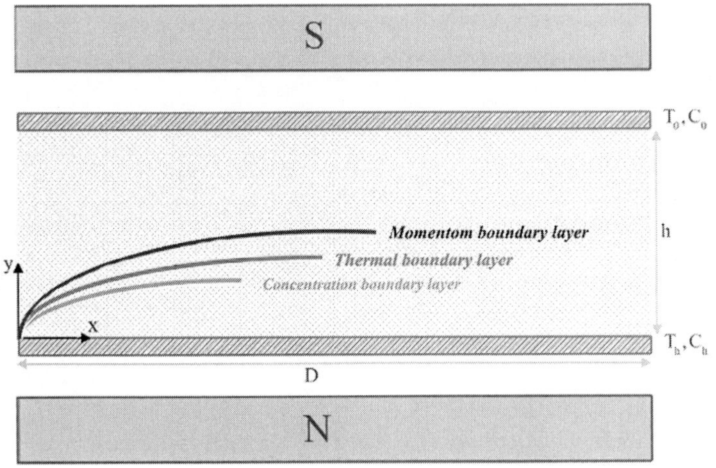

Fig. 4.1.1. The geometry of the problem.

$$\rho_f\left(u\frac{\partial_u}{\partial_x}+v\frac{\partial_u}{\partial_y}\right)=-\frac{\partial_p}{\partial_x}+\mu\left(\frac{\partial^2{}_u}{\partial_x{}^2}+\frac{\partial^2{}_u}{\partial_y{}^2}\right)-\sigma B^2{}_0u \tag{4.1.2}$$

$$\rho_f\left(u\frac{\partial_u}{\partial_x}+v\frac{\partial_u}{\partial_y}\right)=-\frac{\partial_p}{\partial_x}+\mu\left(\frac{\partial^2{}_u}{\partial_x{}^2}+\frac{\partial^2{}_u}{\partial_y{}^2}\right) \tag{4.1.3}$$

$$\begin{aligned}\left(u\frac{\partial T}{\partial_x}+v\frac{\partial T}{\partial_y}\right)=a\left(\frac{\partial^2T}{\partial_x{}^2}+\frac{\partial^2T}{\partial_y{}^2}\right)\\+\frac{(\rho C_P)p}{(\rho C_P)c}[D_B\left(\frac{\rho C}{\rho x}.\frac{\rho T}{\rho x}+\frac{\rho C}{\rho y}.\frac{\rho T}{\rho y}\right)\\+(\frac{D_T}{D_C})(\left(\frac{\partial_T}{\partial_x}\right)^2+\left(\frac{\partial_T}{\partial_y}\right)^2)]\end{aligned} \tag{4.1.4}$$

$$u\frac{\partial_C}{\partial_X}+\frac{\partial_C}{\partial_y}=D_B\left(\frac{\partial^2{}_C}{\partial_x{}^2}+\frac{\partial^2{}_C}{\partial_y{}^2}\right)+\left(\frac{D_T}{D_0}\right)+(\frac{\partial^2T}{\partial_x{}^2}+\frac{\partial^2T}{\partial_y{}^2}) \tag{4.1.5}$$

Where u and v are the velocity components in the x and y directions, respectively, and $\rho_f, D_B, \mu, k, C,\ p^*$, and T are base fluid's density, diffusion coefficient of the diffusing species, viscosity, thermal conductivity, specific heat of nanofluid, modified fluid pressure and temperature. The boundary conditions are as follows:

$$\{ @y = 0 \rightarrow u = ax, v = 0, T = T_h, C = C_h$$
$$\{ @y = h \rightarrow u = 0, v = 0, T = T_0, C = C_0 \tag{4.1.6}$$

Consider the non-dimensional variables in the following order:

$$\eta = \frac{y}{h}, u = axf'(\eta), v = -ahf(\eta), \theta(\eta) = \frac{T - T_h}{T_0 - T_h}, \phi(\eta) = \frac{C - C_h}{C_0 - c_h} \tag{4.1.7}$$

So, the dimensionless forms of governing equations are as follows:

$$f^{(4)} - R(f'f'' - ff'') - Mf'' = 0 \tag{4.1.8}$$

$$\theta'' + \mathrm{Pr}Rf\theta' + Nb\phi'\theta' + Nt\theta'^2 = 0 \tag{4.1.9}$$

$$\phi'' + \mathrm{R.Sc}f\phi' + \frac{\mathrm{Nt}}{\mathrm{Nb}}\theta'' = 0 \tag{4.1.10}$$

And the boundary conditions of dimensionless equations are as follows:

$$\{ @\eta = 0 \rightarrow f = 0, f' = 1, \theta = 1, \phi = 1$$
$$\{ @\eta = 1 \rightarrow f = 0, f' = 0, \theta = 0, \phi = 0 \tag{4.1.11}$$

Dimensionless quantities are defined as bellows. R is the viscosity coefficient, M is the magnetic parameter, Pr is the Prandtl number, Sc is the Schmidt number, Nb is the Brownian motion parameter, and Nt is the thermophoretic parameter.

$$R = \frac{ah^2}{\upsilon}, = \frac{\sigma {B_0}^2}{\rho \upsilon}, \mathrm{Pr} = \frac{\mu}{\rho_f \alpha}, Sc = \frac{\mu}{\rho_f D}, Nb = \frac{(\rho C)_p D_B (C_h)}{(\rho C)_f a}, Nt = \frac{(\rho C)_p D_T (T_H)}{(\rho C)_f a T_C} \tag{4.1.12}$$

The Nusselt number Nu along the stretching wall is defined as:

$$Nu = -\theta'$$

Due to the coupled system of nonlinear differential equations and the AGM's basic idea, we rewrite Eqs. (4.1.8)–(4.1.10) in the following order:

$$F(\eta) = f^{iv} - R(f'f'' - ff'') - Mf'' = 0 \tag{4.1.14}$$

$$\vartheta(\eta) = \theta'' + \Pr Rf\theta' + Nb\phi'\theta' + Nt\theta'^2 = 0 \tag{4.1.15}$$

$$\phi(\eta) = \phi + R.Scf\phi' + \frac{Nt}{Nb}\theta = 0 \tag{4.1.16}$$

According to the basic idea of AGM, we have investigated an appropriate trial function as the solution of the considered differential equation, which is a finite series of polynomials with constant coefficients, as follows:

$$F(\eta) = \sum_{i=0}^{12} a_i\eta^i = a_0 + a_1\eta + a_2\eta^2 + a_3\eta^3 + a_4\eta^4 + a_5\eta^5 + a_6\eta^6 + a_7\eta^7 + a_8\eta^8 + a_9\eta^9 + a_{10}\eta^{10} + a_{11}\eta^{11} + a_{12}\eta^{12} \tag{4.1.17}$$

$$\vartheta(\eta) = \sum_{i=0}^{6} b_i\eta^i = b_0 + b_1\eta + b_2\eta^2 + b_3\eta^3 + b_4\eta^4 + b_5\eta^5 + b_6\eta^6 \tag{4.1.18}$$

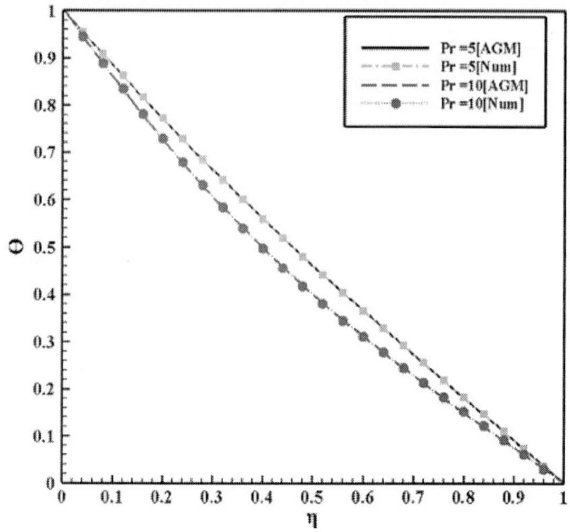

a

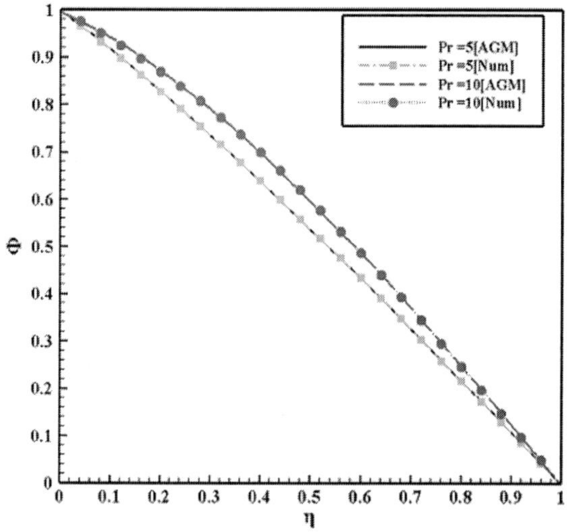

b

Fig. 4.1.2. Comparison of the temperature and concentration profiles between the AGM and the numerical method (Runge-Kutte 4th) when M=1, R=1, Sc=0.1, Nt=0.1, Nb=0.1.

$$\phi(\eta) = \sum_{i=0}^{6} c_i \eta^i = c_0 + c_1\eta + c_2\eta^2 + c_3\eta^3 + c_4\eta^4 + c_5\eta^5 + c_6\eta^6 \tag{4.1.19}$$

In AGM, the boundary conditions are exerted in order to calculate constant coefficients of Eqs. (4.1.17)–(4.1.19) according to the following approaches:

(a) Exerting the boundary conditions on Eqs. (4.1.17)–(4.1.19) are expressed as follows:

$$y = y(B.C) \tag{4.1.20}$$

Where B.C is a compendium of boundary conditions, due to the above explanations, the boundary conditions are exerted on Eqs. (4.1.17)–(4.1.19) in the following order:

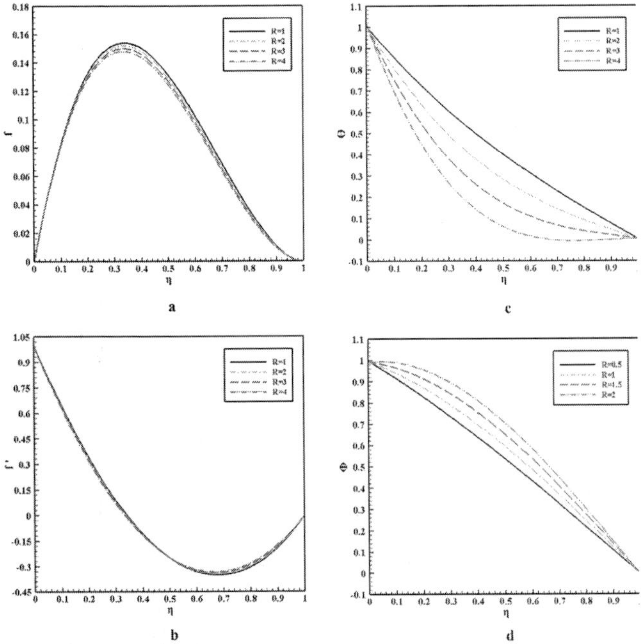

Fig. 4.1.3. The effect of viscosity coefficient on profiles of velocity, temperature, and concentration when Pr=10, M=1, Sc=0.1, Nt=0.1, Nb=0.1.

$$f(0) = 0 \rightarrow a_0 = 0 \tag{4.1.21}$$

$$f(1) = 0 \rightarrow a_0 + a_1 + a_2 + a_3 + a_4 + a_5 + a_6 + a_7 + a_8 + a_9 + a_{10} + a_{11} + a_{12} = 0 \tag{4.1.22}$$

$$f'(0) = 1 \rightarrow a_1 = 1 \tag{4.1.23}$$

$$f'(1) = 0 \rightarrow a_1 + 2a_2 + 3a_3 + 4a_4 + 5a_5 + 6a_6 + 7a_7 + 8a_8 + 9a_9 + 10a_{10} + 11a_{11} + 12a_{12} = 0 \tag{4.1.24}$$

$$\theta(0) = 1 \rightarrow b_0 = 1 \tag{4.1.25}$$

$$\theta(1) = 0 \rightarrow b_0 + b_1 + b_2 + b_3 + b_4 + b_5 + b_6 = 0 \tag{4.1.26}$$

$$\phi(0) = 1 \rightarrow c_0 = 1 \tag{4.1.27}$$

$$\phi(1) = 0 \rightarrow c_0 + c_1 + c_2 + c_3 + c_4 + c_5 + c_6 = 0 \tag{4.1.28}$$

(b) Exerting the boundary conditions on the main differential equations in the present study are Eqs. (4.1.14)–(4.1.16), and also on their derivatives done after substituting Eqs. (4.1.17)–(4.1.19) into the main differential equations in the following form:

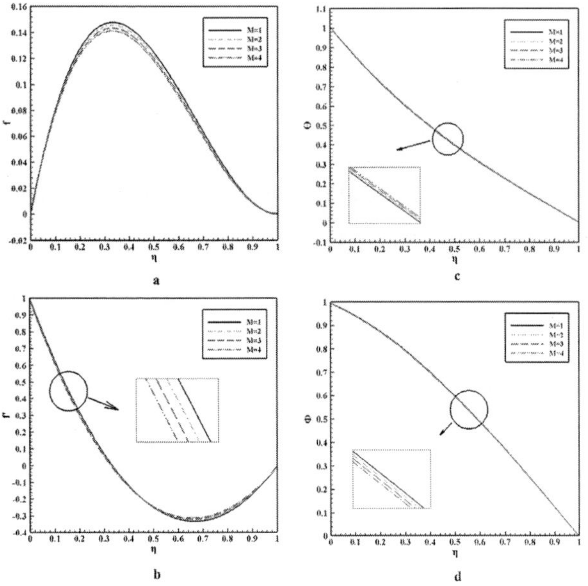

Fig. 4.1.4. The effect of a magnetic parameter on profiles of velocity, temperature, and concentration when Pr=10, R=1, Sc=0.1, Nt=0.1, Nb=0.1.

$$F\big(f(\eta)\big) \to F\big(f(B.C)\big) = 0, F'\big(f(B.C)\big) = 0, \ldots \tag{4.1.29}$$

$$\Theta\big(\theta(\eta)\big) \to \Theta\big(\theta(B.C)\big) = 0, \Theta'\big(\theta(B.C)\big) = 0, \ldots \tag{4.1.30}$$

$$\Phi(\phi(\eta)) \to \Phi(\phi(B.C) = 0, \Phi'(\phi(B.C) = 0, \ldots \tag{4.1.31}$$

The boundary conditions are exerted on the above differential equation. In fact, according to the AGM's superiority to other methods, we can achieve a set of polynomials in the solution that are the same number as the overall used constant coefficients in trial functions. In the proposed problem, we have considered two trial functions containing 27 constant coefficients and just eight equations according to Eqs. (4.1.21)–(4.1.28); we must create 19 additional equations from Eqs. (4.1.29)–(4.1.31) to reach a set of polynomials containing 27 equations and 27 constants. Due to the above explanations, we have created extra equations from Eqs. (4.1.29)–(4.1.31) as follows:

A. Nine equations have been created by calculating equations obtained from:

$$F(0) = 0, F(1) = 0, F'(0) = 0, F'(1) = 0, F''(0)=0, F''(1) = 0, F'''(0) = 0, F'''(1) = 0, F^{(4)}(1) = 0$$

B. 5 equations have been created by calculating equations obtained from:

$$\Theta(0) = 0, \Theta(1) = 0, \Theta'(1) = 0, \Theta''(0) = 0, \Theta''(1) = 0$$

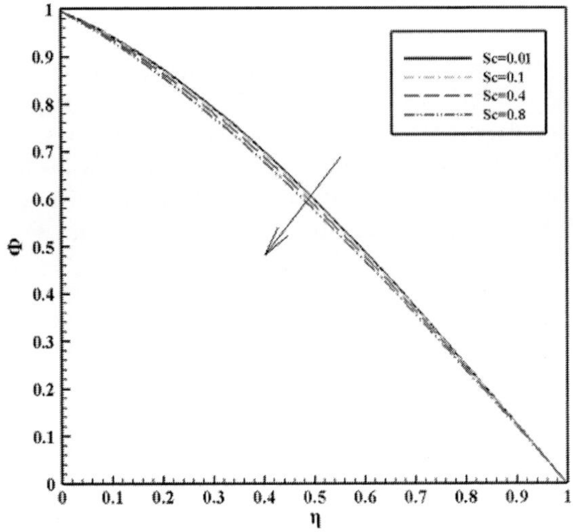

a

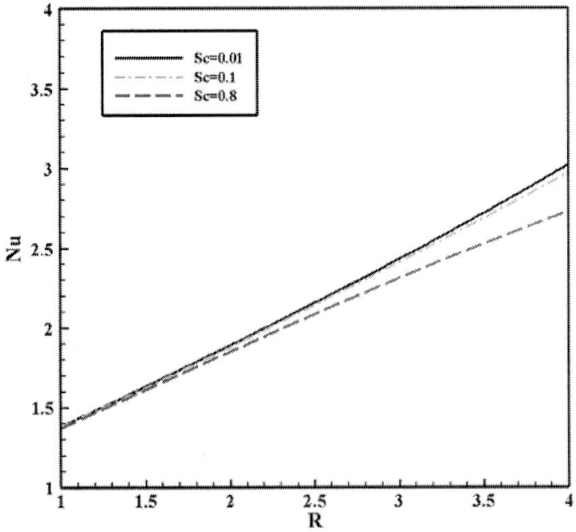

b

Fig. 4.1.5. The effect of the Schmidt number on the Nusselt number and concentration profile when Pr=10, M=1, Nt=0.1, Nb=0.1.

C. Five equations have been created by calculating equations from:

$$\Phi(0) = 0, \Phi(1) = 0, \Phi'(1) = 0, \Phi''(1) = 0, \Phi'''(0) = 0$$

By using the above actions, we have received a set of polynomials containing 27 equations and 27 constants that, by solving them, we can reach Eqs. (4.1.17)–(4.1.19). For example, when Nb=0.1, Nt=0.1, R=1, M=1, Pr=10, Sc=0.1, and substituting constant coefficients from actions in Eqs. (4.1.17)–(4.1.19) can be yielded as follows:

$$\begin{aligned} f(\eta) = 1.03116811 * 10^{-9}\eta^{12} + 1.424738541 \\ * 10^{-7}\eta^{11} - 8.030473083 * 10^{-7}\eta^{10} \\ + 0.00001815343807\eta^{9} \\ - 0.00007562389332\eta^{8} \\ + 0.00130938364\eta^{7} \\ - 0.00423562052\eta^{6} \\ + 0.5499411288\eta^{5} \\ - 0.1270686157\eta^{4} + 1.099882258\eta^{3} \\ - 2.024823388\eta^{2} + \eta \end{aligned} \tag{4.1.32}$$

$$\begin{aligned} \theta(\eta) = -0.0871257746\eta^{6} + 0.7954320201\eta^{5} \\ - 2.010948956\eta^{4} + 1.826878948\eta^{3} \\ - 0.1393336177\eta^{2} - 1.38490262\eta \\ + 1 \end{aligned} \tag{4.1.33}$$

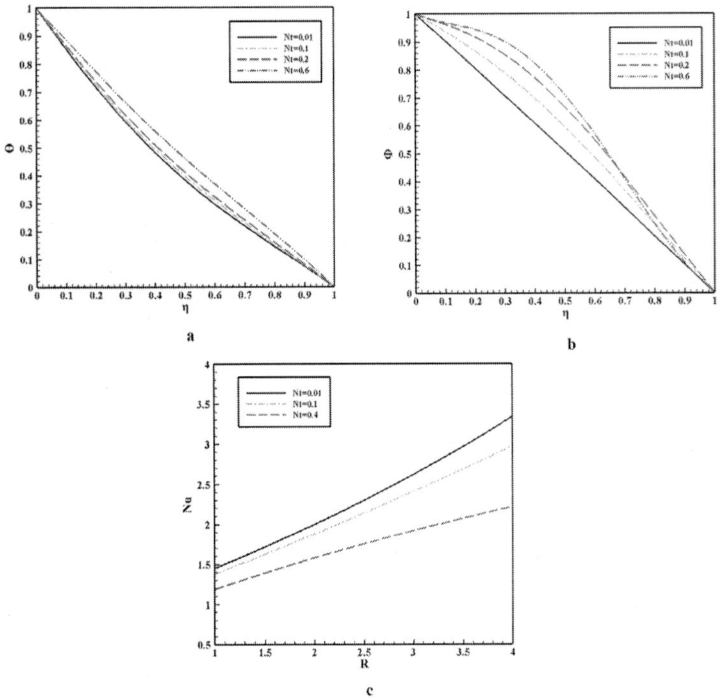

Fig. 4.1.6. The effect of thermophoresis parameter on the Nusselt number, temperature profile, and concentration profile when Pr=10, M=1, Sc=0.1, Nb=0.1.

$$\begin{aligned}\phi(\eta) = 0.08120070963\eta^6 - 0.7625150144\eta^5 \\ + 1.945663591\eta^4 - 1.776406245\eta^3 \\ + 0.1393336177\eta^2 - 0.6272766586\eta \\ + 1\end{aligned} \quad (4.1.34)$$

This study uses AGM to simulate nanofluid flow and heat transfer in the magnetic field. The influences of effective parameters such as viscosity coefficient, magnetic, thermophoretic, and Brownian parameters on flow, heat, and mass transfer characteristics have been depicted as graphs. In order to verify the correctness of the present AGM code, we have compared the results for the velocity, temperature, and concentration profiles with the numerical method (Runge-Kutte 4th). Fig. 4.1.2 proves the precision of AGM. Figs. 4.1.3 and 4.1.4 illustrate the influence of magnetic and viscosity coefficients on profiles of velocity, temperature, and concentration. As shown in Figs. 4.1.3 and 4.1.4, the temperature profile increases with an increase in the magnetic parameter while it reduces with the augmentation of the viscosity coefficient. Also, as the magnetic parameter increases, the concentration profile reduces while the opposite trend is observed for the viscosity coefficient. The effect of the Schmidt number on the concentration profile and the Nusselt number is depicted in Fig. 4.1.5. Both reduce with the rise of the Schmidt number. Figs. 4.1.6 and 4.1.7 show the effect of thermophoresis and the Brownian parameters on temperature, concentration, and the Nusselt number.

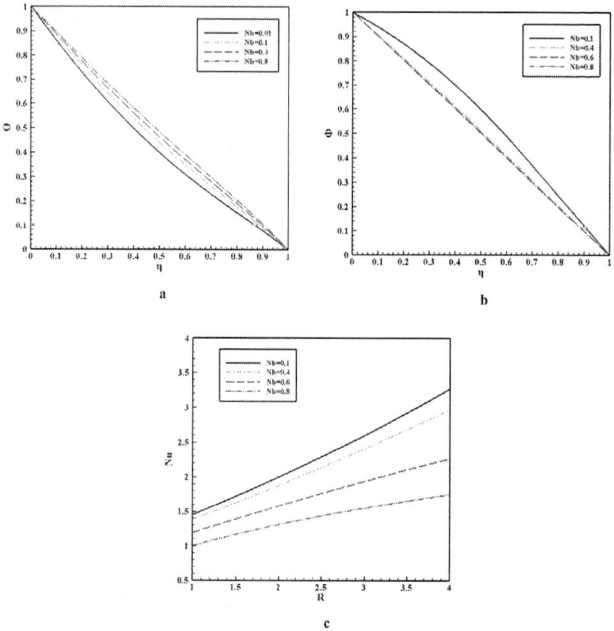

Fig. 4.1.7. The effect of the Brownian parameter on the Nusselt number, the temperature profile, and the concentration profile when Pr=10, M=1, Sc=0.1, Nb=0.1.

Thermophoresis is created due to the Brownian motion of nanoparticles in fluids with a constant gradient of temperature externally sustained. An analysis of the heat convection and mass diffusion shows that particle dispersion and Brownian force are great when the fluid temperature is higher. Small particles disperse faster in the hotter domain and slower in the colder domain because of a temperature gradient in the flow region of the suspension. Migration from a hotter to a colder region of the fluid domain is the collective impact of the dispersion of the nanoparticles. In other words, with the help of the temperature gradient, nanoparticles motion on average against this gradient. The average motion of this nanoparticle is known as thermophoresis. Research shows that thermophoretic and Brownian motion on the mass diffusion is completely reversed. Nanoparticle concentration reduces for higher Brownian motion. However, it increases with a stronger thermophoretic of the nanofluid, enhancing its thermal behaviour. Increasing thermophoretic and Brownian parameters augment the temperature profile, but the Nusselt number has a reverse relationship with these parameters. Also, the concentration profile increases with an increase in the thermophoretic parameter, but it reduces with the rise of the Brownian parameter.

Nanofluid flow, heat, and mass transfer between two parallel plates in a uniform magnetic field are investigated. The effects of thermophoretic and Brownian are considered. Governing equations are solved utilising AGM. A comparison between AGM and the numerical method (Runge-Kutte 4th) is done. The Schmidt number, thermophoresis, Brownian, Magnetic, and viscosity coefficient roles are demonstrated as graphs. The results show that the Nusselt number decreases with increased thermophoretic and Brownian parameters, while the opposite trend is observed for the viscosity coefficient. The concentration of nanofluid decreases when Brownian motion increases while it enhances with a rise in the thermophoretic parameters.

4.2. A Hydrothermal Analysis of MHD Nanofluid

The nanofluid flow between two stretchable and rotating disks with homogeneous and heterogeneous reactions and Joule heating is investigated in a magnetic field. The thermal radiation effect has been considered in the energy equation.

This analysis considers two continuously stretching rotating disks with the nanofluid flow and titanium dioxide (TiO_2) and gallic oxide (GO) as nanoparticles with water (H_2O) as the base. The thermos-physical properties of the nanofluid are given in Table 4.2.1.

The lower disk at $z = 0$ is rotating with angular velocity Ω_1 around the axial direction, and its stretching rate is a_1 while the upper disk (at a distance h apart) rotates with angular velocity Ω_2 and the stretching rate a_2 (Fig. 4.2.1).

The heat transfer contains radiation and Joule heating. The magnetic field of strength b_0 is applied in the z-direction. Homogeneous-heterogeneous reactions are considered. The homogeneous reaction of cubic autocatalysis is:

$$B + 2C = 3C, \omega = k_c\, bc^2 \tag{4.2.1}$$

And the first isothermal reaction on the surface of the catalyst is:

$$B \rightarrow C, \omega = k_s\, b \tag{4.2.2}$$

In the reactions, k_c, k_s, B and C are the rate constants and chemical species, respectively, whereas the b and c are concentrations. We adopted cylindrical coordinates (r, θ, z) with velocity $(\hat{u}, \hat{v}, \hat{w})$ to construct the velocity, temperature, and concentration equation as follows:

$$\frac{du}{dr} + \frac{u}{r} + \frac{dw}{dz} = 0 \tag{4.2.3}$$

$$\hat{u}\frac{\partial \hat{u}}{\partial r} + \hat{w}\frac{\partial \hat{u}}{\partial z} - \frac{\hat{v}^2}{r} = -\frac{1}{\rho_{nf}}\frac{\partial \hat{p}}{\partial r} + \upsilon_{nf}\left(\frac{\partial^2 \hat{u}}{\partial r^2} + \frac{1}{r}\frac{\partial \hat{u}}{\partial r} + \frac{\partial^2 \hat{u}}{\partial z^2} - \frac{\hat{u}}{r^2}\right) - \frac{\sigma_{nt}}{\rho_{nf}}B_0^2 \hat{u} \tag{4.2.4}$$

$$\hat{u}\frac{\partial \hat{v}}{\partial r} + \hat{v}\frac{\partial \hat{v}}{\partial z} - \frac{\hat{u}\hat{v}}{r} = \upsilon_{nf}\left(\frac{\partial^2 \hat{v}}{\partial r^2} + \frac{1}{r}\frac{\partial \hat{v}}{\partial r} + \frac{\partial^2 \hat{v}}{\partial z^2} - \frac{\hat{v}}{r^2}\right) - \frac{\sigma_{nf}}{\rho_{nf}}B_0^2 \hat{v} \tag{4.2.5}$$

$$\hat{w}\frac{\partial \hat{w}}{\partial z} + \hat{u}\frac{\partial \hat{w}}{\partial r} = -\frac{1}{\rho_{nf}}\frac{\partial \hat{p}}{\partial z} + \upsilon_{nf}\left(\frac{\partial^2 \hat{w}}{\partial r^2} + \frac{1}{r}\frac{\partial \hat{w}}{\partial r} + \frac{\partial^2 \hat{w}}{\partial z^2}\right) \tag{4.2.6}$$

$$(\rho c_p)_{nf}(\hat{u}\frac{\partial \hat{T}}{\partial r}+\hat{v}\frac{\partial \hat{T}}{\partial z}) = (k_{nf}+\frac{16\sigma^o \hat{T}_2^3}{3k^o})(\frac{1}{r}\frac{\partial \hat{T}}{\partial r}+\frac{\partial^2 \hat{T}}{\partial r^2} + \frac{\partial^2 \hat{T}}{\partial z^2}) + \sigma_{nf}B_o^2(\hat{u}^2+\hat{v}^2) \tag{4.2.7}$$

$$\hat{u}\frac{\partial b}{\partial r}+\hat{w}\frac{\partial b}{\partial z} = D_B(\frac{1}{r}\frac{\partial b}{\partial r}+\frac{\partial^2 b}{\partial r^2}+\frac{\partial^2 b}{\partial z^2}) - k_c bc^2 \tag{4.2.8}$$

$$\hat{u}\frac{\partial c}{\partial r}+\hat{w}\frac{\partial c}{\partial z} = D_C(\frac{1}{r}\frac{\partial c}{\partial r}+\frac{\partial^2 c}{\partial r^2}+\frac{\partial^2 c}{\partial z^2}) + k_c bc^2 \tag{4.2.9}$$

$$\hat{u} = ra_1, \hat{v} = r\Omega_1, \hat{w} = 0, \hat{T} = \hat{T}_1,$$

$$D_B\frac{\partial b}{\partial z} = k_s b, D_C\frac{\partial c}{\partial z} = -k_s b \ @ \ z = 0 \tag{4.2.10}$$

$$\hat{u} = ra_2, \hat{v} = r\Omega_2, \hat{w} = 0, \hat{T} = \hat{T}_2, b \to b_o, c \to 0 \ @ \ z = h$$

In which $\hat{p}$ indicates pressure, $\hat{T}$ the temperature, $\hat{T}_1$ and $\hat{T}_2$ temperatures at lower and upper disks, respectively, σ^o the Stefan-Boltzmann constant, and k^o the mean absorption coefficient.

The effective nanofluid density ρ_{nf}, thermal conductivity K_{nf}, heat capacitance $(\rho c_p)_{nf}$, dynamic viscosity μ_{nf} and electrical conductivity σ_{nf}, can be introduced as follows:

$$\rho_{nf} = \rho_f(1-\phi)+\rho_s\phi \tag{4.2.11}$$

$$\frac{k_{nf}}{k_f} = \frac{k_s+2k_f-2\phi(k_f-k_s)}{k_s+2k_f+2\phi(k_f-k_s)} \tag{4.2.12}$$

$$(\rho c_p)_{nf} = (\rho c_p)_f(1-\phi)+(\rho c_p)_s\phi \tag{4.2.13}$$

$$\mu_{nf} = \frac{\mu_f}{(1-\phi)^{2.5}} \tag{4.2.14}$$

$$\frac{\sigma_{nf}}{\sigma_f}=1+\frac{3(\frac{\sigma_s}{\sigma_f}-1)\phi}{(\frac{\sigma_s}{\sigma_f}+2)-(\frac{\sigma_s}{\sigma_f}-1)\phi} \tag{4.2.15}$$

In the above expressions, subscript *nf* represents the thermophysical properties of nanofluid, s nano-solid particles, f base fluid and Ø denotes the nanoparticle's solid volume fraction.

With using Karman transformation, it is considered:

$$\hat{u}=r\Omega_1 f'(\xi), \hat{v}=r\Omega_1 g(\xi), w=-2h\Omega_j f(\xi), \hat{\theta}=\frac{\hat{T}-\hat{T}_2}{\hat{T}_1-\hat{T}_2} \tag{4.2.16}$$

$$p=\rho_f\Omega_1 v_f(P(\xi)+\frac{1}{2}\frac{r^2}{h^2}\varepsilon), b=b_o\tilde{\phi}, c=b_o\tilde{l}, \xi=\frac{z}{h}$$

The continuity equation is satisfied, and the other equations (4.2.4-4.2.10) are transformed to:

$$\frac{1}{(1-\phi)^{2.5}(1-\phi+\frac{\rho_s}{\rho_f}\phi)}f''+\mathrm{Re}\,(2ff''-f'^2+g^2-\frac{M}{1-\phi+\frac{\rho_s}{\rho_f}\phi}\frac{\sigma_{nf}}{\sigma_f}f')-\frac{\varepsilon}{1-\phi+\frac{\rho_s}{\rho_f}\phi}=0 \tag{4.2.17}$$

$$\frac{1}{(1-\phi)^{25}(1-\phi+\frac{\rho_s}{\rho_f}\phi)}g''+\mathrm{Re}\,(2fg'-2f'g-\frac{M}{1-\phi+\frac{\rho_s}{\rho_f}\phi}\frac{\sigma_{nf}}{\sigma_f}g)=0 \tag{4.2.18}$$

$$\frac{1}{1-\phi+\frac{\rho_s}{\rho_f}\phi}P' = -4\mathrm{Reff}' - \frac{2}{(1-\phi)^{25}(1-\phi+\frac{\rho_s}{\rho_f}\phi)}f'' \tag{4.2.19}$$

$$\frac{1}{\mathrm{Pr}}(\frac{k_m}{k_f}+R)\theta'' + 2\mathrm{Re}\,(1-\phi+\frac{(\rho c_p)_s}{(\rho c_p)_f}\phi)f\theta' + \mathrm{Re}\;M\frac{\sigma_m}{\sigma_f}Ec(f'^2+g^2) = 0 \tag{4.2.20}$$

$$\frac{1}{\mathrm{Re}}\frac{1}{Sc}\phi'' + 2f\phi' - k_1\phi l^2 = 0 \tag{4.2.21}$$

$$\frac{1}{\mathrm{Re}}\frac{\delta}{Sc}l'' + 2fl' - k_1\phi l^2 = 0 \tag{4.2.22}$$

$$f(0)=0, f(1)=0, f'(0)=A_1, f'(1)=A_2, g(0) = 1, g(1)=\Omega, \theta(0)=1,$$
$$\phi'(0)=K_2\phi(0) \tag{4.2.23}$$
$$\phi(1)=1, \delta l'(0) = -K_2\phi(0), l(1)=0, P(0)=0$$

$$\mathrm{Re}=\frac{\Omega_1 h^2}{\nu_f}, \mathrm{Pr}=\frac{(\rho c_p)_f \nu_f}{k_f}, M=\frac{B_0^2\sigma_f}{\rho_f\Omega_1},$$
$$Ec=\frac{r^2\Omega_1^2}{c_p(\hat{T}_1-\hat{T}_2)}, Sc=\frac{\nu_f}{D_B}, \delta=\frac{D_C}{D_B} \tag{4.2.24}$$
$$A_1=\frac{a_1}{\Omega_1}, A_2=\frac{a_2}{\Omega_1}, \tau=\frac{\Omega_2}{\Omega_1}, k_1=\frac{k_c b_0^2}{\Omega_1}, k_2=\frac{k_s h}{D_B}$$

Where Re denotes the Reynolds number, Pr the Prandtl number, M the Hartmann number, Ec the Eckert number, Sc the Schmidt number, δ ratio of diffusion coefficient, A_1 and A_2 the scaled stretching parameters, τ rotation number, and k_1 and k_2 the measure of the strength of homogeneous and heterogeneous reactions. The diffusion coefficients of chemical species B and C are of comparable size.

$$\phi(\xi) + l(\xi) = 1 \tag{4.2.25}$$

From expressions (4.2.21) and (4.2.22), one obtains:

$$\frac{1}{\text{Re}}\frac{1}{Sc}\phi'' + 2f\phi' - k_1\phi(1-\phi)^2 = 0 \tag{4.2.26}$$

$$\phi'(0) = K_2\phi(0), \phi(1) = 1 \tag{4.2.27}$$

For making the simpler form of Eq. (4.2.17) and removing ε, it can be differentiated with respect to ξ

$$\frac{1}{(1-\phi)^{25}(1-\phi+\frac{\rho_s}{\rho_f}\phi)}f'' + \text{Re}\,(2ff''' - 2gg' - \frac{M}{1-\phi+\frac{\rho_s}{\rho_f}\phi}\frac{\sigma_{nf}}{\sigma_f}f'') = 0 \tag{4.2.28}$$

At the lower disc, the shear stress in the radial and tangential direction is described by:

$$\tau_{zr} = \mu_{nf}\frac{\partial \hat{u}}{\partial z}|_{z=0} = \frac{\mu_f r\Omega_j f''(0)}{(1-\phi)^{2,5}h}, \tau_{z\theta} = \mu_{nf}\frac{\partial \hat{v}}{\partial z}|_{z=0} = \frac{\mu_f r\Omega_1 g'(0)}{(1-\phi)^{2.5}h} \tag{4.2.29}$$

Therefore the total stress is defined as:

$$\tau_w = \sqrt{\tau_{zr}^2 + \tau_{z\theta}^2} \tag{4.2.30}$$

The skin friction coefficients Cf_1 and Cf_2 in the lower and upper disks are illustrated by:

$$Cf_1 = \frac{\tau_w|_{z=0}}{\rho_f(r\Omega_1)^2} = \frac{1}{Re_r(1-\phi)^{25}}[(f''(0))^2 + (g'(0))^2]^{1/2}$$
$$Cf_2 = \frac{\tau_w|_{z=h}}{\rho_t(r\Omega_1)^2} = \frac{1}{Re_y(1-\phi)^{25}}[(f''(1))^2 + (g'(1))^2]^{'/2} \tag{4.2.31}$$

in which the local Reynolds number is $\text{Re}_r = \frac{r\Omega_1 h}{v_f}$.

The Nusselt numbers for lower and upper disks can be expressed as:

$$Nu_{x1} = \frac{hq_w}{k_f(\hat{T}_1 - \hat{T}_2)}|_{z=0}, Nu_{x2} = \frac{hq_w}{k_f(\hat{T}_1 - \hat{T}_2)}|_{z=h} \quad (4.2.32)$$

In which wall heat flux is:

$$\begin{aligned} q_w|_{z=0} &= -k_{nf}\frac{\partial \hat{T}}{\partial z} + q_r|_{z=0} \\ &= -\frac{(\hat{T}_1 - \hat{T}_2)}{h}(k_{nf} + \frac{16\sigma^o \hat{T}_2^3}{3k^o})\theta'(0) \end{aligned} \quad (4.2.33)$$

$$\begin{aligned} q_w|_{z=h} &= -k_{nf}\frac{\partial \hat{T}}{\partial z} + q_r|_{z=h} \\ &= -\frac{(\hat{T}_1 - \hat{T}_2)}{h}(k_{nf} + \frac{16\sigma^o \hat{T}_2^3}{3k^o})\theta'(1) \end{aligned} \quad (4.2.34)$$

Therefore the Nusselt number can be written as:

$$Nu_{x1} = \frac{hq_w}{k_f(\hat{T}_1 - \hat{T}_2)}|_{z=0}, Nu_{x2} = \frac{hq_w}{k_f(\hat{T}_1 - \hat{T}_2)}|_{z=h} \quad (4.2.35)$$

The general forms of the equation with its boundary conditions are as follows:

$$p_k: f(u, u', u'', .. u^{(m)}) = 0, u = u(x) \quad (4.2.36)$$

$$\begin{cases} u(x) = u_0, u(x) = u_1, u^{(m-1)}(x) = u_{m-1} \quad \text{at} \quad x = 0 \\ u(x) = u_{L_0}, u(x) = u_{L_1}, u^{(m-1)}(x) = u_{L_{m-1}} \quad \text{at} \quad x = L \end{cases} \quad (4.2.37)$$

It is assumed that the solution to this equation is as follows:

$$u(x) = \sum_{i=0}^{n} a_i x^i = a_0 + a_1 x^1 + a_2 x^2 + \cdots + a_n x^n \quad (4.2.38)$$

The larger the *n*, the higher the accuracy of the solution. Due to boundary conditions of Eq. (4.2.37), a set of $n+1$ equations which consists of $n+1$ unknown coefficients (a_0 to a_n), is obtained. The differential Eq. (4.2.36) regarding the series from degree (n) can be solved by solving the set of equations.

First, it introduced the residuals:

$$PP = \frac{1}{(1-\phi)^{25}(1-\phi+\frac{\rho_s}{\rho_f}\phi)} f'' + \text{Re}\,(2ff'' - 2gg' - \frac{M}{1-\phi+\frac{\rho_s}{\rho_f}\phi}\frac{\sigma_m}{\sigma_f} f'') = 0 \quad (4.2.39)$$

$$QQ = \frac{1}{(1-\phi)^{25}(1-\phi+\frac{\rho_s}{\rho_f}\phi)} g'' + \text{Re}\,(2fg' - 2f\hat{g} - \frac{M}{1-\phi+\frac{\rho_s}{\rho_f}\phi}\frac{\sigma_{nf}}{\sigma_f} g) = 0 \quad (4.2.40)$$

$$RR = \frac{1}{1-\phi+\frac{\rho_s}{\rho_f}\phi} P' + 4\text{Reff}' + \frac{2}{(1-\phi)^{2.5}(1-\phi+\frac{\rho_s}{\rho_f}\phi)} f'' = 0 \quad (4.2.41)$$

$$ZZ = \frac{1}{\text{Pr}}(\frac{k_{nf}}{k_f} + R)\theta'' + 2\text{Re}\,(1-\phi+\frac{(\rho c_p)_s}{(\rho c_p)_f}\phi) f\theta' + \text{Re}\,M\frac{\sigma_{nf}}{\sigma_f}\text{Ec}\,(f'^2 + g^2) = 0 \quad (4.2.42)$$

$$XX = \frac{1}{\text{Re}}\frac{1}{Sc}\phi'' + 2f\phi' - k_1\phi(1-\phi)^2 = 0 \quad (4.2.43)$$

It is assumed that the solutions to these equations are as follows:

$$f = \sum_{i=0}^{7} a_i\xi^i,\ g = \sum_{i=0}^{7} b_i\xi^i,\ P = \sum_{i=0}^{7} c_i\xi^i,\ \theta = \sum_{i=0}^{7} d_i\xi^i,\ \phi = \sum_{i=0}^{7} e_i\xi^i \quad (4.2.44)$$

According to the below equations, all constant parameters can be obtained.

$$f = f(B.C.), g = g(B.C.), P = P(B.C.), \theta = \theta(B.C.), \phi = \phi(B.C.)$$

$$PP(BC.) = 0, PP'(BC.) = 0, PP''(B.C.) = 0, PP'''(B.C.) = 0, QQ(BC.) = 0, QQ'(B.C.) = 0 \quad (4.2.45)$$

$$QQ''(BC.) = 0, RR(B.C.) = 0, RR'(B.C.) = 0, ZZ(B.C.) = 0, ZZ'(B.C.) = 0, ZZ''(B.C.) = 0$$

$$XX(B.C.) = 0, XX'(B.C.) = 0$$

The impact of boundary conditions in radial, axial, and tangential velocities is described in the figures. Figs. 4.2.2a and 4.2.2b show that with the increase of the values of A_1, the radial and axial velocities enhance near the lower disk (the lower value of ξ). Also, the increase of these velocities decays near the upper disk. In Fig. 4.2.2a, the negative values near the upper disk (the more value of ξ) describe that the velocity of the lower disk is more than that of the upper disk. As shown in Fig. 4.2.2c, the tangential velocity is decreased with an increment in the value of A_1. Fig. 4.2.2d describes the effect of A_2 on the chemical specie concentration. As can be seen, the concentration is increased for larger values of A1. The effect of the values of A_1 on the temperature is shown in Fig. 4.2.2e. As shown in this figure, the temperature increase for larger values of A_1.

As shown in Fig. 4.2.3a, the axial velocity is decreased with an increment of the value of A_2. In Fig. 4.2.3b, the radial velocity decays near the lower disc with increasing A_2 while this velocity is enhanced near the upper disk. The increase of radial velocity for higher A_2 near the upper disk is the higher stretching rate of the upper disk. The effects of the variation of the value of A_2 on the tangential velocities are shown in Fig. 4.2.3c. Here this velocity is more for higher A_2. Fig. 4.2.3d displays the effect of A_2 on chemical concentration $\emptyset(\xi)$. As shown in this figure, for larger values of A_2, the fluid concentration enhances. Fig. 4.2.3e shows that the temperature slightly enhances with the increase of A2.

Figs. 4.2.4a, 4.2.4b, and 4.2.4c show the impact of the Reynolds number on radial, axial, and tangential velocities. It is shown in Figs. 4.2.4a and 4.2.4b that with the increase in Re, since inertial forces directly relate to Re, the magnitude of radial and axial velocities decreases near the lower disk. Fig. 4.2.4c displays the effect of the Reynolds number on tangential velocity. In this figure, $g(\xi)$ is an increasing function of Re near the lower disk while it is a decreasing function near the upper one. The influence of the Reynolds number on concentration and temperature is described in Figs. 4.2.4d and 4.2.4e. It is noted that concentration is a decreasing function of Re while the temperature is increasing.

Figs. 4.2.5a, 4.2.5b, and 4.2.5c show the impact of stretching parameters (A_1, A_2) and the Reynolds number on the skin friction coefficients on the lower and upper disks. It is illustrated that skin friction coefficients are increasing the functions of the stretching parameters while showing decreasing behaviour for the Reynolds number.

The effect of the stretching parameters (A_1) on the Nusselt number on the lower and upper disks is shown in Fig. 4.2.6a. As can be seen, with the increase of the value of A_1, the Nusselt number at the upper disk enhances while the Nusselt number is nearly constant at the lower one. The behaviour of the values of (A_1) and Re on the Nusselt number at lower and upper disks is illustrated in Figs. 4.2.6b and 4.2.6c. It is noted that the Nusselt number is increased for larger stretching parameters and the Reynolds number at the upper disk while it shows decreasing behaviour at the lower one.

The present study discusses the impact of stretching parameters on velocity, temperature, and concentration in MHD nanofluid flow between two stretchable and rotating disks with homogeneous and heterogeneous reactions and Joule heating. The AGM could solve the PDE problems with accurate answers and minimum error. The results show that with the increase of the values of A_1, the radial and axial velocities enhance near the lower disk. Also, with an increment of A_1, the temperature and chemical specie concentration increases while tangential velocity decreases behaviour. The impact of the increase of the value of A_2 is opposite A_1 on the axial, radial, and tangential velocities. The chemical concentration and temperature are increasing functions of A_2. The radial and axial velocities decrease near the lower disk with an increasing Reynolds number.

Skin friction coefficients increase the function of stretching parameters while decreasing the Reynolds number. The Nusselt number is increased for larger stretching parameters and the Reynolds number at the upper disk while it shows decreasing behaviour at the lower one.

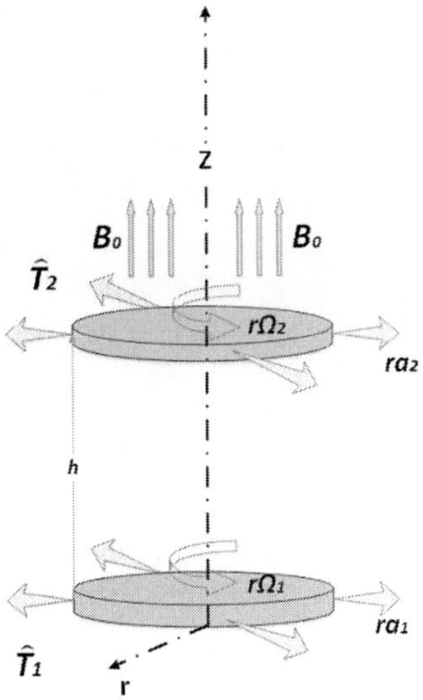

Fig. 4.2.1. Flow geometry

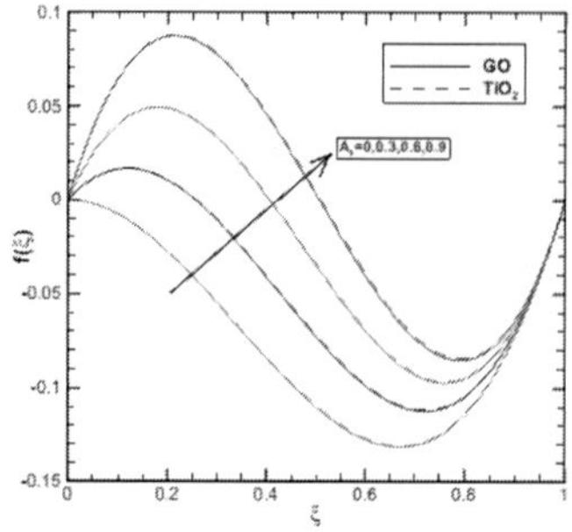

Fig.2a

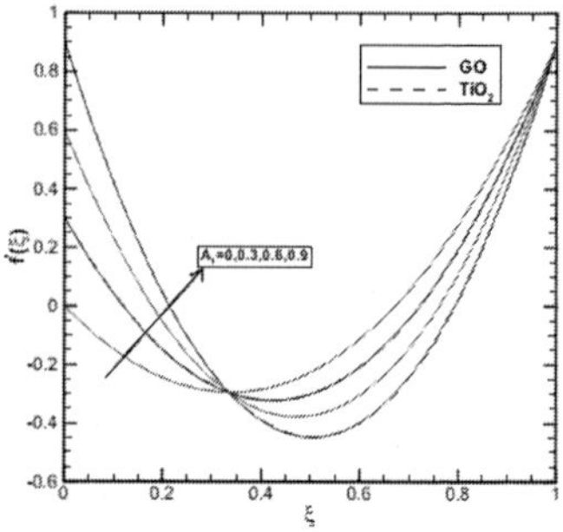

Fig.2b

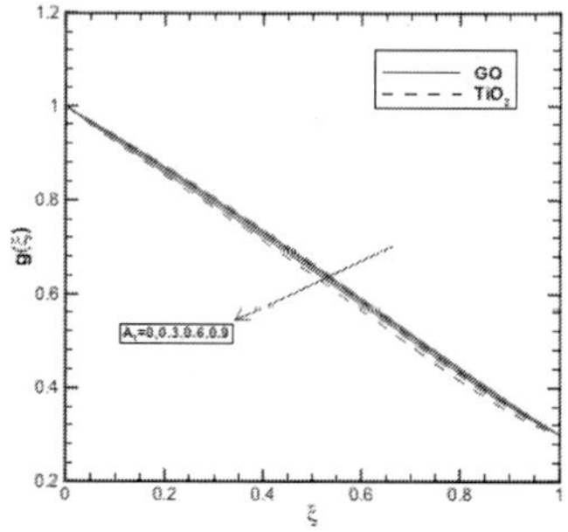

Fig.2c

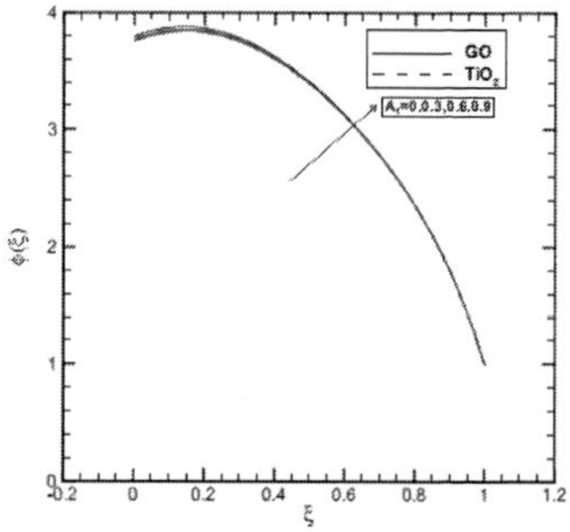

Fig.2d

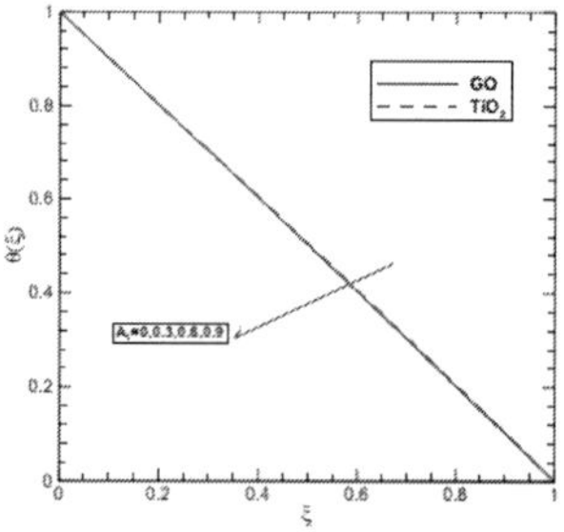

Fig.2e

Fig. 4.2.2. Effect of A_1 on $f(\xi), f'(\xi), g(\xi), \theta(\xi), \phi(\xi)$

$$A_2 = 0.9, M = 0.1, \tau = 0.3, \phi = 0.2, R = Ec = \mathrm{Re} = 0.7$$

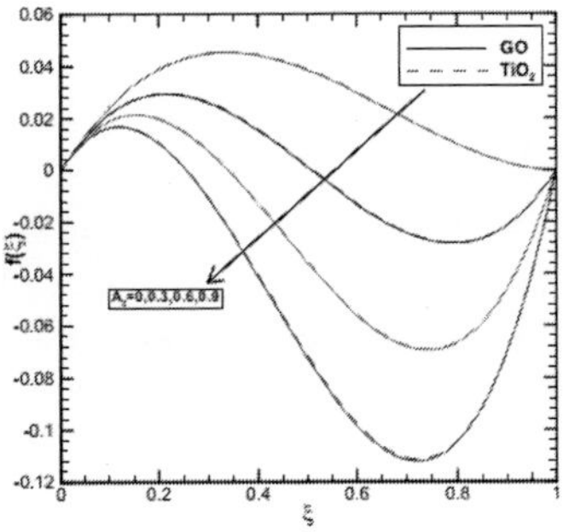

Fig.3a

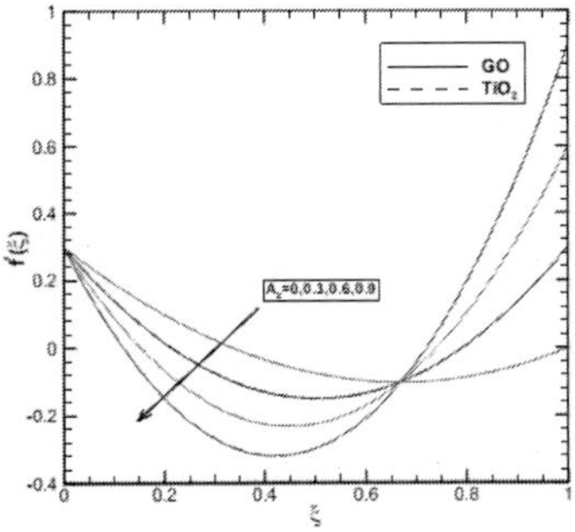

Fig.3b

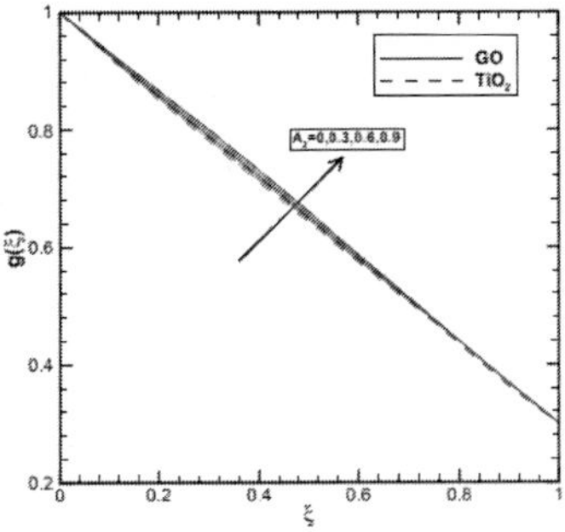

Fig.3c

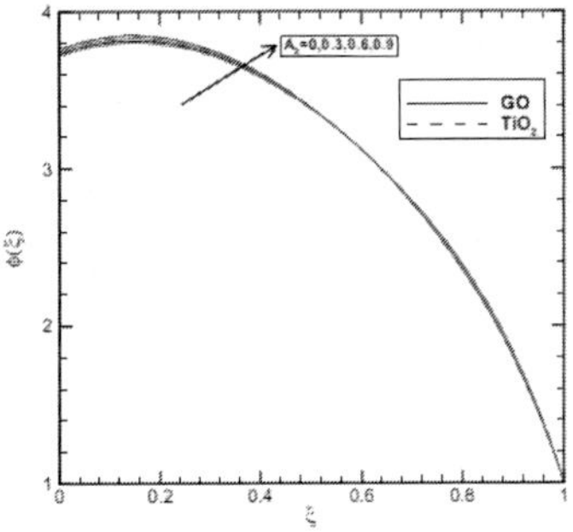

Fig.3d

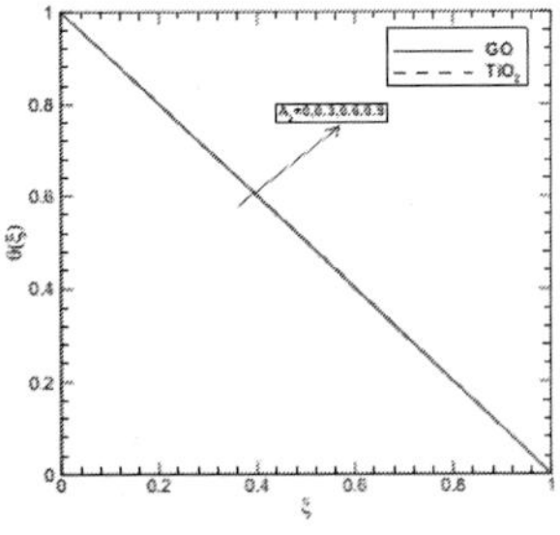

Fig.3e

Fig. 4.2.3. Effect of A_2 on $f(\xi), f'(\xi), g(\xi), \theta(\xi), \phi(\xi)$

$$A_1 = 0.3, M = 0.1, \tau = 0.3, \phi = 0.2, R = Ec = Re = 0.7$$

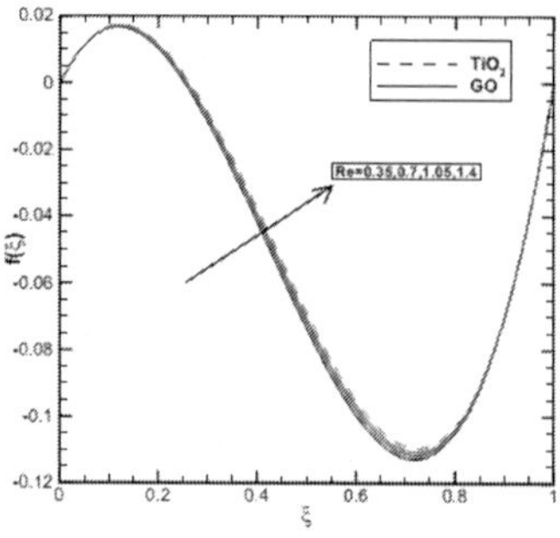

Fig. 4a

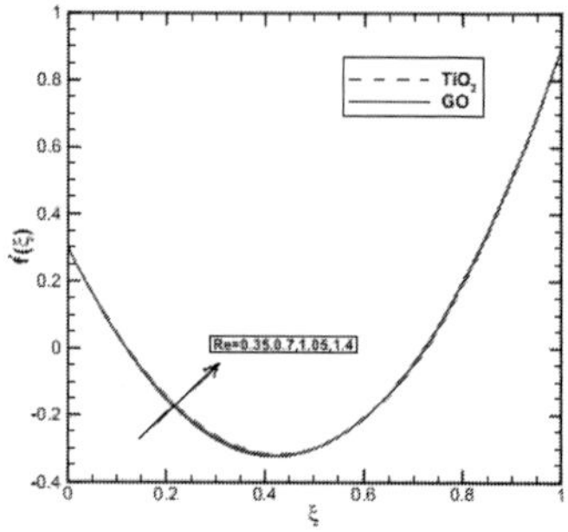

Fig. 4b

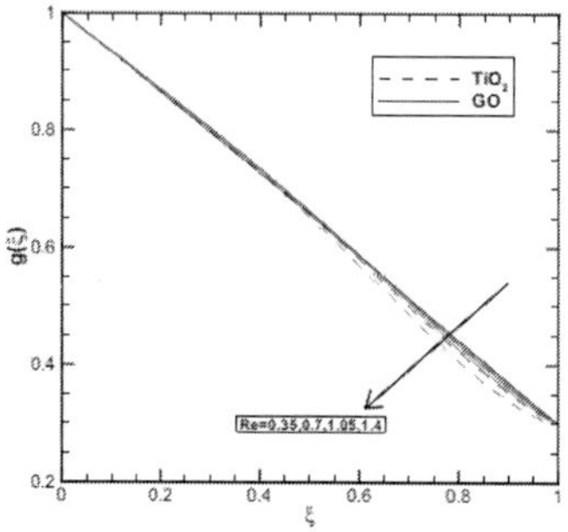

Fig. 4c

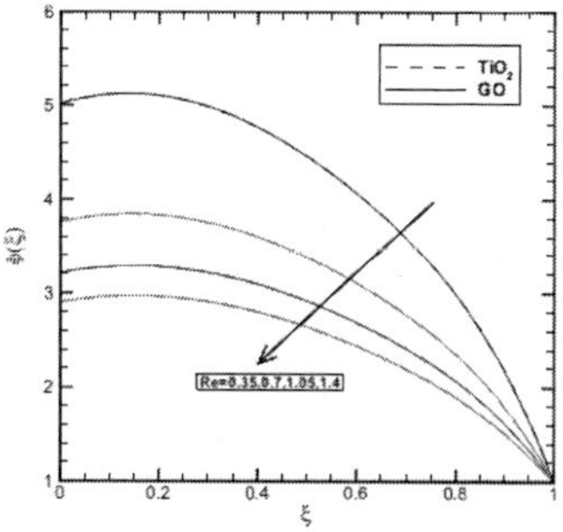

Fig. 4d

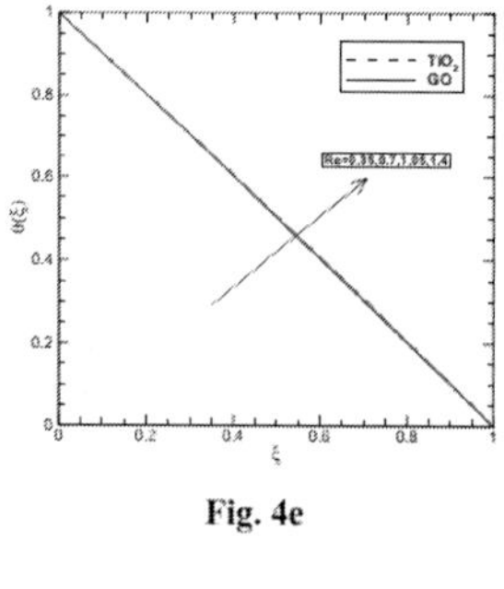

Fig. 4e

Fig. 4.2.4. Effect of Re on $f(\xi), f'(\xi), g(\xi), \theta(\xi), \phi(\xi)$

$A_1 = 0.3, A_2 = 0.9, M = 0.1, \tau = 0.3, \phi = 0.2, R = Ec = 0.7$

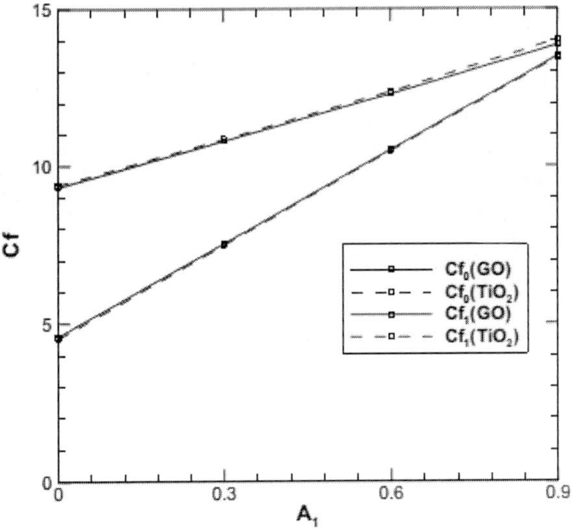

Fig. 4.2.5a. Cf vs. A1

$A_2 = 0.9, M = 0.1, \tau = 0.3, \phi = 0.2, R = Ec = \text{Re} = 0.7$

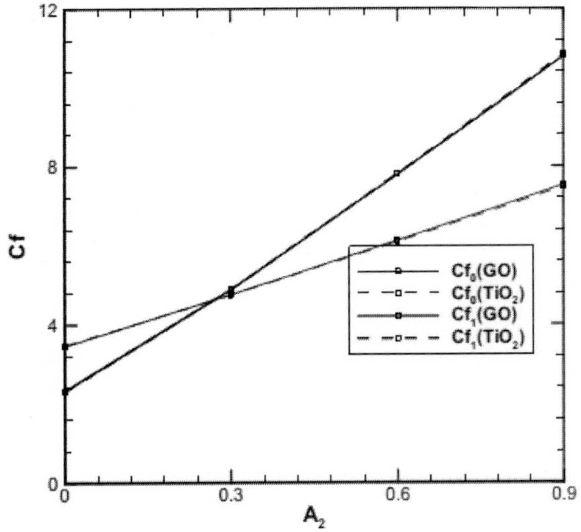

Fig. 4.2.5b. Cf vs. A2

$A_1 = 0.3, M = 0.1, \tau = 0.3, \phi = 0.2, R = Ec = \text{Re} = 0.7$

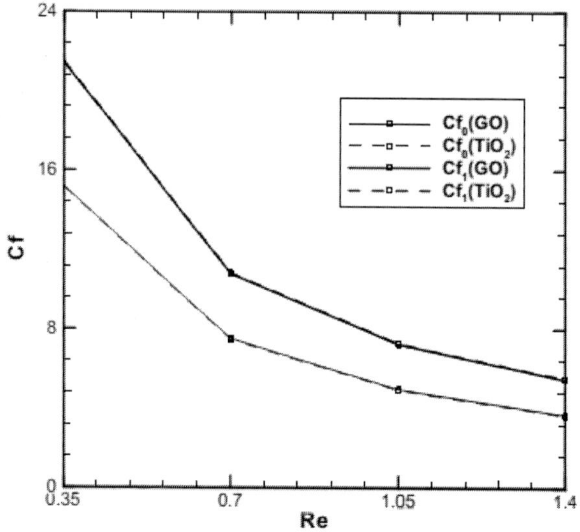

Fig. 4.2.5c. Cf vs. Re

$$A_1 = 0.3, A_2 = 0.9, M = 0.1, \tau = 0.3, \phi = 0.2, R = Ec = 0.7$$

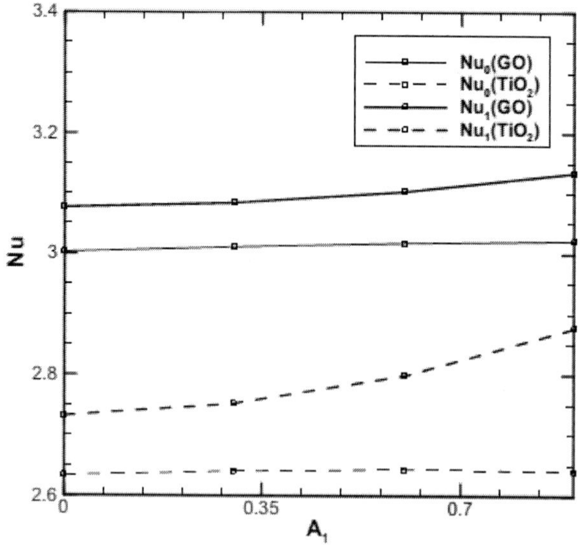

Fig. 4.2.6a. Nu vs. A_1

$$A_2 = 0.9, M = 0.1, \tau = 0.3, \phi = 0.2, R = Ec = \mathrm{Re} = 0.7$$

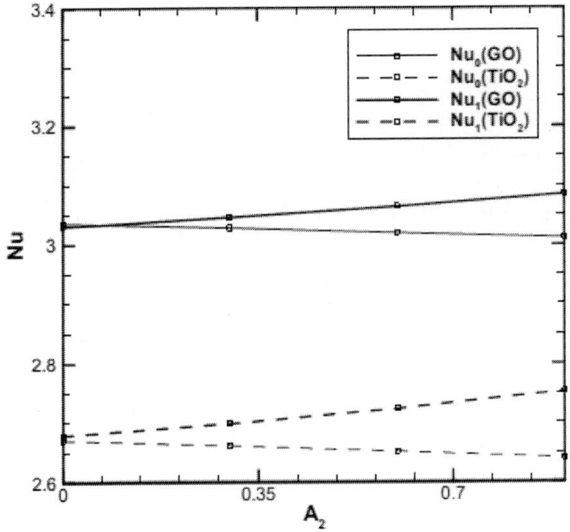

Fig. 4.2.6b. Nu vs. A2

$$A_1 = 0.3, M = 0.1, \tau = 0.3, \phi = 0.2, R = Ec = \mathrm{Re} = 0.7$$

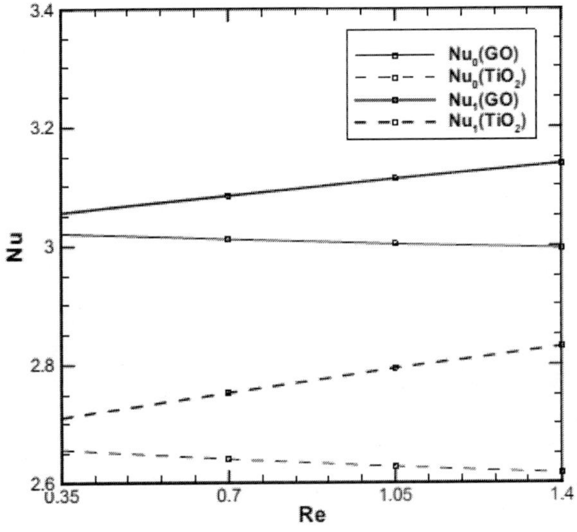

Fig. 4.2.6c. Nu vs. Re

$$A_1 = 0.3, A_2 = 0.9, M = 0.1, \tau = 0.3, \phi = 0.2, R = Ec = 0.7$$

Table 4.2.1. Thermophysical properties of Titanium dioxide and Graphene oxide and water.

	k (W/mK)	ρ (kg/m³)	σ (Um)⁻¹	c_p (J/kgK)
Titanium dioxide	8.9538	4250	2.6e6	686.2
Graphene oxide	5000	1800	1.1e-5	717
Water	0.613	997.1	0.05	4179

4.3. Magnetic Field Effect on Nanofluid Flow Between Two Circular Cylinders

In this problem, uniform magnetic field impact on the nanofluid flow between two circular cylinders is investigated analytically using AGM. A two-phase model is applied for nanofluid.

Laminar 2D flow in cylindrical coordinates is studied (Fig. 4.3.1). The governing formulae are:

$$\mu(\frac{\partial^2 v}{\partial r^2}-\frac{v}{r^2}+\frac{1}{r}\frac{\partial v}{\partial r})-\sigma v B_0^2=\rho_f v\frac{\partial v}{\partial r} \tag{4.3.1}$$

$$\frac{k}{r}\frac{\partial}{\partial r}(r\frac{\partial T}{\partial r})+\mu(\frac{\partial v}{\partial r}-\frac{v}{r})^2+(\rho C_p)_p[D_B\frac{\partial C}{\partial r}\frac{\partial T}{\partial r}+\frac{D_T}{T_1}(\frac{\partial T}{\partial r})^2] \\ -(\rho C_p)_f\frac{\partial T}{\partial r}v=0 \tag{4.3.2}$$

$$-v\frac{\partial C}{\partial r}+\frac{D_B}{r}\frac{\partial}{\partial r}(r\frac{\partial C}{\partial r})=-\frac{D_T}{T_1}\frac{\partial}{\partial r}(r\frac{\partial T}{\partial r}) \tag{4.3.3}$$

$$r=r_1: T=T_1, v(r)=\Omega_1 r_1, C=C_1 \\ r=r_2: T=T_2, v(r)=0, C=C_2 \tag{4.3.4}$$

The dimensionless forms of the above equations are as follows:

$$\frac{\partial^2 v^*}{\partial r^{*2}}-v^*(Ha^2(1-\eta)^{-2}+r^{*-2})-\mathrm{Re}v^*\frac{\partial v^*}{\partial r^*}+\frac{1}{r^*}\frac{\partial v^*}{\partial r^*} \\ =0 \tag{4.3.5}$$

$$\mathrm{EcPr}\left(\frac{\partial v^*}{\partial r^*}-\frac{v^*}{r^*}\right)^2+\frac{1}{r^*}\frac{\partial}{\partial r^*}\left(r^*\frac{\partial \theta}{\partial r^*}\right)-\mathrm{PrRev}^*\frac{\partial \theta}{\partial r^*}+\frac{\partial \theta}{\partial r^*}Nt\frac{\partial \theta}{\partial r^*}+\frac{\partial \phi}{\partial r^*}Nb\frac{\partial \theta}{\partial r^*}=0 \tag{4.3.6}$$

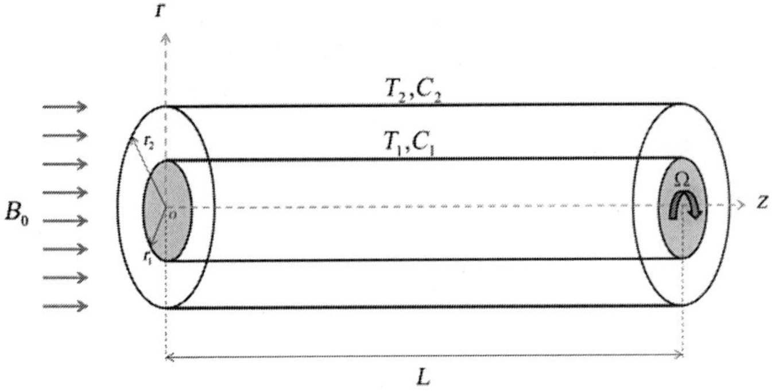

Fig. 4.3.1. The geometry of the problem.

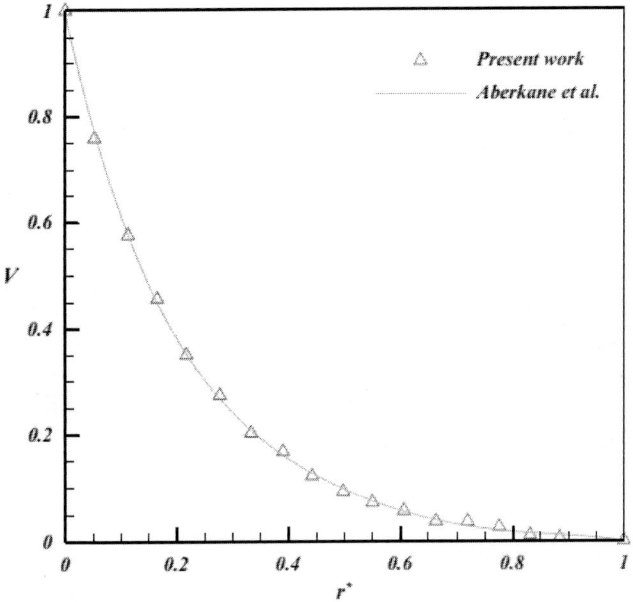

Fig. 4.3.2. Comparison of Aberkane et al. and present velocity profile results, for η= 0.5, Ha = 4.

$$\frac{Nt}{Nb}\left(\frac{1}{r^*}\frac{\partial\theta}{\partial r^*}+\frac{\partial^2\theta}{\partial r^{*2}}\right)-\mathrm{ScRev}^*\frac{\partial\phi}{\partial r^*} = -\frac{1}{r^*}\frac{\partial}{\partial r^*}\left(r^*\frac{\partial\phi}{\partial r^*}\right) \tag{4.3.7}$$

$$\begin{aligned} r^* &= 1: \phi = 0, \theta = 0, v^*(r^*) = 0 \\ r^* &= \eta: \phi = 1, \theta = 1, v^*(r^*) = 1 \end{aligned} \tag{4.3.8}$$

Where

$$\begin{aligned} &\eta = \frac{r_1}{r_2}, v^* = \frac{|v}{\Omega_1 r_1}, r^* = \frac{r}{r_2}, Ha = B_0(r_2 - r_1)\sqrt{\frac{\sigma}{\mu}}, \\ &\theta = \frac{T - T_2}{T_1 - T_2}, \phi = \frac{C - C_2}{C_1 - C_2}, Re = \frac{\rho_f \Omega r_1 r_2}{\mu} \\ &\alpha = \frac{k}{(\rho C_p)_f}, Pr = \frac{\mu}{\rho_f \alpha}, Nb = \frac{(\rho C_p)_p D_B \Delta C}{(\rho C_p)_f \alpha}, \\ &Ec = \frac{\rho_f (\Omega_1 r_1)^2}{(\rho C_p)_f \Delta T}, Nt = \frac{(\rho C_p)_p D_T \Delta T}{(\rho C_p)_f \alpha T_1}, Sc = \frac{\mu}{\rho_f D_B} \end{aligned} \tag{4.3.9}$$

Nu over the hot cylinder is as follows:

$$Nu = -\frac{\partial\theta}{\partial r^*}|_{r^*=\eta} \tag{4.3.10}$$

The physics of the problems in different fields of engineering sciences leads to a set of linear or nonlinear differential equations as its governing equations. According to the physics of these problems and their obtained mathematical formulation, sufficient boundary or initial conditions should be applied to achieve solutions for considered problems. Since applying analytical methods for obtaining the solution to linear and nonlinear differential equations are not an exception to mentioned fact, we could recognize the importance of these boundary and initial conditions in determining the accuracy and efficiency of analytical methods in achieving acceptable solutions due to the physics of problems. In this method, the whole process is clearly stated with boundary conditions; the general form of a differential equation is as follows:

$$p_k: f(u, u', u'', \dots, u^{(m)}) = 0; u = u(x) \tag{4.3.11}$$

The nonlinear differential equation of p, which is a function of u, the parameter u, which is a function of x, and their derivatives are considered as follows:

Boundary conditions:

$$\begin{cases} u(x) = u_0, u'(x) = u_1, \ldots, u^{(m-1)}(x) = u_{m-1} & \text{at } x = 0 \\ u(x) = u_{L_0}, u'(x) = u_{L_1}, \ldots, u^{(m-1)}(x) = u_{L_{m-1}} & \text{at } x = L \end{cases} \tag{4.3.12}$$

To solve the first differential equation with respect to the boundary conditions in x = L in Eq. (4.3.12), the series of letters in the nth order with constant coefficients, which we assume

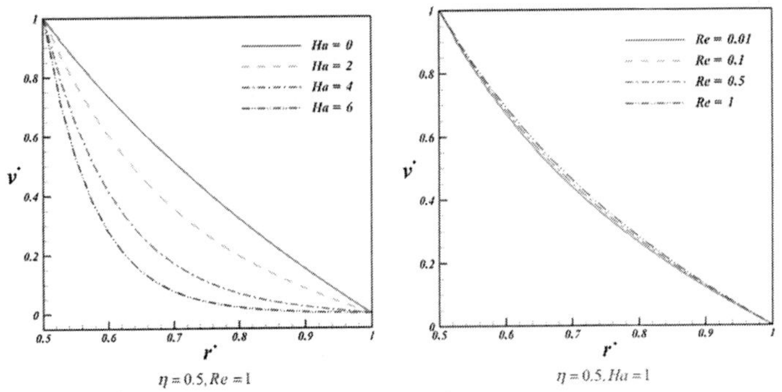

Fig. 4.3.3. Influence of Ha, Re on the velocity profile

As the solution of the first differential equation is considered as follows:

$$u(x) = \sum_{i=0}^{n} a_i x^i = a_0 + a_1 x^1 + a_2 x^2 + \cdots + a_n x^n \tag{4.3.13}$$

The more choice of series sentences from Eq. (4.3.13) causes a more precise solution for Eq. (4.3.11). For obtaining the solution of differential Eq. (4.3.11) regarding the series from degree (n), there are (n+1) unknown coefficients that need (n+1) equations to be specified. The boundary conditions of Eq. (4.3.12) are used to solve a set of equations consisting of (n+1).

Applying the boundary conditions:

(a) The application of the boundary conditions for the answer of differential Eq. (4.3.13) is in the form of the following:

When X=0:

$$\begin{cases} u(0) = a_0 = u_0 \\ u'(0) = a_1 = u_1 \\ u''(0) = a_2 = u_2 \\ \vdots \end{cases} \tag{4.3.14}$$

And when X=L

$$\begin{cases} u(L) = a_0 + a_1 L + a_2 L^2 + \cdots + a_n L^n = u_{L_0} \\ u'(L) = a_1 + 2a_2 L + 3a_3 L^2 + \cdots + n a_n L^{n-1} = u_{L_1} \\ u''(L) = 2a_2 + 6a_3 L + 12a_4 L^2 + \cdots + n(n-1) a_n L^{n-2} = u_{L_{m-1}} \\ \vdots \end{cases} \tag{4.3.15}$$

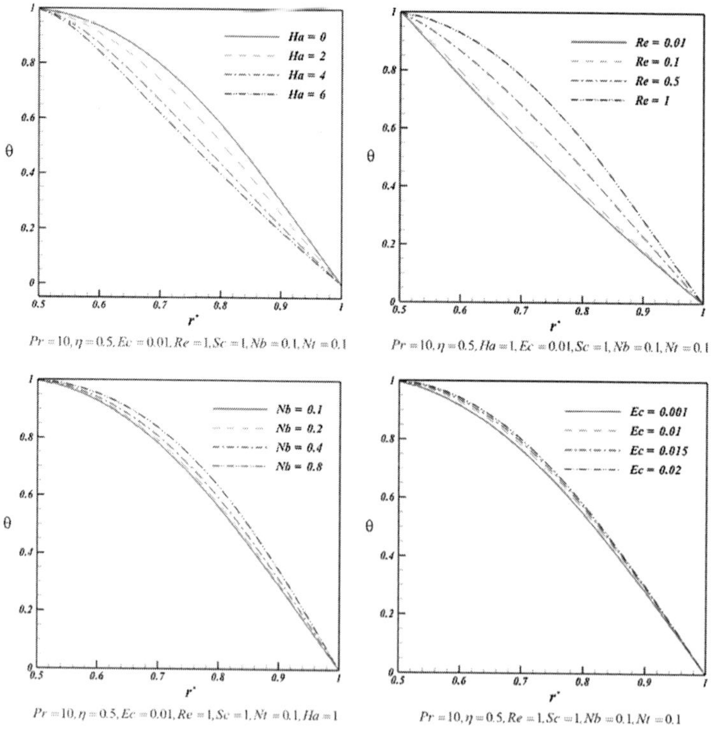

Fig. 4.3.4. Influence of Ha, Re, Nb, Ec on the temperature profile.

(b) After substituting Eq. (4.3.15) into Eq. (4.3.11), the application of the boundary conditions on differential Eq. (4.3.11) is done according to the following procedure:

$$\begin{aligned} p_0&: \quad f(u(0), u'(0), u''(0), \ldots, u^{(m)}(0)) \\ p_1&: \quad f(u(L), u'(L), u''(L), \ldots, u^{(m)}(L)) \end{aligned} \tag{4.3.16}$$

Concerning the choice of n, (n<m) sentences from Eq. (4.3.13) and to make a set of equations consisting of n+1 equations and N+1 unknowns, we confront with some additional unknowns which are indeed the same coefficients of Eq. (4.3.20). Therefore, to remove this problem we should derive m times from Eq. (4.3.18) according to the additional unknowns in the sets above of differential equations and then apply the boundary conditions on them.

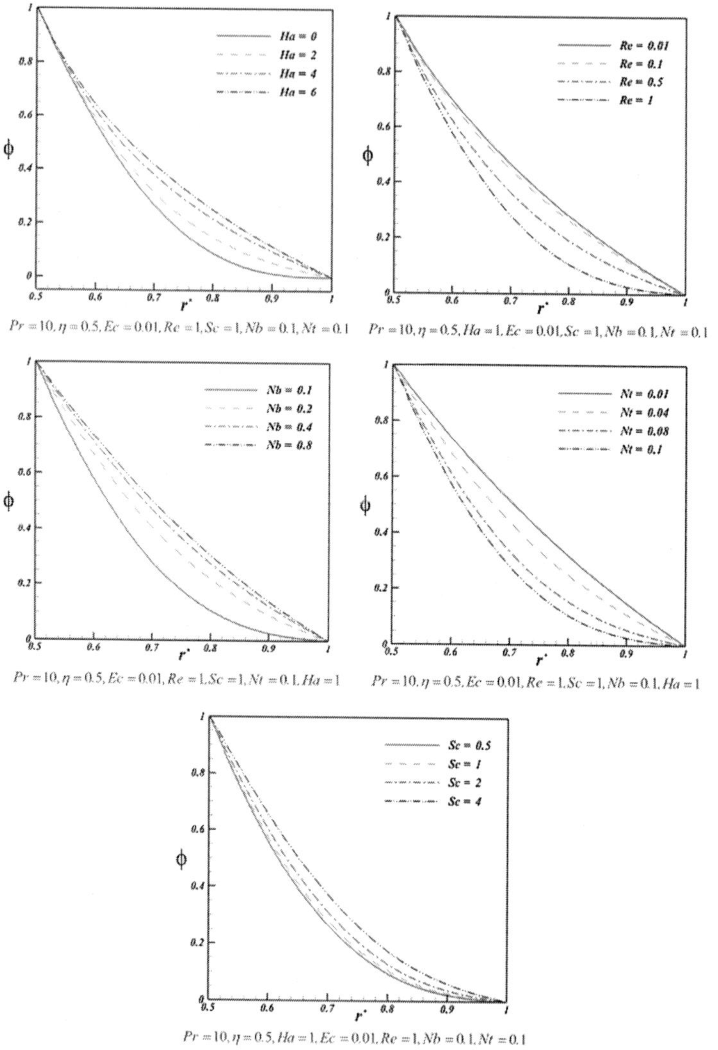

Fig. 4.3.5. Influence of Ha, Re, Nb, Nt, Sc on concentration profile.

$$
\begin{aligned}
&p_k' : f\left(u', u'', u''', \ldots, u^{(m+1)}\right) \\
&p_k'' : f\left(u'', u''', u^{(IV)}, \ldots, u^{(m+2)}\right)
\end{aligned}
\tag{4.3.17}
$$

(c) Application of the boundary conditions on the derivatives of the differential equation p_k, in Eq. (4.3.17) is done in the form of the following:

$$p'_k:\begin{cases} f(u'(0),u''(0),u'''(0),\dots,u^{(m+1)}(0)) \\ f(u'(L),u''(L),u'''(L),\dots,u^{(m+1)}(L)) \end{cases} \tag{4.3.18}$$

$$p''_k:\begin{cases} f(u''(0),u'''(0),\dots,u^{(m+2)}(0)) \\ f(u''(L),u'''(L),\dots,u^{(m+2)}(L)) \end{cases} \tag{4.3.19}$$

(n+1) equations can be made from Eqs. (4.3.14)–(4.3.19) so that (n+1) unknown coefficients of Eq. (4.3.13) such as $a_0, a_1, \dots, a_n$, will be computed. The solution of the nonlinear differential Eq. (4.3.11) will be gained by determining the coefficients of Eq (4.3.13). To comprehend the procedures of applying the following explanation, we have presented the relevant process step by step in the following part.

First, we introduce the residuals:

$$\begin{aligned} F &= \frac{\partial^2 v^*}{\partial r^{*2}} - v^*(Ha^2(1-\eta)^{-2} + r^{*-2}) - \mathrm{Rev}^* \frac{\partial v^*}{\partial r^*} + \frac{1}{r^*}\frac{\partial v^*}{\partial r^*} = 0 \\ G &= EcPr(\frac{\partial v^*}{\partial r^*} - \frac{v^*}{r^*})^2 + \frac{1}{r^*}\frac{\partial}{\partial r^*}(r^*\frac{\partial \theta}{\partial r^*}) - \mathrm{Pr\ Rev}^* \frac{\partial \theta}{\partial r^*} \\ &\quad + \frac{\partial \theta}{\partial r^*} Nt \frac{\partial \theta}{\partial r^*} + \frac{\partial \phi}{\partial r^*} Nb \frac{\partial \theta}{\partial r^*} = 0 \\ R &= \frac{Nt}{Nb}(\frac{1}{r^*}\frac{\partial \theta}{\partial r^*} + \frac{\partial^2 \theta}{\partial r^2}) - ScRev^* \frac{\partial \phi}{\partial r^*} + \frac{1}{r^*}\frac{\partial}{\partial r^*}(r^*\frac{\partial \phi}{\partial r^*}) = 0 \end{aligned} \tag{4.3.20}$$

We assume that the solutions to these equations are as follows:

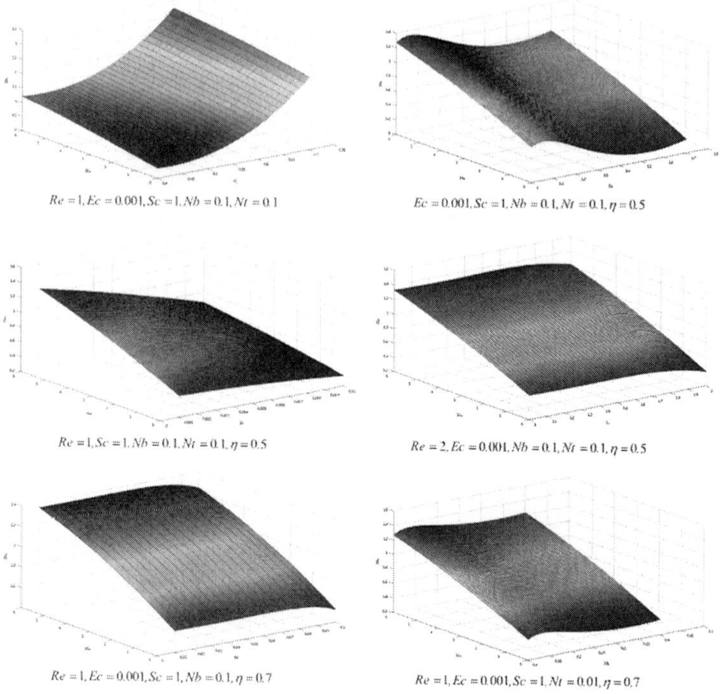

Fig. 4.3.6. Influence of Re, Ec, Sc, Nb, Nt, η, Ha on Nusselt number.

$$v^* = \sum_{i=0}^{7} a_i (r^*)^i, \theta = \sum_{i=0}^{7} b_i (r^*)^i, \phi = \sum_{i=0}^{7} c_i (r^*)^i \tag{4.3.21}$$

According to the below equations, all constant parameters can be obtained.

$$\begin{gathered} v^* = v^*(B.C), \theta = \theta(B.C), \phi = \phi(B.C) \\ F(B.C) = 0, F'(B.C) = 0 \\ G(B.C) = 0, G'(B.C) = 0 \\ R(B.C) = 0, R'(B.C) = 0 \end{gathered} \tag{4.3.22}$$

Al_2O_3-H_2O nanofluid two-phase modelling in an annulus is investigated analytically using AGM. The horizontal magnetic field impact is taken into account. The influence of the effective parameters is depicted as a graph. Fig. 4.3.2 proves the accuracy of AGM.

Fig. 4.3.3 illustrates the influence of Ha; Re on the velocity profile as an inertial force dominates the viscous forces and velocity augments. As electromagnetic forces dominate the viscous force, velocity decreases. Velocity increases with the rise of Re but reduces with the rise of Ha. The impact of Ha, Re, Nb, and Ec on the temperature profile is illustrated in Fig. 4.3.4. Lorentz forces reduce the velocity and generate secondary flow; the temperature reduces with the rise of the Hartmann number. The temperature gradient near the inner pipe reduces with the rise of the magnetic field but increases with the rise of Re; Nb. A higher Ec means a higher viscous dissipation, so the temperature rises with an increased Ec. Fig. 4.3.5 demonstrates the influence of Ha, Re, Nb, Nt, and Sc on the concentration profile. The concentration of Al_2O_3-H_2O nanofluid increases with the rise of Ha, Nb, and Sc, but it decreases with the rise of Re, Nt. Fig. 4.3.6 shows the impact of Re, Ec, Sc, Nb, Nt, η, Ha on the heat transfer rate. The Nusselt number enhances with an increase of Ha but decreases with the rise of Re, Ec, Sc, Nb, Nt. The distance between the two pipes reduces with increases of η, so the heat transfer rate increases.

Magnetohydrodynamic Al_2O_3-H_2O nanofluid heat transfer between two pipes is examined. The governing formulae are solved by employing AGM. The roles of the aspect ratio, the Reynolds number, the Eckert number, the Hartmann number, thermophoresis, the Brownian parameters, and the Schmidt number are demonstrated as graphs. The results revealed that temperature enhances with a rise in the Brownian parameter, the Eckert number, and the Reynolds number but decreases with the rise in the Hartmann number. The concentration increases with the rise of the Lorentz forces and the Brownian motion but decreases with the rise in thermophoresis forces.

4.4. The Jeffery-Hamel flow with a High Magnetic Field and Nanoparticles

This study studies the effects of magnetic field and nanoparticles on the Jeffery-Hamel flow using two powerful analytical methods, HPM and a simple and innovative approach, which we named AGM.

Consider a system of cylindrical polar coordinates (r, θ, z) in which a steady two-dimensional flow of an incompressible conducting viscous fluid from a source or sink at channel walls lie in planes and intersects in the z-axis. Assuming purely radial motion means no change in the flow parameter along the $z-$direction. The flow depends on r and θ; we assume no magnetic field in the $z-$direction. The reduced forms of continuity, Navier-Stokes, and Maxwell's equations are:

$$\frac{\rho_{nf}}{r}\frac{\partial(ru(r,\theta))}{\partial r}(ru(r,\theta)) = 0 \quad (4.4.1)$$

$$u(r,\theta)\frac{\partial u(r,\theta)}{\partial r} = -\frac{1}{\rho_{nf}}\frac{\partial P}{\partial r} + v_{nf}(\frac{\partial^2 u(r,\theta)}{\partial r^2} + \frac{1}{r}\frac{\partial u(r,\theta)}{\partial r} + \frac{1}{r^2}\frac{\partial^2 u(r,\theta)}{\partial \theta^2} - \frac{u(r,\theta)}{r^3}) - \frac{\sigma B_0^2}{\rho_{nf} r^2}u(r,\theta) \quad (4.4.2)$$

$$\frac{1}{\rho_{nf}}\frac{\partial P}{\partial \theta} - \frac{2v_{nf}}{r^2}\frac{\partial u(r,\theta)}{\partial \theta} = 0 \quad (4.4.3)$$

Where B_0 is the electromagnetic induction, σ is the conductivity of the fluid, $u(r)$ is the velocity along the radial direction, P is the fluid pressure, v_{nf}, is the coefficient of kinematic viscosity, and p_{nf}, is defined as fluid density. The effective density ρ_{nf}, the effective dynamic viscosity μ_{nf}, and the kinematic viscosity v_{nf} of the nanofluid are given as:

$$\rho_{nf} = \rho_f(1-\phi) + \rho_x\phi \quad (4.4.4)$$

$$\mu_{nf} = \frac{\mu_f}{(1-\phi)^{2.5}} \quad (4.4.5)$$

$$v_{nf} = \frac{v_f}{\rho_{nf}} \quad (4.4.6)$$

Here ϕ is the solid volume fraction. Considering $\frac{\partial u(r,\theta)}{\partial \theta}$, for the purely radial flow, we can define the velocity parameter as:

$$f(\theta) = ru(r)$$

Introducing $\eta = \frac{\theta}{\alpha}$, as the dimensionless degree, the dimensionless form of the velocity parameter can be obtained by dividing its maximum value as:

$$f(\eta) = \frac{f(\eta)}{f_{max}} \quad (4.4.8)$$

Substituting Eq. (4.4.7) into Eq. (4.4.2) and Eq. (4.4.3) and eliminating *P*, one can obtain the ordinary differential equation for the normalised function profile as:

$$f'''(\eta) + 2\alpha \mathrm{Re}\ A^{*}(1-\phi)^{2.5} f(\eta) f'(\eta) + (4-(1-\phi))^{1.25} Ha)\alpha^{2} f'(\eta) = 0 \tag{4.4.9}$$

Where A^* is a parameter, the Reynolds number is *Re,* and the Hartmann number *Ha* is based on the electromagnetic parameter as follows:

$$A* = (1-\phi) + \frac{\rho_s}{\rho_f}\phi \tag{4.4.10}$$

$$Re = \frac{f_{max}\alpha}{\upsilon_f} = \frac{U_{max} r\alpha}{\upsilon_f} \tag{4.4.11}$$

$$Ha = \sqrt{\frac{\sigma B_0^2}{\rho_f \upsilon_f}} \tag{4.4.12}$$

With the following reduced form of boundary conditions:

$$f(0) = 1, f'(0) = 0, f(1) = 0$$

Physically, these boundary conditions mean that maximum velocity values are observed at the centre line $\eta = 0$, as shown in Fig. 4.4.1, and we consider fully developing the velocity profile. Thus, the rate of velocity is zero at $\eta = 0$. Also, in fluid dynamics, the no-slip condition for fluid states that at a solid boundary, the fluid will have zero velocity relative to the boundary. The fluid velocity at all fluid-solid boundaries is equal to that of the solid boundary, so we can see that value of velocity is zero at $\eta = 1$.

To illustrate the basic concepts of the HPM, consider the following differential equation:

$$Lu + Nu = g(t) \tag{4.4.14}$$

Where *L* is a linear operator, a nonlinear operator, and *g*(*t*) is an inhomogeneous term. So, the homotopy perturbation structure is shown in the following equation:

$$v = v_0 + pv_1 + p^2 v_2 + p^3 v_3 + p^4 v_4 + \cdots \tag{4.4.15}$$

Where p is an embedding parameter and u_0 shows the initial guess. It should satisfy the boundary condition. The process of changes in p from zero to unity is that of $v(r, p)$ changing from u_0 to $u(r)$. We consider v as follows:

$$v = v_0 + pv_1 + p^2 v_2 + p^3 v_3 + p^4 v_4 + \cdots \tag{4.4.16}$$

And the best approximation for the solution is:

$$u = \lim_{p \to 1} v = v_0 + v_1 + v_2 + v_3 + v_4 + \cdots \tag{4.4.17}$$

By applying HPM into Eq. (4.4.9) and substituting (see Eq. (4.4.16)) and rearranging it based on the powers of terms, it can be obtained:

$$\begin{gathered} p^0 : \frac{d^3}{d\eta^3} f_0(\eta) \\ p^1 : 2\alpha \mathrm{Re}\ A^*(1-\phi)^{2.5} f_0(\eta)(\frac{d}{d\eta} f_0(\eta)) \\ -\alpha^2 (1-\phi)^{1.25} \mathrm{Ha}\ (\frac{d}{d\eta} f_0(\eta)) \\ +4\alpha^2 (\frac{d}{d\eta} f_0(\eta)) + \frac{d^3}{d\eta^3} f_1(\eta) \\ p^2 : 4\alpha^2 (\frac{d}{d\eta} f_1(\eta)) + 2\alpha \mathrm{Re}\ A^*(1-\phi)^{2.5} f_1(\eta)(\frac{d}{d\eta} f_0(\eta)) \\ +\frac{d^3}{d\eta^3} f_2(\eta) + 2\alpha ReA^*(1-\phi)^{2.5} f_0(\eta)(\frac{d}{d\eta} f_1(\eta)) \\ -\alpha^2 (1-\phi)^{1.25} \mathrm{Ha}\ (\frac{d}{d\eta} f_1(\eta)) \end{gathered} \tag{4.4.18}$$

And, etc.

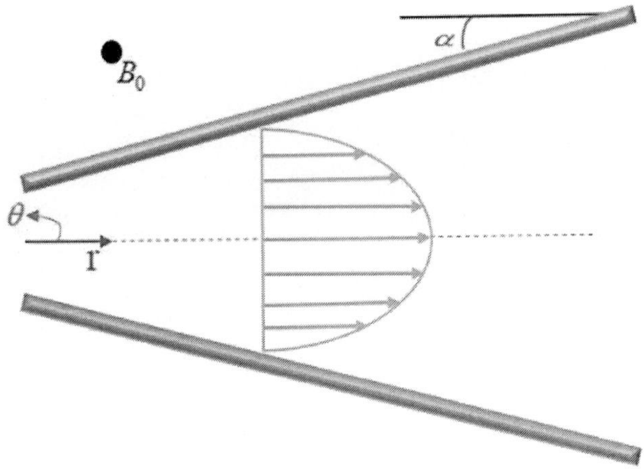

Fig. 4.4.1. Schematic diagram of the physical model.

Therefore,

$$f_0(\eta) = -\eta^2 + 1$$

$$\begin{aligned} f_1(\eta) = -2\alpha(&\frac{1}{120}(2A^*\mathrm{Re}\ \phi^2\sqrt{1-\phi} - 4A^*\mathrm{Re}\ \sqrt{1-\phi} \\ &+2A\sqrt{1-\phi}Re)\eta^6 + \frac{1}{24}(2A^*\mathrm{Re}\ \phi^2\sqrt{1-\phi} \\ &+4A^*\mathrm{Re}\ \phi\sqrt{1-\phi} - Ha(1-\phi)^{\frac{1}{4}}\alpha\phi - 2A^*\sqrt{1-\phi}Re \\ &Ha(1-\phi)^{\frac{1}{4}}\alpha - 4\alpha)\eta^4 \\ &-\frac{1}{60}\frac{1}{-1+\phi}((1-\phi)^{\frac{1}{4}}\alpha(-24A^*(1-\phi)^{\frac{1}{4}}\mathrm{Re}\ \phi^2 \\ &+8A^*(1-\phi)^{\frac{1}{4}}\mathrm{Re}\ \phi^3 + 24A^*(1-\phi)^{\frac{1}{4}}\mathrm{Re}\ \phi \\ &-8A^*(1-\phi)^{\frac{1}{4}}Re + 5Ha\alpha\phi^2\alpha - 10Ha\phi\alpha + 5Ha\alpha \\ &-20(1-\phi)^{\frac{3}{4}}\alpha)\eta^2) \end{aligned} \tag{4.4.19}$$

And so on.

The solution of this equation when $p \to 1$ would be as follows:

$$f(\eta) = f_0(\eta) + f_1(\eta) + f_2(\eta) + f_3(\eta) + f_4(\eta) + \cdots \quad (4.4.20)$$

For example, by substituting $A^* = 1.1$, $Ha = 500$, $Re = 50$, $\alpha = 5°$, $\theta = 0.1$ in Eq. (4.4.9) the following solution would be obtained by HPM,

$$\begin{aligned} f = \; & -1.233827541\eta^2 + 1 - 0.2220652426\eta^6 \\ & +0.4123741072\eta^4 - 0.01007475689\eta^{10} \\ & +0.05359343297\eta^8 \end{aligned} \quad (4.4.21)$$

Boundary and initial conditions are required for analytical methods of each linear and nonlinear differential equation according to the physics of the problem. Therefore, we can solve every differential equation with any degree. To comprehend the given method in this problem, two differential equations governing engineering processes will be solved in this new manner. Under the boundary conditions, the general manner of a differential equation is as follows: the nonlinear differential equation of p which is a function of u, the parameter u, which is a function of x, and their derivatives are considered as follows:

$$p_k: f(u, u', u'', \dots, u^{(m)}) = 0; u = u(x) \quad (4.4.22)$$

Boundary conditions:

$$\begin{cases} u(0) = u_0, & u'(0) = u_1, & \dots, & u^{(m-1)}(0) = u_{m-1} \\ u(L) = u_{L_0}, & u'(L) = u_{L_1}, & \dots, & u^{(m-1)}(L) = u_{L_{m-1}} \end{cases} \quad (4.4.23)$$

To solve the first differential equation with respect to the boundary conditions in $x = L$ in Eq. (4.4.23), the series of letters in the nth order with constant coefficients, which is the answer of the first differential equation, is considered as follows:

$$u(x) = \sum_{i=0}^{n} a_i x^i = a_0 + a_1 x^1 + a_2 x^2 + \cdots + a_n x^n \quad (4.4.24)$$

The more precise the answer to Eq. (4.4.22), the more choice of series sentences from Eq. (4.4.24). Approximately five or six sentences from the series are enough to solve nonlinear differential equations in applied problems. In answer to differential Eq. (4.4.24) regarding the series from degree (n), there are (n+1) unknown coefficients that need (n+1) equations to be specified. The boundary conditions of Eq. (4.4.23) are used to solve a set of equations consisting of (n+1) ones. The boundary conditions are applied to the functions as follows:

(a) The application of the boundary conditions for the answer of differential Eq. (4.4.24) is in the form of:

If X=0:

$$\begin{cases} u(0) = & a_0 = u_0 \\ u'(0) = & a_1 \quad u_1 \\ u''(0) = & a_2 \quad u_2 \\ \vdots & \vdots \quad \vdots \end{cases} \tag{4.4.25}$$

And when X=L

$$\begin{cases} u(L) = a_0 + a_1 L + a_2 L^2 + \cdots + a_n L^n = uL_0 \\ u'(L) = a_1 + 2a_2 L + 3a_3 L^2 + \cdots + n a_n L^{n-1} = uL_1 \\ u''(L) = 2a_2 + 6a_3 L + 12 a_4 L^2 + \cdots + n(n-1) a_n L^{n-2} = uL_{m-1} \end{cases} \tag{4.4.26}$$

(b) After substituting Eq. (4.4.26) into Eq. (4.4.22), the application of the boundary conditions on differential Eq. (4.4.22) is done according to the following procedure:

$$\begin{matrix} p_0 & : f(u(0), & u'(0), & u''(0), & \dots & u^{(m)}) \\ p_1 & : f(u(L), & u'(L), & u''(L), & \dots & u^{(m)}(L)) \end{matrix} \tag{4.4.27}$$

Concerning the choice of n; ($n < m$) sentences from Eq. (4.4.24) and to make a set of equations which consist of ($n + 1$) equations and ($n + 1$) unknowns, we confront some additional unknowns, which are indeed the same coefficients of Eq. (4.4.24). Therefore, to remove this problem, we should derive m times from Eq. (4.4.22) according to the additional unknowns in those above set differential equations. Then this is the time to apply the boundary conditions of Eq. (4.4.11).

$$\begin{matrix} p'_k & : f(u', & u'', & u''' & \dots & u^{(m+1)}) \\ p''_k & : f(u'', & u''', & u^{(IV)}, & \dots & u^{(m+2)}) \end{matrix} \tag{4.4.28}$$

(c) The application of the boundary conditions on the derivatives of the differential equation p_k in Eq. (4.4.28) is done in the form of:

$$p'_k : \begin{cases} f(u'(0), u''(0), u'''(0), \dots, u^{(m+1)}(0)) \\ f(u'(L), u''(L), u'''(L), \dots, u^{(m+1)}(L)) \end{cases} \tag{4.4.29}$$

$$p''_k : \{ f(u''(0), \quad u'''(0), \quad \dots, \quad u^{(m+2)}(0)) \tag{4.4.30}$$

$(n + 1)$ equations can be made from Eq. (4.4.25) to Eq. (4.4.29) so that $(n + 1)$ unknown coefficients of Eq. (4.4.24) take for example $a_0, a_1, a_2, \dots, a_n$, will be computed. The answer to the nonlinear differential Eq. (4.4.22) will be gained by determining the coefficients of Eq. (4.4.24). Note that according to Eq. (4.4.9) we consider $g(\eta)$ as follows:

$$g(\eta)\colon f'''(\eta) + 2\alpha\text{Re}\ A^*(1-\phi)^{2.5} f(\eta) f'(\eta) \\ +(4 - 1(1-\phi)^{2.5} Ha)\alpha^2 f'(\eta) = 0 \tag{4.4.31}$$

By considering variables k_1 and k_2 as follows:

$$k_1 = 2\alpha ReA^*(1-\phi)^{2.5}, k_2 = (4 - (1-\phi)^{2.5} Ha)\alpha^2 \tag{4.4.32}$$

Eq. (4.4.31) will convert to the following form:

$$f(\eta)\colon \frac{d^3}{d\eta^3} f(\eta) + k_1 f \frac{d}{d\eta} f(\eta) + k_2 \frac{d}{d\eta} f(\eta) = 0 \tag{4.4.33}$$

In AGM, the answer to the differential equations is considered a finite series of polynomials with constant coefficients as follows:

$$f = \sum_{k=0}^{6} a_k \eta^k = a_0 + a_1\eta + a_2\eta^2 + a_3\eta^3 + a_4\eta^4 \\ + a_5\eta^5 + a_6\eta^6 \tag{4.4.34}$$

It is notable that in the above equations, the constant coefficients of a_0 to a_6, are obtained by applying the introduced boundary conditions.

In AGM, the boundary conditions are applied to compute constant coefficients of Eq. (4.4.34) in two ways as follows:

(4) Applying the boundary conditions on Eq. (4.4.34) is expressed as follows:

f=f (B.C.)

Notably, B.C. is the abbreviation of boundary conditions. According to the above explanations, applying the boundary conditions on Eq. (4.4.34) is done in the following form:

$$f(0) = 1 \rightarrow a_0 = 1 \tag{4.4.36}$$

$$f(1) = 0 \rightarrow a_6 + a_5 + a_4 + a_3 + a_2 + a_1 + a_0 = 0 \tag{4.4.37}$$

$$f'(0) = 0 \rightarrow a_1 = 0 \tag{4.4.38}$$

(b) Applying the boundary conditions to the main differential equation, which in this case study is Eq. (4.4.33) and its derivative is done after substituting Eq. (4.4.34) for the main differential equation Eq. (4.4.33) as follows:

$$g(f(\eta)) \rightarrow g(f(BC)) = 0, g'(f(\eta)) \rightarrow g'(f(BC)) = 0 \tag{4.4.39}$$

Based on the above formulae, the boundary conditions are applied to the achieved differential equation as follows:

$$g(f(0)): 6a_3 + k_1 a_0 a_1 + k_2 a_1 = 0 \tag{4.4.40}$$

$$\begin{aligned} g(f(1)):& \, 6a_3 + 24a_4 + 60a_5 + 120a_6 + \\ & k_1(a_0 + a_1 + a_2 + a_3 + a_4 + a_5 + a_6) \\ & \cdot (a_1 + 2a_2 + 3a_3 + 4a_4 + 5a_5 + 6a_6) \\ & + k_2(a_1 + 2a_2 + 3a_3 + 4a_4 + 5a_5 + 6a_6) = 0 \end{aligned} \tag{4.4.41}$$

Then, the boundary conditions are applied to the derivative of the acquired differential equation in the following form:

$$g'(f(0)): 24a_4 + k_1 a_1^2 + 2k_1 a_0 a_2 + 2k_2 a_2 = 0 \tag{4.4.42}$$

$$\begin{aligned} g'(f(1)):& \, 24a_4 + 120a_5 + 36a_6 \\ & + k_1(a_1 + 2a_2 + 3a_3 + 4a_4 + 5a_5 + 6a_6)^2 \\ & + k_1(a_0 + a_1 + a_2 + a_3 + a_4 + a_5 + a_6) \\ & \cdot (2a_2 + 6a_3 + 12a_4 + 20a_5 + 30a_6) \\ & + k_2(2a_2 + 6a_3 + 12a_4 + 20a_5 + 30a_6) = 0 \end{aligned} \tag{4.4.43}$$

Solving a set of seven algebraic equations with seven unknowns from Eq. (4.4.36-4.4.43), the constant coefficients of Eq. (4.4.34) can easily be gained.

For simplicity, the following new variables are considered as follows:

$$
\begin{aligned}
\psi = {} & 518400k_1 + 155520k_1^2 - k_2^5 - 7632k_1^2k_2 - 2880k_1^3 \\
& +233280k_1k_2 - 2160k_1k_2^2 + 60k_2^4 - 2k_2^4k_1 \\
& -31104000 - k_1^2k_2^3 + 518400k_2 - 48k_1^2k_2^2 + 12k_2^3k_1 \\
& +(-3.224862720 \cdot 10^{13}k_1 - 3.224862720 \cdot 10^{13}k_2 \\
& -1.397440512 \cdot 10^{13}k_1k_2 + 2.687385600 \cdot 10^{11}k_2^2 \\
& +1.845338112 \cdot 10^{12}k_1^2k_2 - 1.155575808 \cdot 10^{13}k_1^2 \\
& +3.881779200 \cdot 10^{10}k_2^3k_1 - 3.981312000 \cdot 10^{8}k_2^4k_1 \\
& +4.367001600 \cdot 10^{10}k_1^2k_2^2 - 3.565762560 \cdot 10^{9}k_1^3k_2^2 \\
& -3.670272000 \cdot 10^{9}k_1^2k_2^3 - 6.136197120 \cdot 10^{10}k_1^3k_2 \\
& -1.150848000 \cdot 10^{7}k_2^4k_1^2 - 2.488320000 \cdot 10^{7}k_2^5k_1 \\
& +1.866240000 \cdot 10^{6}k_2^5k_1^2 + 7.948800000 \cdot 10^{5}k_2^6k_1 \\
& -3.732480000 \cdot 10^{9}k_2^4 + 1.244160000 \cdot 10^{8}k_2^5 \\
& +1.021206528 \cdot 10^{12}k_1^3 - 1.036800000 \cdot 10^{6}k_2^6 \\
& +1.700352000 \cdot 10^{6}k_2^4k_1^3 - 27792k_2^6k_1^2 - 5760k_2^7k_1 \\
& -72k_2^7k_1^2 - 264k_2^8k_1 + 2.53808640010^{8}k_1^4k_2 \\
& +6.801408000 \cdot 10^{7}k_1^3k_2^3 - 29644k_2^5k_1^3 + 6k_2^8k_1^2 \\
& +4k_2^9k_1 + 6.985958400 \cdot 10^{7}k_1^4k_2^2 + 4.216320000 \cdot 10^{5}k_1^4 \\
& +7.464960000 \cdot 10^{6}k_1^5k_2 - 2.073600000 \cdot 10^{5}k_1^5k_2^2 \\
& -5760k_1^5k_2^3 - 16992k_1^4k_2^4 + 96k_1^4k_2^5 + 168k_1^3k_2^6 \\
& +k_1^4k_2^6 + 4k_1^3k_2^7 + 3600k_2^8 - 120k_2^9 + k_2^{10} \\
& -2.956124160 \cdot 10^{10}k_1^4 + 2.687385600 \cdot 10^{8}k_1^5 \\
& +1.791590400 \cdot 10^{10}k_1k_2^2 + 9.674588160 \cdot 10^{4})^{0.5} \\
\Delta_1 = {} & k_1^2 - 120k_1 - 120k_2 + k_2^2 + 2k_1k_2 + 3600 \\
\Delta_2 = {} & k_1k_2 + 36k_1 + k_2^2 - 720
\end{aligned}
\tag{4.4.44}
$$

With regards to the above variables, the constant coefficients of Eq. (4.4.34) are achieved in the forms of:

$$
\begin{aligned}
a_0 &= 1 \\
a_1 &= 0 \\
a_2 &= \frac{12}{\Delta_2}\{\frac{\psi}{k_1\Delta_2}(0.1041 + 0.00173) - 5k_2 - 60\} \\
a_3 &= 0 \\
a_4 &= \frac{-1}{k_1\Delta_1\Delta_2}\{5k_1^2\Delta_1 - 0.00173\psi k_1 k_2 - 0.1042\psi k_1 \\
&\quad +5\Delta_1 k_1 k_2^2 - 0.001736\psi k_2^2 - 0.10416\psi k_2\} \\
a_5 &= \frac{2}{k_1\Delta_1\Delta_2}\{2k_2^2 k_1\Delta_1 - 0.00173\psi k_2^2 \\
&\quad -0.001736\psi k_1 k_2 + 2\Delta_1 k_2 k_1^2 - 0.0834\psi k_1 \\
&\quad -0.0416\psi k_2 + 12\Delta_1 k_1^2 + 1.25\psi\} \\
a_6 &= \frac{0.001736\psi}{k_1\Delta_1}
\end{aligned}
\qquad (4.4.45)
$$

Table 4.4.1. The values obtained by the numerical solution.

η	0	0.2	0.4	0.6	0.8	1
$f(\eta)$	1.000	0.949064	0.805313	0.589751	0.321003	0.000
$f'(\eta)$	0.000	-0.501233	-0.917467	-1.221339	-1.464184	-1.770954

By substituting the constant values of Eq. (4.4.45) into Eq. (4.4.34), the differential equation is answered in the following form:

$$
\begin{aligned}
f(\eta) = 1 + \frac{12}{\Delta_2}\{\frac{\psi}{k_1\Delta_2}(0.1041 + 0.00173) - 5k_2 - 6(\\
\frac{-1}{k_1\Delta_1\Delta_2}\{5k_1^2\Delta_1 - 0.00173\psi k_1 k_2 - 0.1042\psi k_1 + 5\Delta\\
-0.001736\psi k_2^2 - 0.10416\psi k_2\}\eta^4\\
\frac{2}{k_1\Delta_1\Delta_2}\{2k_2^2 k_1\Delta_1 - 0.00173\psi k_2^2 - 0.001736\psi k_1\\
+2\Delta_1 k_2 k_1^2 - 0.0834\psi k_1 - 0.0416\psi k_2 + 12\Delta_1 k\\
+1.25\psi\}\eta^5 + (\frac{0.001736\psi}{k_1\Delta_1})\eta^6
\end{aligned}
\qquad (4.4.46)
$$

By selecting the following physical quantities:

$$Ha = 500, Re = 50, \alpha = 5°, \phi = 0.1 \qquad (4.4.47)$$

According to Eq. (4.4.10) $A*$ is defined as follows:

$$A^* = (1 - \phi) + \frac{\rho_s}{\rho_f}\phi \qquad (4.4.48)$$

By substituting $\rho_s = 2\rho_f$, with Eq. (4.4.48), the value of A* would be obtained as follows:

A* = 1.1

With the physical values of Eq. (4.4.47) and Eq. (4.4.49), the answer function is obtained as follows:

$$
\begin{aligned}
f(\eta) = 1 - 1.22477\eta^2 + 0.4153\eta^4 - 0.1148\eta^5\\
- 0.07572\eta^6
\end{aligned}
\qquad (4.4.50)
$$

Concerning the given physical values of Eq. (4.4.47) and Eq. (4.4.49) and the introduced domain of $\eta \in \{1,0\}$, the numerical solutions to the problem are obtained as follows (see Tables 4.4.1–Table 4.4.3 and Figs. 4.4.2–Fig. 4.4.8). The following charts show the difference between the results based on the yielded solution from Eq. (4.4.50) by AGM and the results of Table 4.4.1 by the numerical method.

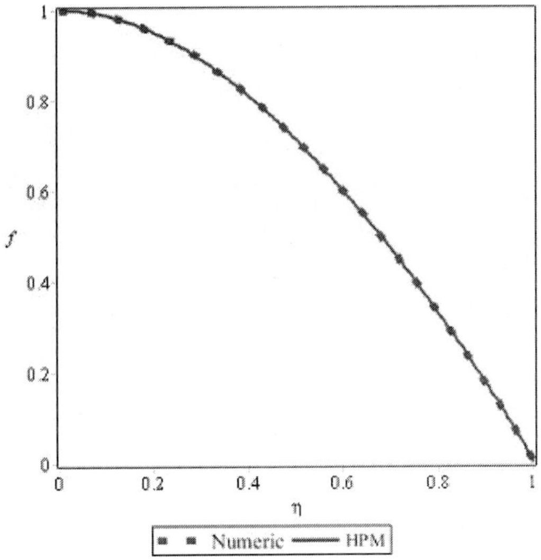

Fig. 4.4.2. A comparison between the solutions for f(η) by HPM and the numerical method.

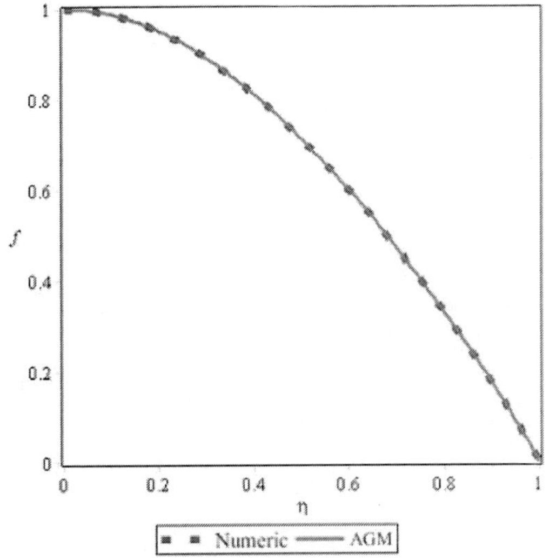

Fig. 4.4.3. A comparison between the solutions for f(η) by AGM and the numerical method.

Table 4.4.2. Comparing the values for f(η) by HPM and AGM with the numerical solution.

η	HPM AGM	AGM	Numeric	percentage of HPM	percentage of AGM
0	1.000	1.000	1.000	—	—
0.2	0.951292	0.951632	0.949064	0.23	0.27
0.4	0.812268	0.813182	0.805313	0.86	0.97
0.6	0.599744	0.600445	0.589751	1.69	1.81
0.8	0.328955	0.328783	0.321003	2.47	2.42
1	0	0.000	0.000	—	—

Table 4.4.3. Comparing the values for f(η) by HPM and AGM with the numerical solution

η	HPM AGM	AGM	Numeric	percentage of HPM	percentage of AGM
0	0.000	0.000	0.000	—	—
0.2	- 0.480755	- 0.477682	- 0.501233	4.08	4.70
0.4	- 0.894461	- 0.892847	- 0.917467	2.51	2.68
0.6	- 1.216921	- 1.220631	- 1.221339	0.36	0.06
0.8	- 1.489787	- 1.493100	- 1.464184	1.74	1.97
1	- 1.822550	- 1.816706	- 1.770954	2.91	2.58

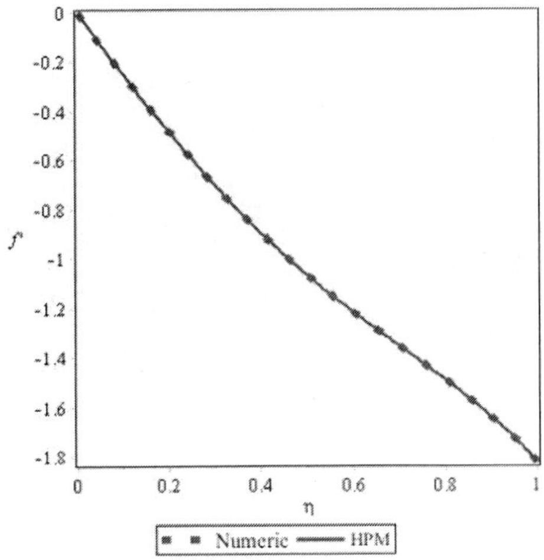

Fig. 4.4.4. A comparison between the solutions for f' (η) by HPM and the numerical method.

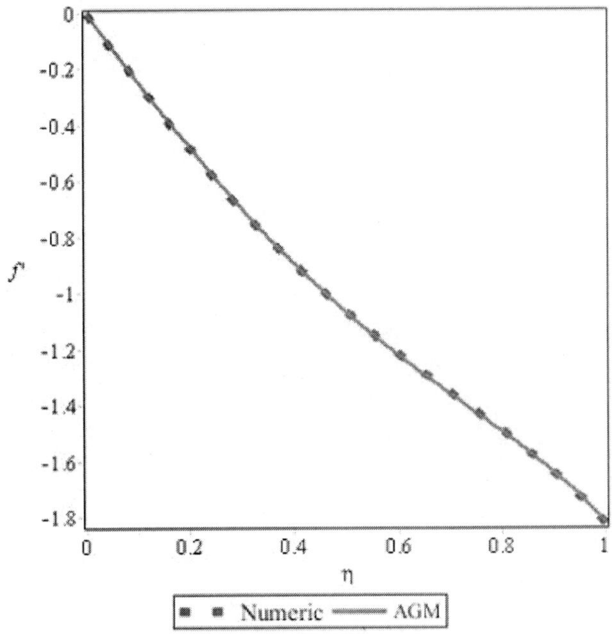

Fig. 4.4.5. A comparison between the solutions for f'(η) by AGM and the numerical method.

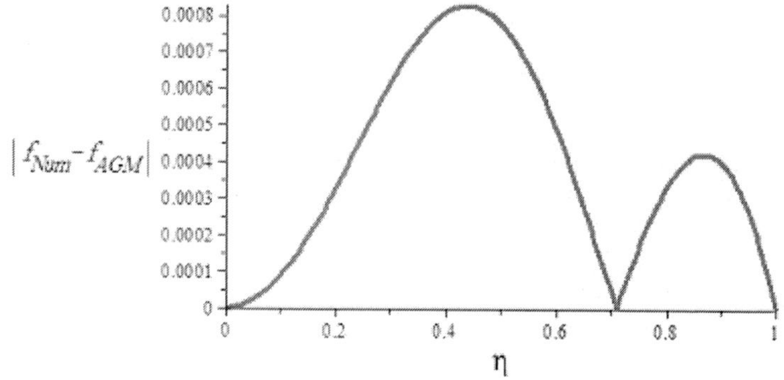

Fig. 4.4.6. The absolute difference between the solutions obtained by AGM and the numerical method for $f(\eta)$.

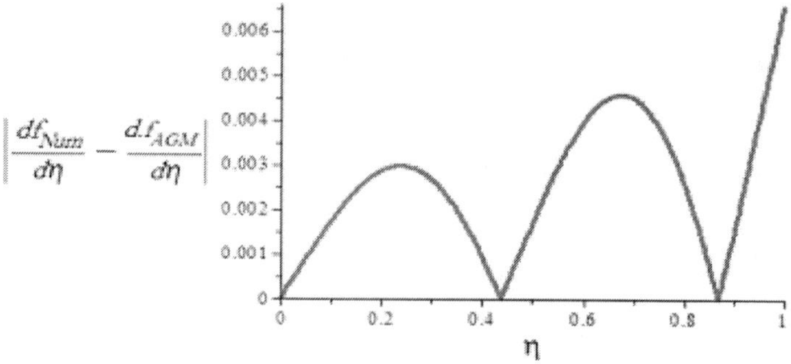

Fig. 4.4.7. The absolute difference between the solutions obtained by AGM and the numerical method for $f'(\eta)$.

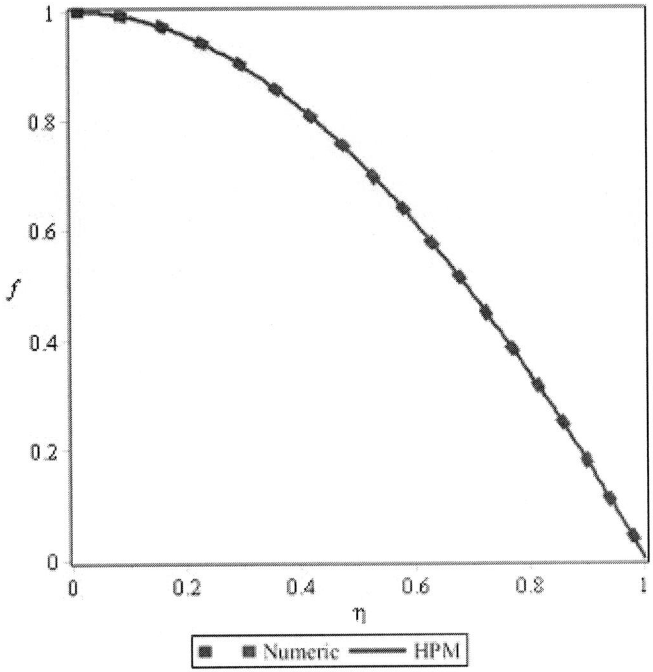

Fig. 4.4.8. Comparison between the solutions obtained for $f(\eta)$ by HPM and the numerical method

($A^*= 1$, $Ha = 0$, $Re = 50$, $\alpha = 1^{\circ}$, $\Phi= 0$)

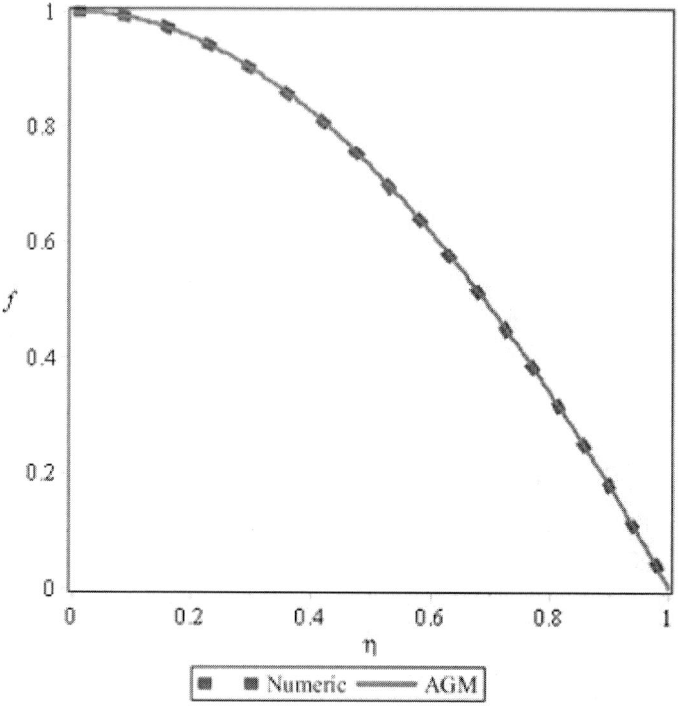

Fig. 4.4.9. Comparison between the solutions obtained for $f(\eta)$ by AGM and the numerical method.

The graphs for the different values of the Reynolds number and α are presented to show the accuracy of HPM and AGM. As is obvious, both methods are accurate and applicable.

Now we present graphs to investigate the effect of different parameters. Fig. 4.4.18 shows the effect of the Reynolds number at constant α and the Hartmann number ($\alpha = 1^o, Ha = 0$). Fig. 4.4.19 and Fig 4.4.20 show the effect of the Hartmann number at constant α and the Reynolds number ($\alpha = 1^o, 5^o, Re = 200$). Fig. 4.4.19 shows that with small amounts of α, $f(\eta)$ is almost independent of the Hartmann number, but Fig. 4.4.20 shows that with greater amounts of α, $f(\eta)$ is dependent on the Hartmann number. Fig. 4.4.21 shows the effect of α for constant Reynolds and Hartmann numbers (Re = 200, Ha = 0). Fig. 4.4.22 and Fig. 4.4.23 show the effect of Φ for constant α, the Reynolds number, and the Hartmann number ($\alpha = 5^o, Re =$

$200, Ha = 0$). Note that A^* is dependent on Φ at first; we chose $2\rho_f = \rho_s$ (Fig. 4.4.22) and then we chose ρ_s= 8933 (density of Cu.) and $\rho_s = 997.1$ (density of water). Therefore, Fig. 4.4.23 shows $f(\eta)$ for Cu-water.

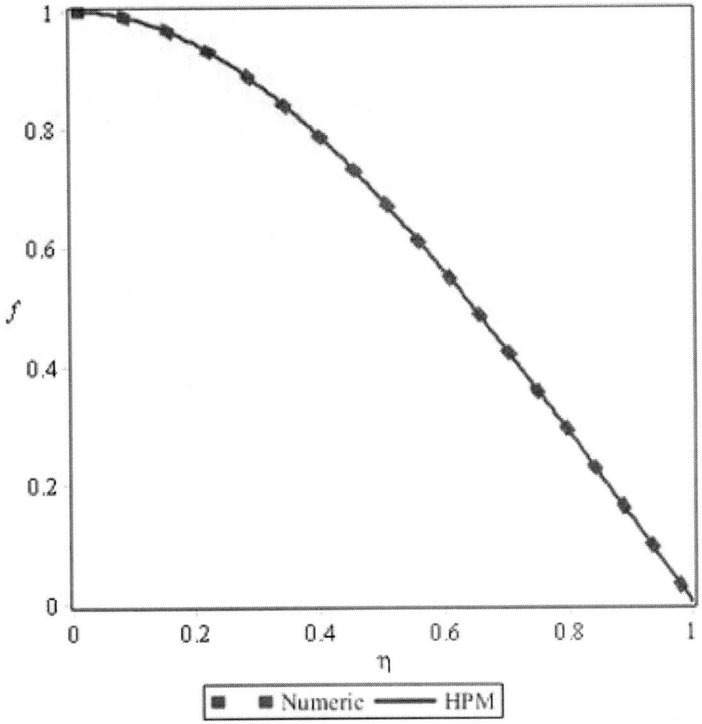

Fig. 4.4.10. Comparison between the solutions obtained for $f(\eta)$ by HPM and the numerical method

(A^*= 1, Ha = 0, Re = 150, α = 1∘,Φ= 0)

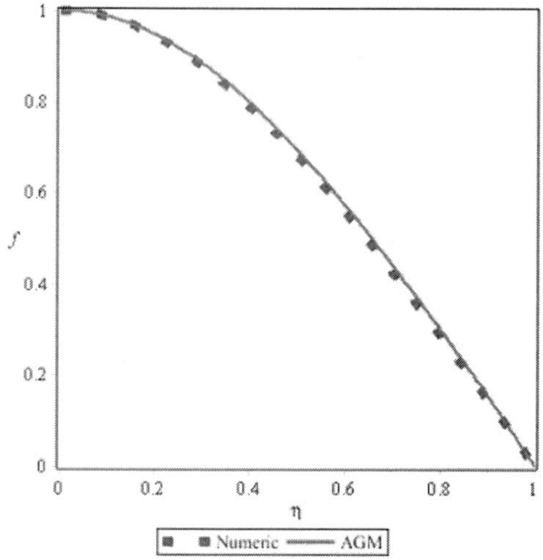

Fig. 4.4.11. Comparison between the solutions obtained for $f(\eta)$ by HPM and the numerical method

(A*= 1, Ha = 0, Re = 150, α = 1◦,Φ= 0)

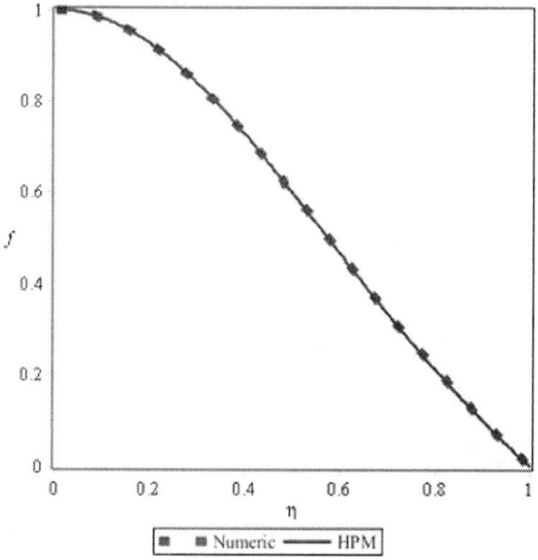

Fig. 4.4.12. Comparison between the solutions obtained for $f(\eta)$ by HPM and the numerical method

(A*= 1, Ha = 0, Re = 150, α = 1◦,Φ= 0)

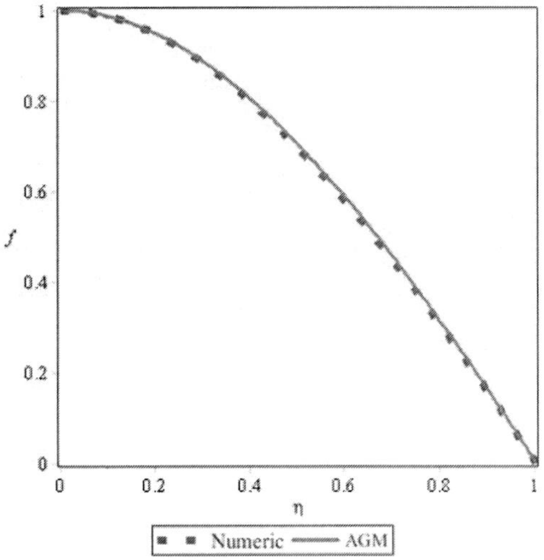

Fig. 4.4.13. Comparison between the solutions obtained for $f(\eta)$ by AGM and the numerical method

($A^*= 1$, $Ha = 0$, $Re = 150$, $\alpha = 1^\circ$, $\Phi= 0$)

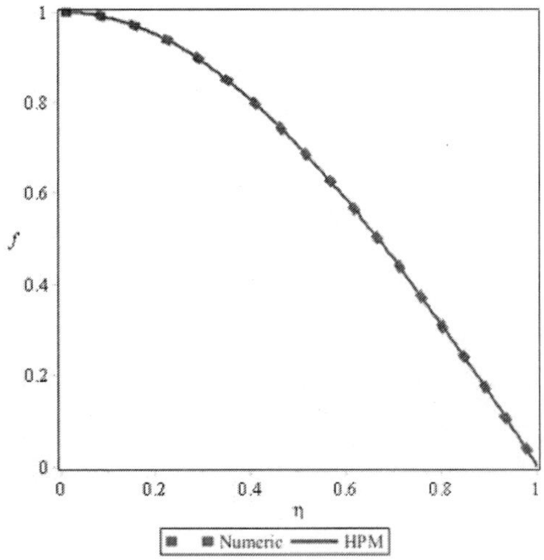

Fig. 4.4.14. Comparison between the solutions obtained for $f(\eta)$ by HPM and the numerical method

($A^*= 1$, $Ha = 0$, $Re = 50$, $\alpha = 1^\circ$, $\Phi= 0$)

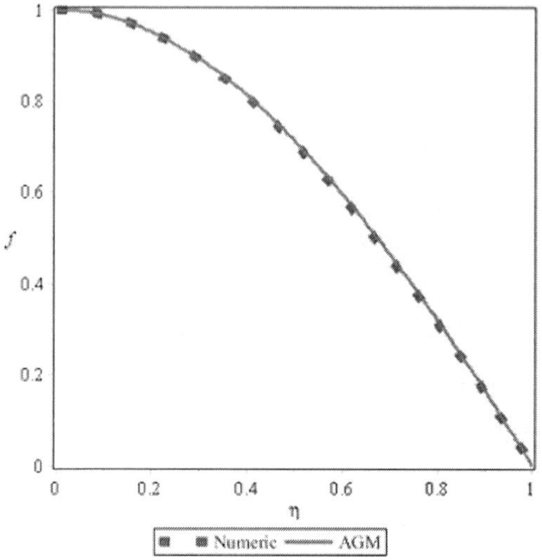

Fig. 4.4.15. Comparison between the solutions obtained for $f(\eta)$ by AGM and the numerical method

(A^*= 1, Ha = 0, Re = 50, α = 2◦,Φ= 0)

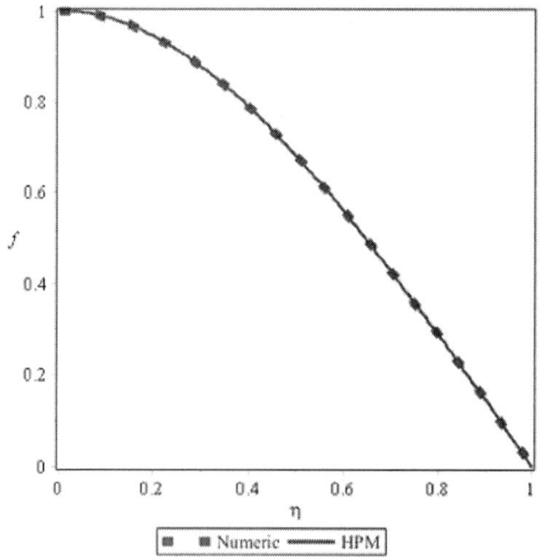

Fig. 4.4.16. Comparison between the solutions obtained for $f(\eta)$ by HPM and the numerical method

(A^*= 1, Ha = 0, Re = 50, α = 3◦,Φ= 0)

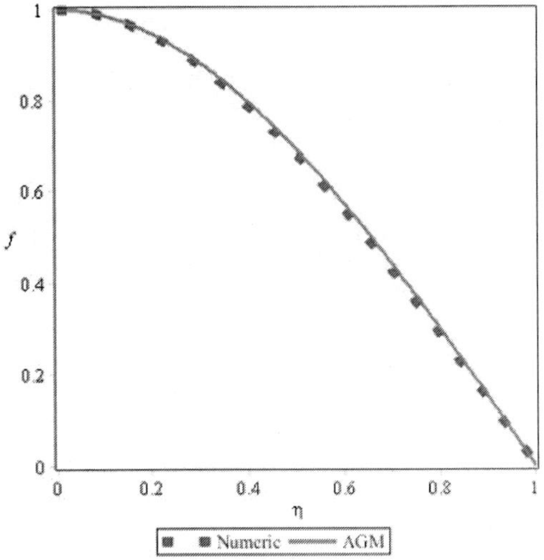

Fig. 4.4.17. Comparison between the solutions obtained for $f(\eta)$ by HPM and the numerical method

($A^*= 1$, $Ha = 0$, $Re = 50$, $\alpha = 3°$, $\Phi= 0$)

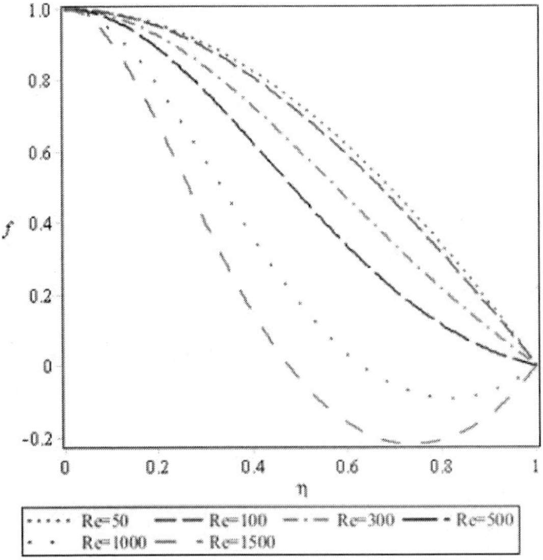

Fig. 4.4.18. $f(\eta)$ for different values of *Re* ($A^*= 1$, $Ha = 0$, $\alpha = 1°$, $\Phi= 0$)

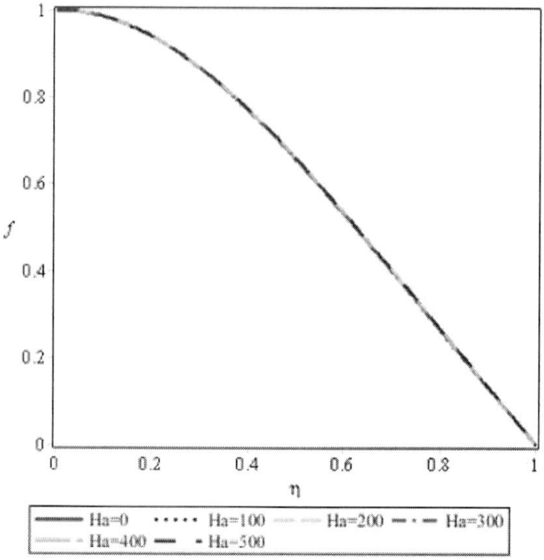

Fig. 4.4.19. ***f*****(η) for different values of** ***Ha (A*= 1, Ha = 0, α = 1◦,Φ= 0, R=200)***

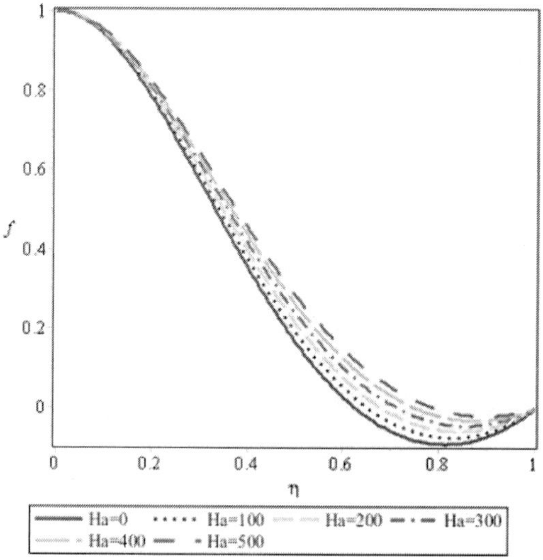

Fig. 4.4.20. f(η) for different values of Ha (A*= 1, Ha = 0, *α* =5◦,*Φ*= 0, R=200)

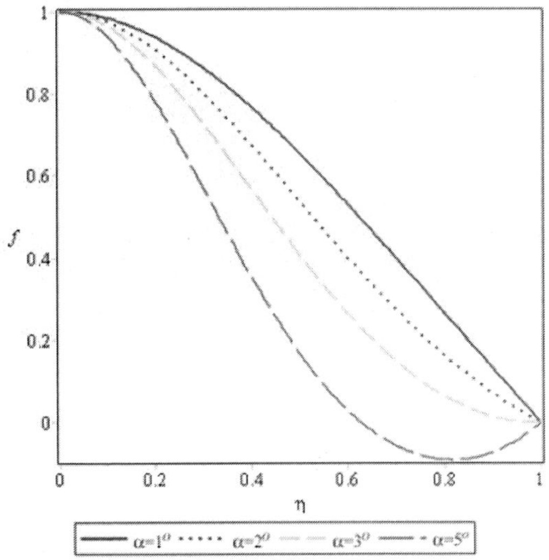

Fig. 4.4.21. f(η) for different values of α (A*= 1, Ha = 0, α =5◦,Φ= 0, R=200)

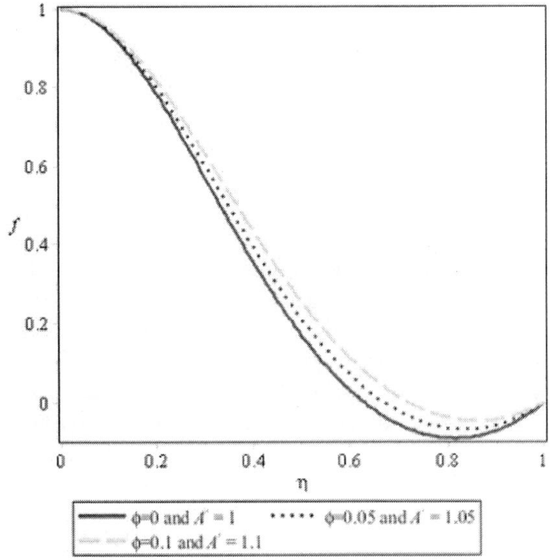

Fig. 4.4.22. *f*(η) for different values of ϕ and *A *(Ha = 0, Re = 200 Ha = 0, α = 5)***

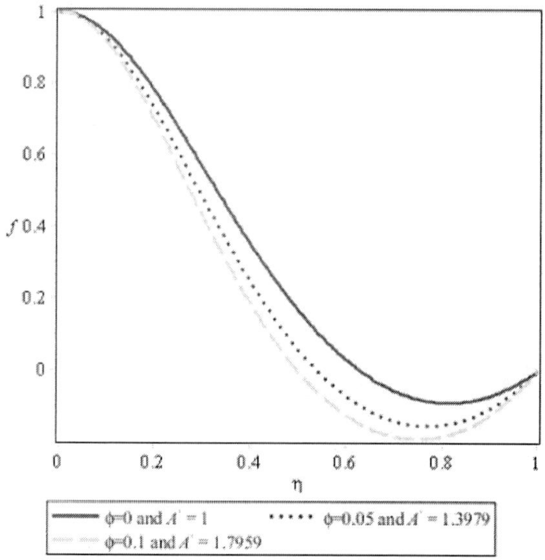

Fig. 4.4.23. f(η) for different values of φ and A* (Ha = 0, Re = 200, α = 5°).

This case study offers two methods to solve a nonlinear differential equation governing the Jeffery-Hamel flow with a high magnetic field and nanoparticles: the HPM and AGM. The above has been done to indicate the ability of HPM and AGM to solve a broad range of differential equations. We showed that both methods are efficient and accurate. In AGM, in contrast to most analytical methods, which can solve differential equations only by introducing dimensionless parameters, a wide variety of complicated differential equations are easily solvable directly, meaning there is no need to convert variables into new ones. In AGM, considering three or four sentences of the series of polynomials, and the answer to differential equations, is sufficient to obtain reliable solutions. Increasing the number of sentences makes the solution closer to the exact solution. In contrast, the error might increase or decrease in most other methods.

4.5. Adding Nanoparticles to the Blood Flow in the Presence of a Magnetic Field

The present study examines numerical and analytical approaches to the fluid flow of blood containing nanoparticles in a porous media affected by a magnetic field.

For this problem, a steady flow in a porous blood vessel in a magnetic field is examined for an incompressible, non-Newtonian nanofluid. The schematic model of the problem is presented in Fig. 4.5.1. Obviously, the blood flow is known as 3D matter. Despite the 3D feature, at present, 3D modelling of blood flow is not practical due to the far-reaching computational time required to model the vessel channels with consideration of fluid-solid interactions between the blood and artery walls. Assuming blood flow as a one-dimensional stream is useful for analysing the hemodynamics and dynamics related to cardiovascular diseases. As a lateral reason for assuming one-dimensional blood flow, it has been proved that pressure flow caused by the heart produces long wavelengths, and the transversal flows are negligible.

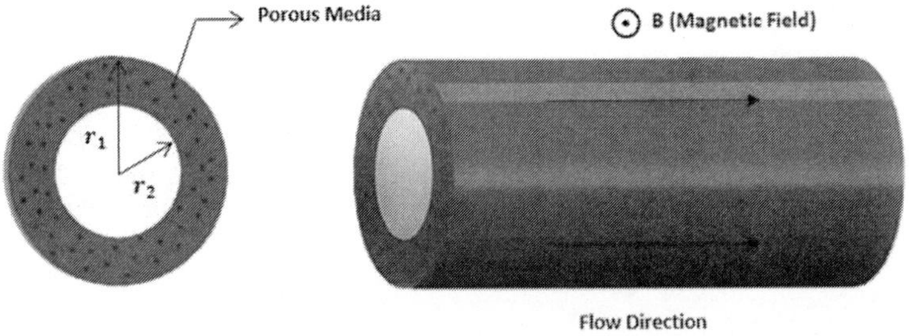

Fig. 4.5.1. Schematic of blood flow through a porous artery in the presence of a magnetic field.

Table 4.5.1. Some fundamental operations of differential transformation method.

Original function	Transformed function
$x(t)$	$X(k) = \alpha F(k) \pm \beta G(k)$
$x(t) = \frac{d^m f(t)}{dt^m}$	$X(k) = \frac{(k+m)! F(k+m)}{K!}$
$x(t) = f(t) g(t)$	$X(k) = \sum_{l=0}^{k} F(l) G(k-l)$
$x(t) = t^m$	$X(k) = \delta(k-m) = \begin{cases} 1, & \text{if } k = m \\ 0, & \text{if } k \neq m \end{cases}$
$x(t) = exp(t)$	$X(k) = \frac{1}{k!}$
$x(t) = \sin(\omega t + \alpha)$	$X(k) = \frac{\omega^k}{k!} \sin\left(k\frac{\pi}{2} + \alpha\right)$
$x(t) = \cos(\omega t + \alpha)$	$X(k) = \frac{\omega^k}{k!} \cos\left(k\frac{\pi}{2} + \alpha\right)$

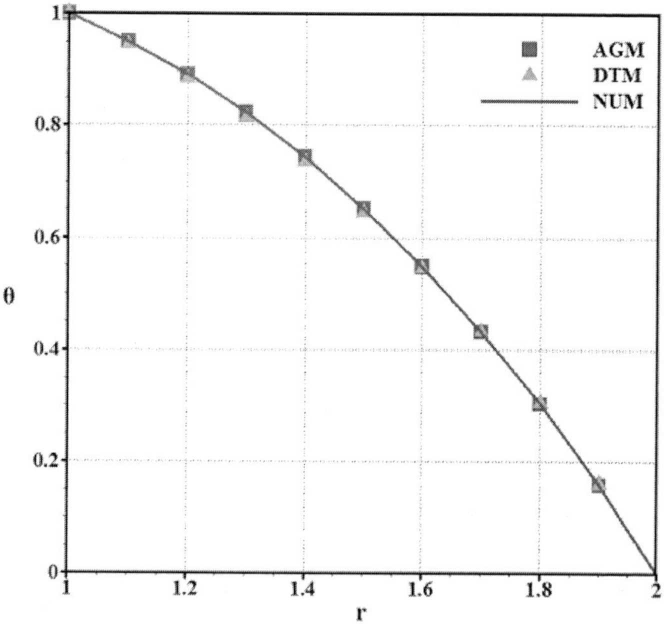

Fig. 4.5.2. Comparison of AMG, DTM versus the numerical method for(θ) considering. Nb = Nt =α= ∧ = P = M = C = Gr = Br = 1, $\mu(r) = 1$

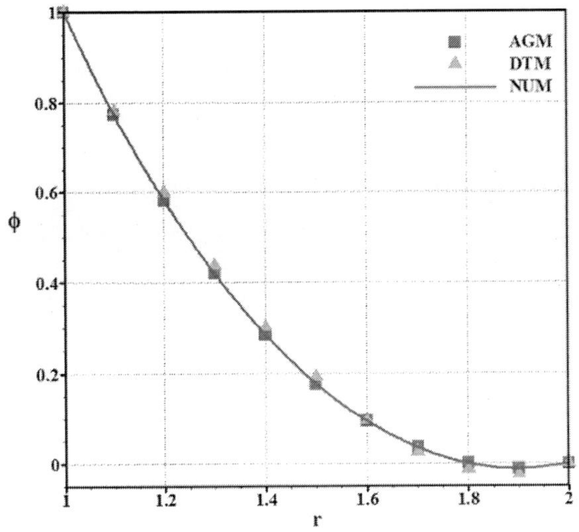

Fig. 4.5.3. Comparison of AMG, DTM versus the numerical method for(θ) considering. Nb = Nt =α= ∧ = P = M = C = Gr = Br = 1, μ(r)=1

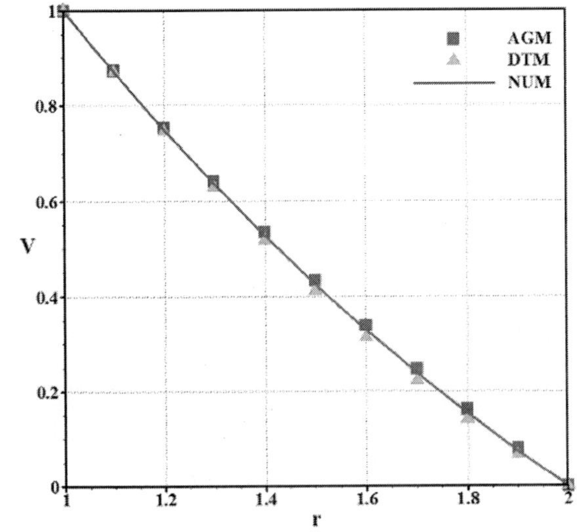

Fig. 4.5.4. Comparison of AMG, DTM versus the numerical method for(θ) considering. Nb = Nt =α= ∧ = P = M = C = Gr = Br = 1, $\mu(r) =$ 1

Table 4.5.2. Comparison of DTM and AGM with HPM and the numerical method for the case of Nb = Nt =α= ∧ = P = M = C = Gr = Br = 1, $\mu(r) = 1$

r	θ (HPM)	θ (NUM)	θ (DTM)	θ (AGM)
1.0	1.000000	1.0000000000	1.0000000001	0.9999999998
1.1	0.980296	0.9504559134	0.9456133828	0.9504561032
1.2	0.842245	0.8916480651	0.8840660114	0.8916486117
1.3	0.710272	0.8228344628	0.8144196733	0.8228355669
1.4	0.588083	0.7432801843	0.7356409200	0.7432820809
1.5	0.474328	0.6522567719	0.6465958502	0.6522597214
1.6	0.367917	0.5490417180	0.5460448916	0.5490459651
1.7	0.267960	0.4329180235	0.4326375823	0.4329236588
1.8	0.281307	0.3031738135	0.3049073522	0.3031804065
1.9	0.084572	0.1591020031	0.1612663049	0.1591077556
2.0	0.000000	0.0000000000	-0.0000000008	-0.000000001

Table 4.5.3. Comparison of DTM and AGM with HPM and the numerical method for the case of Nb=Nt=α=ʌ=P=M=C=Gr=Br=1,μ(r)=1.

r	ϕ (HPM)	ϕ (NUM)	ϕ (DTM)	ϕ (AGM)
1.0	1.000000	1.0000000000	1.0000000000	0.9999999692
1.1	0.774537	0.7745370395	0.7801431600	0.7745390936
1.2	0.582283	0.5822831235	0.5986760830	0.5822860047
1.3	0.420142	0.4201422909	0.4381670180	0.4201476844
1.4	0.285866	0.2858661617	0.3016702660	0.2858718363
1.5	0.177818	0.1778182270	0.1927689170	0.1778242505
1.6	0.094814	0.0948144721	0.0956175670	0.094817906
1.7	0.036012	0.0360124842	0.0269850590	0.036010508
1.8	0.000832	8.3237335873	0.013700000	0.00084804
1.9	0.000011	0.011100841	0.024300000	0.01109970
2.0	0.000000	0.0000000000	0.0000000000	0.00000546

Table 4.5.4. Comparison of DTM and AGM with HPM and the numerical method for the case of Nb=Nt=α=ʌ=P=M=C=Gr=Br=1, μ(r)=1.

r	v (HPM)	v (NUM)	v (DTM)	v (AGM)
1.0	1.000000	1.0000000000	1.000000000	1.000000000
1.1	0.87048	0.8717659804	0.8708987908	0.873468310
1.2	0.778394	0.7504530614	0.7470026760	0.753863413
1.3	0.673639	0.6356758515	0.6287848450	0.640802355
1.4	0.572138	0.5271461028	0.5167184901	0.533963063
1.5	0.473287	0.4246545624	0.4112768008	0.433031890
1.6	0.376522	0.3280599001	0.3129329673	0.337651155
1.7	0.281307	0.2372829329	0.2221601803	0.247366682
1.8	0.187133	0.1523052745	0.1394316289	0.161575345
1.9	0.093517	0.0731721880	0.0652205057	0.079472590
2.0	0.000000	0.0000000000	0.000000000	0.000000000

Table 4.5.5. The absolute error of DTM for the case of Nb=Nt=α=ʌ=P=M=C=Gr=Br=1, μ(r)=1.

r	θ	ϕ	v
1	0.0000000001	0.0000000000	0.0000000000
1.1	0.0048425306	0.0065052497	0.0008671896
1.2	0.0075820537	0.0001411157	0.0034503854
1.3	0.0084147895	0.0059010518	0.0068910065
1.4	0.0076392643	0.0073082620	0.0104276127
1.5	0.0056609217	0.0040774908	0.0133777616
1.6	0.0029968264	0.0018511750	0.0151269328
1.7	0.0002804412	0.0078259750	0.0151227526
1.8	0.0017335387	0.000537627	0.0128736456
1.9	0.0021643018	0.0092178440	0.0079516823
2	0.0000000008	0.0000000000	0.0000000000

Table 4.5.6. The absolute error of AGM for the case of Nb=Nt=α=ʌ=P=M=C=Gr=Br=1, μ(r)=1.

r	θ	ϕ	v
1	0.0000000002	0.0000000308	0.0000000000
1.1	0.0000001898	0.0000020541	0.0017023296
1.2	0.0000005466	0.0000028812	0.0034103516
1.3	0.0000011041	0.0000053935	0.0051265035
1.4	0.0000018966	0.0000056746	0.0068169602
1.5	0.0000029495	0.0000060235	0.0083773276
1.6	0.0000042471	0.0000034339	0.0095912549
1.7	0.0000056353	0.0000019762	0.0100837491
1.8	0.0000065930	0.000015667	0.0092700705
1.9	0.0000057525	0.0000011410	0.0063004020
2	0.0000000010	0.0000054600	0.0000000000

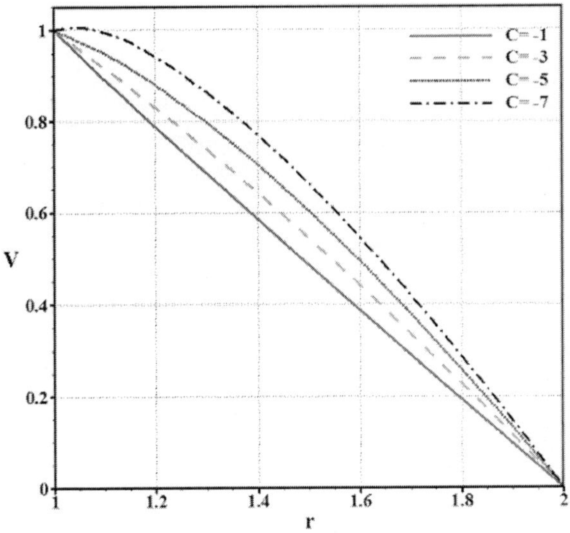

Fig. 4.5.5. The effect of a pressure gradient on the velocity profile

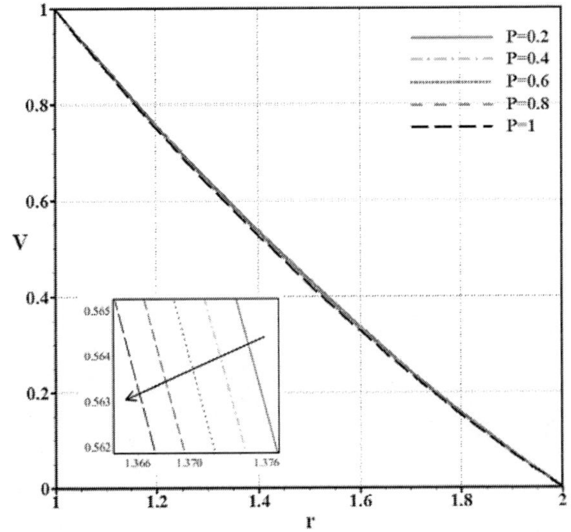

Fig. 4.5.6. The effect of the porosity parameter on the velocity profile.

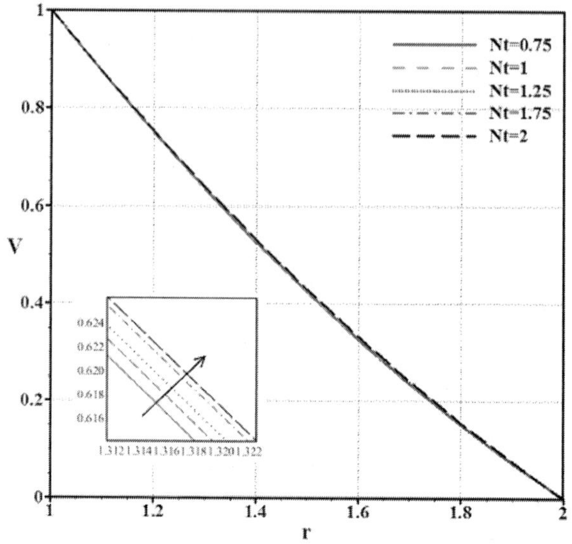

Fig. 4.5.7. The effect of a thermophoresis parameter on the velocity profile.

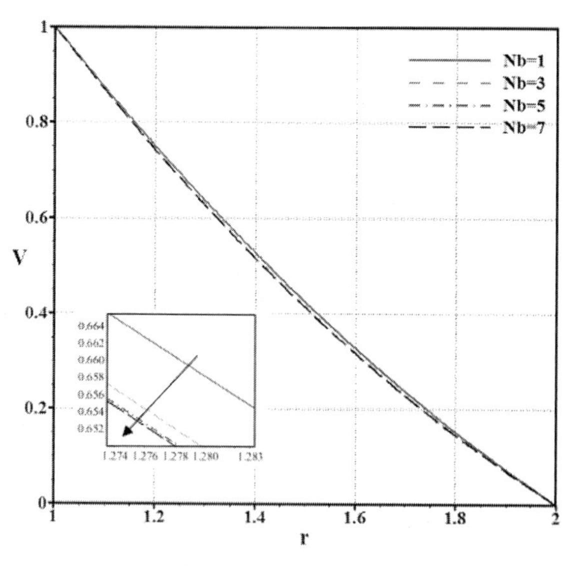

Fig. 4.5.8. The effect of a Brownian motion parameter on the velocity profile.

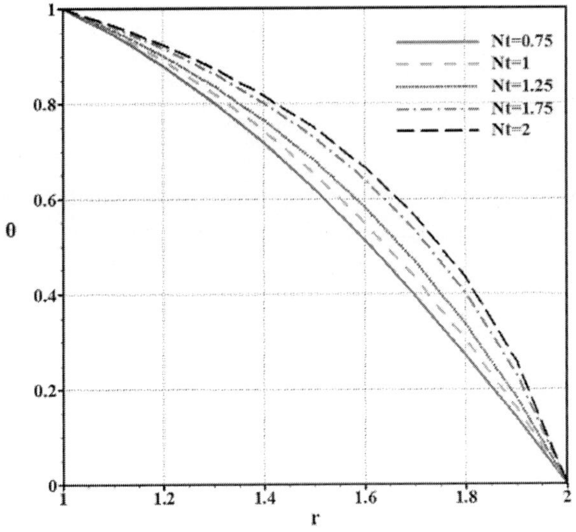

Fig. 4.5.9. The effect of a thermophoresis parameter on the temperature profile.

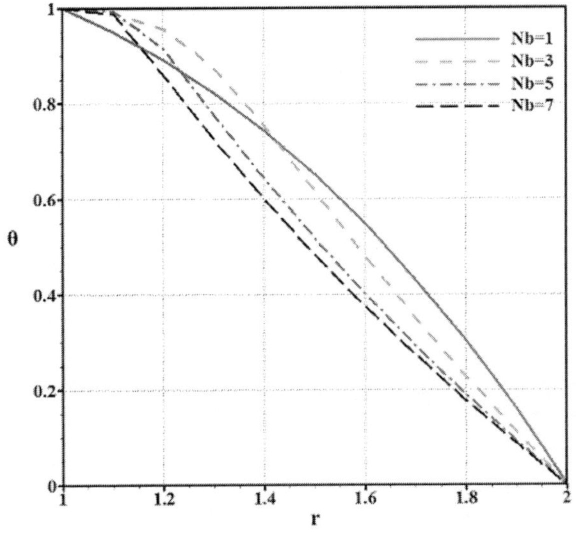

Fig. 4.5.10. The effect of a Brownian motion parameter on the temperature profile.

For a nanofluid with the stated characteristics, the density (ρ) is presented by:

$$\rho = \phi\rho_p + (1-\phi)\rho_{f_0} \\ \cong \phi\rho_p + (1-\phi)\{\rho_f(1-\beta_T(\theta-\theta_w))\} \tag{4.5.1}$$

In the above formula, ρ_{f_0}, is the density of the nanofluid, θ_w, is the reference temperature ρ_f, is the nanofluid's density at the reference temperature, ρ_f is the density of the nanoparticles, and the volumetric coefficient of expansion (the bulk coefficient) is β_T. The total density of the fluid with nanoparticles(ρ), would be:

$$\rho \cong \phi\rho_p + (1-\phi)\{\rho_f(1-\beta_T(\theta-\theta_w))\} \tag{4.5.2}$$

By the incompressibility assumption, the conservation of momentum, total mass, thermal energy, and nanoparticles equations become:

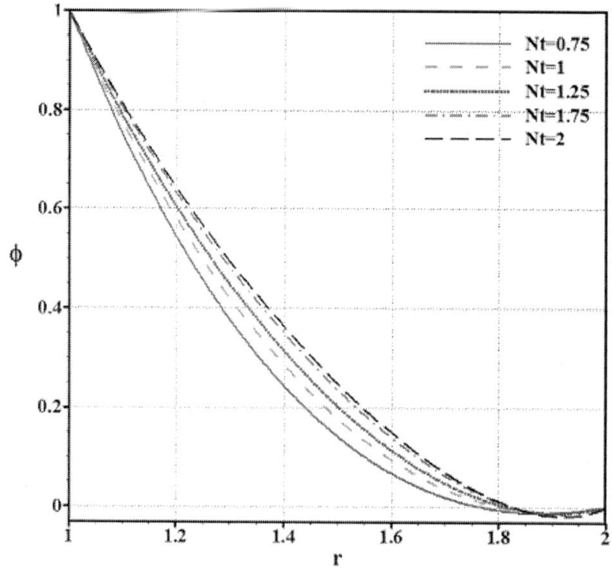

Fig. 4.5.11. The effect of thermophoresis parameters on the nanoparticles concentration profile.

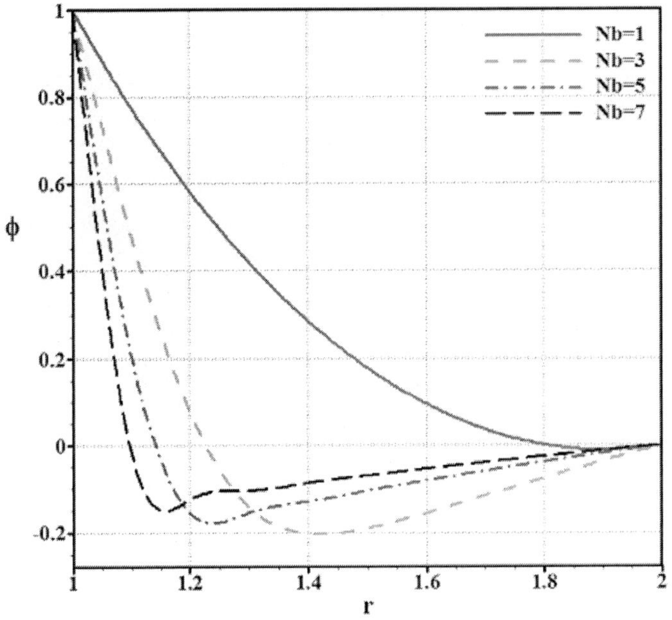

Fig. 4.5.12. The effect of the Brownian motion parameter on the nanoparticles concentration profile.

$$\rho_f(\frac{\partial V}{\partial t} + V.\nabla V) = \text{div } T - \frac{\mu\varphi}{k}(1 + \lambda_r\frac{\partial}{\partial t})V + [\phi\rho_p + (1 - \phi)\rho_f[1 \tag{4.5.3}$$

$$-\beta_T(\theta - \theta_w)]]g + J \times B$$

$$(\rho c)_f(\frac{\partial \theta}{\partial t} + V.\nabla\theta) = k\nabla^2\theta + (\rho c)_p[D_b\nabla\phi \cdot \nabla\theta + \frac{D_T}{\theta_w}\nabla\theta \cdot \nabla\theta] \tag{4.5.4}$$

$$(\frac{\partial \phi}{\partial t} + V.\nabla\phi) = D_b\nabla^2\phi + \frac{D_T}{\theta_w}\nabla^2\theta \tag{4.5.5}$$

In which φ is the porosity of the medium, and ϕ is the nanoparticles mass concentration. In our study, conductive fluid is examined. Therefore, we should consider the last term on the right-hand side of Eq. (4.5.3). Through Ohm's law, we have:

$$J = \sigma(E + V \times B) \tag{4.5.6}$$

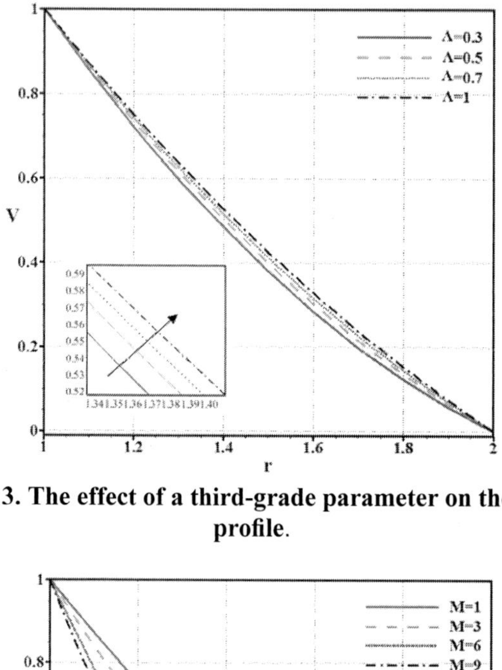

Fig. 4.5.13. The effect of a third-grade parameter on the velocity profile.

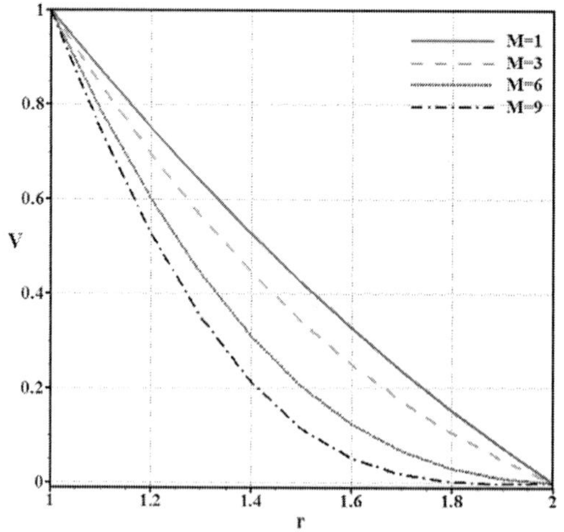

Fig. 4.5.14. The effect of an MHD parameter on the velocity profile.

Where E is the electric field, B=B_0+b indicates the total magnetic field in which B_0 and b are representatives of the uniformly applied magnetic field and variable magnetic quantity, respectively. σ is known as the electrical conductivity, V is the velocity vector, and J represents the electric current density. The induced magnetic field is neglected for small Reynolds' numbers. The simplified equation is:

$$J \times B = -\sigma B_0^2 V \tag{4.5.7}$$

At $r = r_1$ and $r = r_2$ the boundary conditions are as follows:

$$v(r_2) = v_0, v(r_1) = 0, \theta(r_2) = 0_w, \theta(r_1) = 0, \phi(r_2) = \phi_w, \phi(r_1) = 0 \tag{4.5.8}$$

For a third-grade fluid, stress is given by:

$$T = -PI + \mu A_1 + \alpha_1 A_2 + \alpha_2 A_1^2 + \beta_1 A_3 + \beta_2 (A_1 A_2 + A_2 A_1) + \beta_3 (\text{tr } A_1^2) A_1 \tag{4.5.9}$$

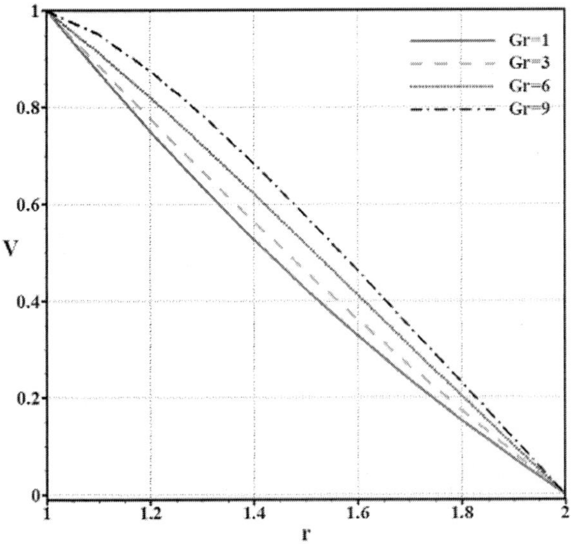

Fig. 4.5.15. The effect of the Grashof number on the velocity profile.

$\alpha_1, \alpha_2, \beta_1, \beta_2, \beta_3$, are the material modules. Commonly, they are functions of temperature. –PI shows the spherical stress caused by restraint of incompressibility and the kinematical tensors A_1, A_2, A_3 can be introduced by the following equations:

$$A_1 = (\nabla V) + (\nabla V)^t \tag{4.5.10}$$

$$A_n = \frac{dA_{n-1}}{dt} + A_{n-1}(\nabla V) + (\nabla V)^t A_{n-1} \Rightarrow n = 2{,}3 \tag{4.5.11}$$

In Eqs. (4.5.10)–(4.5.11), ∇ represents the gradient operator V = [0,0, v(r)] is the velocity field, and D/Dt shows the material time derivative, which has the following form:

$$\frac{D(\cdot)}{Dt} = \frac{\partial(\cdot)}{\partial t} + [\text{grad } (\cdot)]V \tag{4.5.12}$$

It should be noticed that specific Helmholtz free energy is minimum when the local fluid is at rest.

$$\mu \geq 0, \alpha_1 \geq 0, |\alpha_1 + \alpha_2| \leq \sqrt{24\mu\beta_3}, \beta_1 = \beta_2 = 0, \beta_3 \geq 0 \tag{4.5.13}$$

The fluid is assumed to be thermodynamically compatible, so Eq. (4.5.9) will be reduced to:

$$T = -PI + \mu A_1 + \alpha_1 A_2 + \alpha_2 A_1^2 + \beta_3(\text{tr } A_1^2)A_1$$

And by rearranging, we have:

$$T = -PI + [\mu + \beta_3(trA_1^2)]A_1 + \alpha_1 A_2 + \alpha_2 A_1^2 \tag{4.5.15}$$

For equation simplification, we define the non-dimensional parameters as:

$$v = \frac{v}{V_0}, \bar{r} = \frac{r}{R}, \bar{\mu} = \frac{\mu}{\mu_0}, \bar{\theta} = \frac{\theta - \theta_w}{\theta_m - \theta_w}, \bar{\phi} = \frac{\phi - \phi_w}{\phi_m - \phi_w} \tag{4.5.16}$$

Where $V_0, \mu_0, \theta_w, \theta_m$ and ϕ_m, denote the reference velocity, viscosity, pipe temperature, fluid temperature, and mass concentration. By placing Eq. (4.5.15) in the linear momentum equation and applying the non-dimensional quantities given in Eq. (4.5.16) to the governing Eq. (4.5.3)–(4.5.5), the dimensionless forms of the governing equations will be obtained as the following non-dimensional coupled equations:

$$N_b\frac{d\theta}{dr}\frac{d\phi}{dr}+\frac{1}{r}\frac{d\theta}{dr}+\alpha\frac{d^2\theta}{dr^2}+\alpha_1 N_t(\frac{d\theta}{dr})^2=0 \quad (4.5.17)$$

$$N_b(\frac{d^2\theta}{dr^2}+\frac{1}{r}\frac{d\theta}{dr})+N_t(\frac{d^2\phi}{dr^2}+\frac{1}{r}\frac{d\phi}{dr})=0 \quad (4.5.18)$$

$$\frac{d\mu}{dr}\frac{dv}{dr}+\frac{\mu}{r}\frac{dv}{dr}+\mu\frac{d^2v}{dr^2}+\frac{\Lambda}{r}(\frac{dv}{dr})^3+3\Lambda(\frac{dv}{dr})^2\frac{d^2v}{dr^2} = Pv+M^2v+c-G_r\theta-B_r\phi \quad (4.5.19)$$

P and M are porosity and MHD parameters, respectively. Boundary conditions are described as follows:

$$v(r_2)=1, v(r_1)=0, \theta(r_2)=1, \theta(r_1)=0, \phi(r_2) = 1, \phi(r_1)=0 \quad (4.5.20)$$

These boundary conditions are assumed for a hollow blood vessel with an inner and outer wall. Eqs. (4.5.17)–(4.5.20) are defined by consideration of the following relations:

$$\Lambda=\frac{2\beta_3 v_0^2}{\mu_0 R^2}, c=\frac{\frac{\partial p}{\partial z}R^2}{v_0\mu_0}, P=\frac{\varphi}{kR_2^2}, B_r = \frac{(\rho_p-\rho_w)R^2(\phi_m-\phi_w)g}{v_0\mu_0}, M^2 = \frac{\sigma B_0^2 R^2}{\mu_0} \quad (4.5.21)$$

$$N_t=\frac{D_T(\theta_m-\theta_w)}{\theta_w}, G_r = \frac{(\theta_m-\theta_w)\rho_{f,w}R^2(1-\phi_w)g}{v_0\mu_0}, N_b = D_b(\phi_m-\phi_w)$$

As said before, considering blood as a non-Newtonian fluid, it is well known that small-sized nanoparticles are increasingly used in every biomedical application. In other words, nanoparticles, like gold nanoparticles, can be applied to activate or prevent the blood vein's growth. Though there are some drugs for increasing and reducing the blood-capillary growth in specified diseases, they are only efficient for a short time. Researchers have revealed that these nanoparticles could solve some drug administration problems. They showed that these nanoparticles are effective drug-carrying and drug-delivery vehicles since they encapsulate large quantities of therapeutic molecules. In the present problem, the blood containing nanoparticles is considered a non-Newtonian nanofluid. The viscosity of the fluid is determined using constant model Eq. (4.5.22).

$$\mu = 1 \tag{4.5.22}$$

To understand DTM's concept, suppose that x(t) is an analytic function in the certain domain of D and t_i, shows each point of the domain. The function x(t) is represented by power series, which considers t_i, as the central location. The Taylor series expansion function of x(t) is defined below:

$$x(t) = \sum_{k=0}^{\infty} \frac{(t - t_i)^k}{k!} [\frac{d^k x(t)}{dt^k}]_{t=t_i} \qquad \forall t \in D \tag{4.5.23}$$

The Maclaurin series of x(t) can be obtained by taking $t_i = 0$ in Eq. (4.5.23) stated as:

$$x(t) = \sum_{k=0}^{\infty} \frac{t^k}{k!} \left[\frac{d^k x(t)}{dt^k}\right]_{t=0} \qquad \forall t \in D \tag{4.5.24}$$

The differential transformation of the function x(t) is defined as follows:

$$X(k) = \sum_{k=0}^{\infty} \frac{H^k}{k!} [\frac{d^k x(t)}{dt^k}]_{t=0} \tag{4.5.25}$$

X(k) represents the transformed function, also called the T-function, and x(t) represents the original function. The differential spectrum of X(k) is confined to interval $t \in [0, H]$, where H displays a constant value. The differential inverse transform of X(k) is in the following form:

$$x(t) = \sum_{k=0}^{\infty} (\frac{t}{H})^k X(k) \tag{4.5.26}$$

The Taylor series accomplishes the basic concept of the differential transformation method. Under different values of the k argument, the values of the X (k) function would be in discrete forms, i.e., X (0) is known as the zero discrete, X (1) as the first discrete, etc. With more discrete values available, more precision is applied to achieve the desired function. The function x (t) consists of the T-function X (k), whose value is given by summing the T-function with (t/H) k as its coefficient. In real applications, at the correct magnitude of constant H, the larger values of the k argument would reduce the discrete spectrum rapidly. A finite series express the function x (t), and Eq. (4.5.26) can be written as:

$$x(t) = \sum_{k=0}^{n} (\frac{t}{H})^k X(k) \tag{4.5.27}$$

To solve a problem with DTM, first, the DTM transformation form of Eqs. (4.5.17)–(4.5.19) is obtained by using Table 4.5.1.

The transformed form of the equations $\theta(r), \phi(r), v(r)$, is as follows:

$$N_b(\sum_{P=0}^{k} \sum_{L=0}^{P} \delta(L-1)(1+p-L)\theta(1+p-L) \times$$

$$(1+k-p)\phi(1+k-p)) + (1+k)\theta(1+k)$$

$$+a(\sum_{L=0}^{k} \delta(L-1)(1+k-L) \tag{4.5.28}$$

$$(2+k-L)\theta(2+k-L)) + \alpha_1 N_t$$

$$\times(\textstyle\sum_{P=0}^{k} \sum_{L=0}^{P} \delta(L-1)(1+p-L) \times$$

$$\theta(1+p-L)(1+k-p)\theta(1+k-p))=0$$

$$N_b((\sum_{L=0}^{k} \delta(L-1)(1+k-L)(2+k-L) \times$$

$$\theta(2+k-L)) + (1+k)\theta(1+k)) \tag{4.5.29}$$

$$+N_t((\sum_{L=0}^{k} \delta(L-1)(1+k-L)(2+k-L) \times$$

$$\phi(2+k-L)) + (1+k)\phi(1+k))$$

$$(1+k)v(1+k) +$$

$$\mu(\sum_{L=0}^{k} \delta(L-1)(1+k-L)(2+k-L)v(2+k-L))$$

$$+\Lambda(\sum_{P=0}^{k}\sum_{L=0}^{P}(1+L)v(1+L)(1+p-L)v \times$$

$$(1+p-L)(1+k-p)v(1+k-p))$$

$$+3\Lambda(\sum_{j=0}^{k}\sum_{P=0}^{j}\sum_{L=0}^{P}\delta(L-1)(1+p-L)v \times$$

$$(1+p-L)(1+j-p)v(1+j-p)) \tag{4.5.30}$$

$$-P(\sum_{L=0}^{k}\delta(L-1)v(k-L)) -$$

$$M^2(\sum_{L=0}^{k}\delta(L-1)v(k-L)) - C \times \delta(L-1)$$

$$+G_r(\sum_{L=0}^{k}\delta(L-1)\theta(k-L))$$

$$+B_r(\sum_{L=0}^{k}\delta(L-1)\phi(k-L)) = 0$$

The transformed form of boundary conditions, as shown in Eq. (4.5.20), can be written as:

$$\theta(0) = a, \theta(1) = b, \phi(0) = c, \phi(1) = d, v(0) = g, v(1) = q \tag{4.5.31}$$

Where a, b, c, d, g, and q are unknown coefficients that must be determined By solving Eqs. (4.5.28)–(4.5.30) and using the transformed form of boundary conditions in Eq. (4.5.31), the transformed functions at any other discrete number can be found. By doing the procedure until $\theta(3), \phi(5), v(3)$ then $\theta(r), \phi(r), v(r)$ are obtained as follows:

$$\theta(r) = a + br + (-\frac{1}{3}\alpha_1 N_t b^2 - \frac{1}{3}N_b bd)r^3 \tag{4.5.32}$$

$$\begin{aligned}\phi(r) &= c + dr + \frac{1}{3}(N_b b(N_t\alpha_1 b + N_b d)r^3)/ \\ &(N_t) - \frac{1}{2}(N_b \alpha b(N_t\alpha_1 b + N_b d)r^4)/(N_t) \\ &+\frac{1}{5}(N_b b(N_t\alpha_1 b + N_b d) \times \\ &(-2N_t^2\alpha_1 b + N_b^2 b - N_b N_t d + 6N_b\alpha^2)r^5)/(N_t^2)\end{aligned} \tag{4.5.33}$$

$$\begin{aligned}v(r) &= g + qr - \frac{1}{3}\times \\ &((-M^2 g + B_r c + G_r a - Pg - C)r^3)/(3\Lambda q^2 + \mu(r))\end{aligned} \tag{4.5.34}$$

To find the value of a; b; c; d; g; q, we should substitute the boundary conditions Eq. (4.5.20) into each differential Eq. (4.5.32)–(4.5.34), the constant parameters obtained as follow:

$$\begin{aligned}&a = 1.294117647, b = -0.1764705882, \\ &c = 2.956973294, d = -1.988872404 \\ &g = 2.473190396, q = -1.552055462\end{aligned} \tag{4.5.35}$$

To solve differential equations in different areas of mechanical engineering using the AGM technique, a general answer with constant coefficients is needed.

According to the physical form of the problem, boundary and initial conditions are required to achieve the analytical solution for every linear and nonlinear differential equation. Consequently, we can solve differential equations to any degree if the required boundary and initial conditions are available. To help understand the AGM, two differential equations ruling over-engineering actions will be solved with this new method.

Using boundary conditions, the total answer for a differential equation is accomplished as below:

The nonlinear differential equation of the p and u parameters is considered a function of u and x, respectively. The following formula shows the nonlinear differential equation of p in terms of u and its derivatives:

$$p_k: f(u, u', u'', \ldots, u^{(m)}) = 0; u = u(x) \tag{4.5.36}$$

Boundary conditions:

$$\begin{cases} u(x) = u_0, u'(x) = u_1, \ldots, u^{(m-1)}(x) = u_{m-1} \;\; at \;\; x = 0 \\ u(x) = u_{L_0}, u'(x) = u_{L_1}, \ldots, u^{(m-1)}(x) = u_{L_{m-1}} \;\; at \;\; x = L \end{cases} \tag{4.5.37}$$

The first differential equation is solved with attention to boundary conditions, and its answer is defined as a series of the nth order with a constant coefficient, as follows:

$$u(x) = \sum_{i=0}^{n} a_i x^i = a_0 + a_1 x^1 + a_2 x^2 + \cdots + a_n x^n \tag{4.5.38}$$

More precision for Eq. (4.5.39) will be obtained by more choices of series sentences from Eq. (4.5.41). The boundary conditions are applied to the functions as follows:

a) Applying boundary conditions on Eq. (4.5.41), we have:

When x = 0:

$$\begin{cases} u(0) = a_0 = u_0 \\ u'(0) = a_1 = u_1 \\ u''(0) = a_2 = u_2 \\ \quad\quad . \\ \quad\quad . \\ \quad\quad . \end{cases} \tag{4.5.39}$$

and when x= L:

$$\begin{cases} u(L) = a_0 + a_1 L + a_2 L^2 + \cdots + a_n L^n = u_{L_0} \\ u'(L) = a_1 + 2a_2 L + 3a_3 L^2 + \cdots + n a_n L^{n-1} = u_{L_1} \\ u''(L) = 2a_2 + 6a_3 L + 12a_4 L^2 + \cdots + n(n-1) a_n L^{n-2} = u_L \\ \vdots \end{cases} \quad (4.5.40)$$

b) After substitution steps, the effects of the boundary conditions on Eq. (4.5.39) are investigated by the following structure:

$$\begin{cases} p_0: f(u(0), u'(0), u''(0), \dots, u^{(m)}(0)) \\ p_1: f(u(L), u'(L), u''(L), \dots, u^{(m)}(L)) \\ \vdots \end{cases} \quad (4.5.41)$$

More selection of sentences in series is needed to accomplish a more accurate solution for Eq. (4.5.39). Five or six sentences from Eq. (4.5.38) series are normally used to solve the nonlinear differential equations to attain a solution with adequate accuracy. Considering Eq. (4.5.38)'s nth order series, there are still (n+1) unknown coefficients in the answer of Eq. (4.5.41) that need to be determined by (n+1) equations. Therefore, the boundary conditions of Eq. (4.5.40) are used to solve Eq. (4.5.41) completely.

Considering $(n < m)$ and n is the number of sentences from Eq. (4.5.41) equations, (n+1) equations and (n+1) unknowns are obtained from the set of equations, so there are additional unknowns which are the identical coefficients of Eq. (4.5.41). In the following, Eq. (4.5.41) should be derived m times under the number of the additional unknowns in equation sets. Now, the boundary conditions are applied to the derivatives:

$$\begin{cases} p'_k: f(u', u'', u''', \dots, u^{(m+1)}) \\ p''_k: f(u'', u''', u^{(IV)}, \dots, u^{(m+2)}) \\ \vdots \end{cases} \quad (4.5.42)$$

c) Applying the boundary conditions on the derivatives of the differential equation P_k is done, and the resulting equations will be:

$$p'_k:\{\begin{matrix} f(u'(0),u''(0),u'''(0),\dots,u^{(m+1)}(0)) \\ f(u'(L),u''(L),u'''(L),\dots,u^{(m+1)}(L)) \end{matrix} \quad (4.5.43)$$

$$p''_k:\{\begin{matrix} f(u''(0),u'''(0),\dots,u^{(m+2)}(0)) \\ f(u''(L),u'''(L),\dots,u^{(m+2)}(L)) \end{matrix} \quad (4.5.44)$$

(n+1) equations can be made from Eqs. (4.5.40) to Eq. (4.5.44) so that unknown coefficients of Eq. (4.5.39), for example a_0, a_1, $a_2 \dots a_n$, will be computed. The answer to the nonlinear differential Eq. (4.5.37) will be gained by determining the coefficients of Eq. (4.5.39).

In this method, first we rewrite the ruling equation in the following form:

$$F(r):=N_b\frac{d\theta}{dr}\frac{d\phi}{dr}+\frac{1}{r}\frac{d\theta}{dr}+\alpha\frac{d^2\theta}{dr^2}+\alpha_1 N_t(\frac{d\theta}{dr})^2=0 \quad (4.5.45)$$

$$G(r):=N_b(\frac{d^2\theta}{dr^2}+\frac{1}{r}\frac{d\theta}{dr})+N_t(\frac{d^2\phi}{dr^2}+\frac{1}{r}\frac{d\phi}{dr})=0 \quad (4.5.46)$$

$$H(r):=\frac{d\mu}{dr}\frac{dv}{dr}+\frac{\mu}{r}\frac{dv}{dr}+\mu\frac{d^2v}{dr^2}+\frac{\Lambda}{r}(\frac{dv}{dr})^3+3\Lambda(\frac{dv}{dr})^2\frac{d^2v}{dr^2}= Pv+M^2v+C-G_r\theta-B_r\phi \quad (4.5.47)$$

To solve the problem, we consider three finite series of polynomials with constant coefficients as the ruling Eqs. (4.5.17)–(4.5.19):

$$\theta_r=\sum_{i=0}^{9}a_i\cdot r^i=r^9a_9+r^8a_8+r^7a_7+r^6a_6+r^5a_5 +r^4a_4+r^3a_3+r^2a_2+ra_1+a_0 \quad (4.5.48)$$

$$\phi_r=\sum_{i=0}^{15}b_ir^i=r^{15}b_{15}+r^{14}b_{14}+r^{13}b_{13}+r^{12}b_{12} +r^{11}b_{11}+r^{10}b_{10}+r^9b_9+r^8b_8 +r^7b_7+r^6b_6+r^5b_5+r^4b_4 +r^3b_3+r^2b_2+rb_1+b_0 \quad (4.5.49)$$

$$v_r=\sum_{i=0}^{5}c_ir^i=r^5c_5+r^4c_4+r^3c_3+r^2c_2+r^1c_1 +c_0 \quad (4.5.50)$$

It is obvious that by applying the boundary conditions as shown in Eq. (4.5.20), the constant coefficients of Eqs. (4.5.48)–(4.5.50) can be obtained.

At the first step, the boundary conditions (Eq. (4.5.20)) are substituted with Eqs. (4.5.48)–(4.5.50) as follows:

$$\theta(1) = 1 \rightarrow a_9 + a_8 + a_7 + a_6 + a_5 + a_4 + a_3 + a_2 + a_1 + a_0 = 1 \tag{4.5.51}$$

$$\theta(2) = 0 \rightarrow 512a_9 + 256a_8 + 128a_7 + 64a_6 + 32a_5 + 16a_4 + a_3 + a_2 + a_1 + a_0 = 0 \tag{4.5.52}$$

$$\phi(1) = 1 \rightarrow b_0 + b_1 + b_2 + b_3 + b_4 \tag{4.5.53}$$

$$+b_5 + b_6 + b_7 + b_8 + b_9 + b_{10}$$

$$+b_{11} + b_{12} + b_{13} + b_{14} + b_{15} = 1$$

$$\phi(2) = 0 \rightarrow b_0 + 2b_1 + 4b_2 + 8b_3 + 16b_4 + 32b_5 + 64b_6 + 128b_7 + 256b_8 + 512b_9 + 1024b_{10} + 2048b_{11} + 4096b_{12} + 8192b_{13} + 16384b_{14} + 32768b_{15} = 0 \tag{4.5.54}$$

$$v(1) = 1 \rightarrow c_5 + c_4 + c_3 + c_2 + c_1 + c_0 = 1 \tag{4.5.55}$$

$$v(2) = 0 \rightarrow 32c_5 + 16c_4 + 8c_3 + 4c_2 + 2c_1 + c_0 = 1 \tag{4.5.56}$$

After substituting Eqs. (4.5.48)–(4.5.50) for Eqs. (4.5.45)–(4.5.47), we apply the boundary conditions on the equations obtained and their derivatives; constant coefficients $a_0, \ldots, a_9$, $b_0, \ldots, b_{15}$, $c_0, \ldots$, and c_5 create 32 unknown parameters, so 32 equations are needed to find the unknowns. Considering the six equations already achieved (4.5.51)–(4.5.56), 26 more equations are needed. These equations will be obtained by expanding the derivatives of each parameter (F, G, H) until the desired number of equations is obtained. For instance, the derivatives of F will be expanded till the seventh order of F. (F(T1)=0, F′(T1)=0 … … F7(T1) = 0).

The constant coefficients obtained as follow:

By substituting the constant coefficients with Eqs. (4.5.48)–(4.5.50), $\theta(r), \phi(r), v(r)$, are gained as below:

$$
\begin{aligned}
&a_0 = 1.156440588, a_1 = 0.001073927210,\\
&a_2 = -0.009147612675, a_3 = -0.1648512770,\\
&a_4 = 0.02543115181, a_5 = -0.01340488758\\
&a_6 = 0.006067365151, a_7 = -0.001968335703,\\
&a_8 = 0.0003958729576, a_9 = -0.00003679234137,\\
&b_0 = 9.718759072, b_1 = -33.83464356\\
&b_2 = 91.82900059, b_3 = -203.7763879,\\
&b_4 = 349.6762300, b_5 = -464.7752580\\
&b_6 = 482.9152742, b_7 = -394.2950564,\\
&b_8 = 253.0946946, b_9 = -127.0781720\\
&b_{10} = 49.33291291, b_{11} = -14.50655075,\\
&b_{12} = 3.122336260, b_{13} = -0.4638124670\\
&b_{14} = 0.04248076545, b_{15} = -0.001807348805,\\
&c_0 = 2.879101202, c_1 = -2.820390804\\
&c_2 = 1.505200118, c_3 = -0.8078651259,\\
&c_4 = 0.2876716619, c_5 = -0.04371705154
\end{aligned}
\tag{4.5.57}
$$

$$
\begin{aligned}
\theta(r) = -0.00003679234137r^9 + 0.0003958729576r^8\\
- 0.001968335703r^7\\
+ 0.006067365151r^6\\
- 0.01340488758r^5\\
+ 0.02543115181r^4\\
- 0.1648512770r^3\\
- 0.009147612675r^2\\
+ 0.001073927210r + 1.156440588
\end{aligned}
\tag{4.5.58}
$$

$$
\begin{aligned}
&\phi(r) = 9.718759072 - 33.83464356r\\
&+91.82900059r^2 - 203.7763879r^3
\end{aligned}
\tag{4.5.59}
$$

$$+349.6762300r^4 - 464.7752580r^5$$

$$+482.9152742r^6 - 394.2950564r^7$$

$$+253.0946946r^8 - 127.0781720r^9$$

$$+49.33291291r^{10} - 14.50655075r^{11}$$

$$+3.122336260r^{12} - 0.4638124670r^{13}$$

$$+0.04248076545er^{14} - 0.001807348805r^{15}$$

$$v(r) = -0.04371705154r^5 + 0.2876716619r^4 - 0.8078651259r^3 + 1.505200118r^2 - 2.820390804r + 2.879101202 \quad (4.5.60)$$

According to the problem's physics, it is evident that an appropriate numerical method is required to solve the current boundary value problem (BVP). Applying Maple's algebra package, the numerical method of the problem can be solved. In Maple software, two general methods are used. First is the trapezoidal method, which can give efficient answers for typical problems. The second is the midpoint method which helps treat harmless endpoint singularities. Applying the midpoint method, the 4th Runge-Kutta-Fehlberg method increases the Euler method accuracy by adding a midpoint to the step. This algorithm is an accurate way to solve many engineering and mathematical issues. In this study, the midpoint method is a convenient technique for a proper solution.

The present example examines the behaviour of the fluid flow of blood containing nanoparticles in a porous media affected by a magnetic field. At the same time, the blood is considered a third-grade non-Newtonian fluid. Two analytical methods, AGM and DTM, have been conducted to solve the governing equations. Their results are compared with numerical results. Fig. 4.5.1 shows the schematic blood flow in a vein in the presence of a magnetic field. Figs. 4.5.2–4.5.4 compares the methods with the numerical approach for functions such as temperature profile, nanoparticle concentration profile, and velocity profile, respectively. In addition, Tables 4.5.2–4.5.6 show the results obtained from each method, accompanied by the error of each one compared to the numerical approach. It can be seen from the graphs and the results that AGM has the best agreement with the numerical results. And it has the least error and more accurate answers compared to other methods in which multiple unknown parameters are solvable. Fig. 4.5.5 shows the effect of increasing the pressure gradient on the velocity profile; it is shown that an increase in the negative pressure gradient would cause an increase in the velocity profile. This incident would consequently accelerate the blood velocity in the veins due to the high negative pressure gradient and the suction force. Fig. 4.5.6 shows the influence of porosity; as expected, by increasing the porosity of the veins, interactions and the friction between the flow and the blood arterial will be increased so that the velocity profile will be decreased consequently. The effect of the thermophoresis parameter is shown in Fig. 4.5.7.; it is obvious from the graph that an increase in the thermophoresis parameter would cause an increase in the velocity profile and that is because higher temperatures create higher kinetic energy, which leads to an increase in the velocity profile.

Fig. 4.5.8 examines the influence of the Brownian motion parameter on velocity profile; it is interpreted that by increasing this parameter, the velocity profile decreases, which is the result of the random motions of the particles that are increased in response to the increase of the Brownian motion parameter.

Fig. 4.5.9 shows that an increase in the thermophoresis parameter would cause an increase in temperature profile in the whole domain. At the same time, Fig. 4.5.10 indicated that by increasing the Brownian motion parameter, we would have a decrease in the temperature profile. Consequently, the effect of these parameters on the nanoparticle concentration profile has been discussed in Fig. 4.5.11 and Fig. 4.5.12, in which by increasing the thermophoresis parameter and Brownian motion parameter, we would have an increase and a decrease in nanoparticle concentration, respectively. It should be mentioned that by increasing the Brownian motion parameter, the random motions of particles would be increased; therefore, the number of nanoparticles in one specific domain would be decreased, so the nanoparticle concentration at that domain will be decreased in response. According to Fig. 4.5.13, increasing the third-grade parameter would cause an increase in the velocity profile, as already mentioned. The present problem examines fluid flow in the presence of a magnetic field, so it would be convenient to investigate the effect of this parameter on velocity profile. In this case, Fig. 4.5.14 shows that increasing the magnetic field parameter will decrease the velocity profile accordingly. Lastly, the influence of the Grashof number has been discussed in Fig. 4.5.15, in which we increase velocity profile by increasing the Grashof number.

The present study considers blood a third-grade non-Newtonian flow containing nanoparticles. The governing equations have been solved and examined using two analytical methods called AGM and DTM, along with the numerical approach. Interpreting the results indicated that AGM is the most effective method to deal with such problems with the least error from the numerical approach compared to others. AGM could solve many problems with multiple unknown parameters with accurate answers and minimum errors, and as can be seen, it is the best method for dealing with such problems so far. Furthermore, the authors also suggest this method as an excellent method for partial differential equations with infinity boundary conditions.

Following the validation process, the effect of the following parameters is discussed for the case; the pressure gradient, the Brownian motion parameter near the building wall, the thermophoretic diffusion of nanoparticles, and the magnetic nature of blood cells. These outcomes were obtained from the results:

1. Increasing both the Grashof number and the negative pressure gradient will accelerate the velocity of the blood while increasing the magnetic field parameter along with the porosity parameter has the opposite effect on the velocity profile.

2. The Brownian motion parameter and the thermophoresis parameter play an important role in the temperature profile, and the nanoparticles concentration distributes in a way that increases the thermophoresis parameter causing an enhancement for both parameters. In contrast, the Brownian motion parameter has the opposite effect on both parameters.

List of symbols

Ec	Eckret number
P_r	Prandtl number
Ha	Hartmann number
Nu	Nusslet number
Re	Reynolds number
k	Thermal conductivity
σ	Electrical conductivity

Ω	Constant rotation velocity
η	Aspect ratio
ϕ	Nanoparticle volume fraction
ρ	Fluid density
θ	Dimensionless temperature
μ	Dynamic viscosity
C	nanofluid concentration
C_p	specific heat at constant pressure (J/kg.K)
h	distance between the plates (m)
k	thermal conductivity (W/m.K)
p^*	modified fluid pressure (Pa)
R	viscosity coefficient
u, v, w	velocity components along x, y, z axes respectively (m/s)
$u_w(x)$	the velocity of the stretching surface (m/s)
α	thermal diffusivity
ϕ	dimensionless concentration
η	dimensionless variable
υ	kinematic viscosity $\left(\frac{m^2}{s}\right)$
σ	electrical conductivity (S/m)
a_1, a_2	stretching rates of lower and upper disks [1/T]
Ω_1, Ω_2	the rotational velocity of lower and upper disks [1/T]
T_1, T_2, T_f	temperatures of lower and upper disks and mean temperature [K]

B_0	magnetic strength [M/T^{2}A]
ρ_{nf}	density [M/L^3]
ν_{nf}	kinematic viscosity [L^2/T]
k_{nf}	thermal conductivity [ML/T^{3}K]
R	radiation parameter
k	mean absorption coefficient [1/L]
M	Hartmann number
σ°	Stefan-Boltzmann constant [M/T^{3}K^4]
θ, ϕ	temperature and concentration profiles
W_s	suction/injection parameter
A_1	stretching parameter of the lower disk
A_2	stretching parameter of the lower disk
Ω	rotational parameter
α_1, α_2	temperature ratio parameter
Br	Brinkman number
W_0	suction/injection rate [L/T]
D_B	Brownian coefficient
v^*	dimensionless velocity
D_T	thermophoretic coefficient
B	constant applied magnetic field
Nb	thermophoresis parameter
Nt	Brownian parameter
r^*	dimensionless radius

ϕ	nanoparticle volume fraction
α	thermal diffusivity
A, B	Constants in viscosity function
B_r	Brownian diffusion constant
C	pressure gradient parameter
D_b	Brownian diffusion coefficient
D_T	thermophoretic diffusion coefficient
E	electric field
g	gravitational vector
Gr	Grashof number
k	permeability
M	MHD parameter
N_b	Brownian motion parameter
N_T	Thermophoresis parameter
P	porosity
r_1	outer radius
D_b	inner radius
r	Variable radius from r_2 to r_1
V	velocity vectorGreek symbols
α, β	material moduli
β_T	volumetric coefficient of expansion
Λ	third-grade parameter
λ_r	retardation time

ϕ	nanoparticles mass concentration
ϕ_m	mass concentration
φ	porosity
θ	temperature
θ_w	pipe temperature
θ_m	fluid temperature
ρ_f	the density of the base fluid
ρ_p	the density of the nanoparticles
ρ_{f_0}	the density of the nanofluid
σ	electrical conductivity
m	particle mass
u	particle velocity
g	gravity acceleration
D	particle diameter
C_D	drag coefficient
t	time
Re	Reynolds number
n	flow behaviour index
V_0	velocity at t = 0
k	consistency coefficient
$X(n)$	deviation factor
j	number of polynomial statements
u_m	terminal settling velocity

t_m	time of terminal settling velocity
CM	Collocation Method
HPM	Homotopy Perturbation Method
HAM	Homotopy Analysis Method
VIM	Variational Iteration Method
NuM	The numerical method
ρ	fluid density
ρ_s	particle density
P_0	ambient pressure on the film surface
T	temperature
T_0	the temperature on the film surface
T_w	the temperature on the disk
G	gravitational acceleration
t	the thickness of the fluid film
W	velocity of spraying
u, v, w	velocity components along the x, y, and z axes, respectively
Ω	angular velocity
α	thermal diffusivity of the fluid
β	the angle between the horizontal axis and the disk
η	dimensionless variable
δ	normalised thickness
ϑ	kinematic viscosity

References

1. Derakhshan, R., Shojaei, A., Hosseinzadeh, K., Nimafar, M., & Ganji, D. D. (2019). "Hydrothermal analysis of magnetohydrodynamic nanofluid flow between two parallel by AGM." *Case Studies in Thermal Engineering*, 14, 100439.

2. Zangooee, M. R., Hosseinzadeh, K., & Ganji, D. D. (2019). "Hydrothermal analysis of MHD nanofluid (TiO2-GO) flow between two radiative stretchable rotating disks using AGM." *Case Studies in Thermal Engineering*, 14, 100460.

3. Sheikholeslami, M., Jalili, P., & Ganji, D. D. (2018). "Magnetic field effect on nanofluid flow between two circular cylinders using AGM." *Alexandria Engineering Journal*, 57(2), 587–594.

4. Rostami, A., Akbari, M., Ganji, D., & Heydari, S. (2014). "Investigating Jeffery-Hamel flow with high magnetic field and nanoparticle by HPM and AGM." *Open Engineering*, 4(4), 357–370.

5. Ardahaie, S. S., Amiri, A. J., Amouei, A., Hosseinzadeh, K., & Ganji, D. D. (2018). "Investigating the effect of adding nanoparticles to the blood flow in presence of magnetic field in a porous blood arterial." *Informatics in Medicine Unlocked*, 10, 71–81.

6. Abdollahi, S. A., Jalili, P., Jalili, B., Nourozpour, H., Safari, Y., Pasha, P., & Ganji, D. D. (2023). Computer simulation of Cu: AlOOH/water in a microchannel heat sink using a porous media technique and solved by numerical analysis AGM and FEM. *Theoretical and Applied Mechanics Letters*, *13*(3), 100432.

7. Jalili, P., Sadeghi Ghahare, A., Jalili, B., & Domiri Ganji, D. (2023). Analytical and numerical investigation of thermal distribution for hybrid nanofluid through an oblique artery with mild stenosis. *SN Applied Sciences*, *5*(4), 95.

8. Jalili, B., Jalili, P., Sadighi, S., & Ganji, D. D. (2021). Effect of magnetic and boundary parameters on flow characteristics analysis of micropolar ferrofluid through the shrinking sheet with effective thermal conductivity. *Chinese Journal of Physics*, *71*, 136-150.

9. Jalili, B., Sadighi, S., Jalili, P., & Ganji, D. D. (2019). Characteristics of ferrofluid flow over a stretching sheet with suction and injection. *Case Studies in Thermal Engineering*, *14*, 100470.

CHAPTER 5

A COMPARISON BETWEEN AGM AND OTHER METHODS

5.1. Hypocycloid Motion

A hypocycloid is a curve produced by a fixed point on the circumference of a small circle of a specified radius rolling around the inside of a large circle with a bigger radius. In the present problem, a semi-hypocycloid motion of a cylinder has been simulated by extracting the governing nonlinear differential equation of this special motion, and then, the equation obtained has been solved completely.

In general, vibrational equations and their initial conditions are defined for different systems as follows:

$$f(\ddot{u}, \dot{u}, u, F_0, \sin(\omega_0 t)) = 0 \tag{5.1.1}$$

$$\{u(0) = A, \dot{u}(0) = 0\} \tag{5.1.2}$$

A total answer with constant coefficients is required to solve differential equations in AGM. In vibrational systems, with respect to the kind of vibration, it is necessary to choose the answer mentioned in AGM. We divide vibrational systems into two general forms:

1. Vibrational systems without any external force

Differential equations governing this kind of vibrational system are introduced in the following form:

$$f(\ddot{u}, \dot{u}, u) = 0 \tag{5.1.3}$$

Now the answer to this kind of vibrational system is chosen as follows:

$$u(t) = e^{-bt}\{a\cos(\omega t + \varphi)\} \tag{5.1.4}$$

2. Vibrational systems with external force

It is assumed that the external forces exerting on the vibrational systems are defined as:

$$F(t) = F_0 \sin(\omega_0 t) \tag{5.1.5}$$

Therefore, the differential equation governing the vibrational system is expressed as follows:

$$f(\ddot{u}, \dot{u}, u, F_0, \sin(\omega_0 t)) = 0 \tag{5.1.6}$$

The answer to this kind of vibrational system can be written as follows:

$$u(t) = e^{-bt}\{a\cos(\omega t + \varphi)\} + d\cos(\omega t + \emptyset) \tag{5.1.7}$$

To increase the precision of the achieved equation, we can add another term in the form of the Fourier cosine series expansion as follows:

$$u(t) = e^{-bt}\{a\cos(\omega t + \varphi_1) + c\cos(2\omega t + \varphi_2)\} + d\cos(\omega_0 t + \emptyset) \tag{5.1.8}$$

The exact solution of all vibration differential equations can be written as follows:

$$u(t) = e^{-at}\left\{\sum_{k=1}^{\infty} b_k \cos(k\omega t + \varphi_k)\right\} + d\cos(\omega_0 t + \emptyset) \tag{5.1.9}$$

The constant coefficients of Eqs. (5.1.4), (5.1.7), which are $\{a, b, \omega, \varphi, d, \emptyset\}$, will be computed in AGM by applying the initial conditions [Eq. (5.1.2)].

In AGM, the application of initial conditions is done in the two following forms:

(a) Applying the initial conditions on the answer of the differential equation.

$$u(t) = u(IC) \tag{5.1.10}$$

Notably, IC is the abbreviation of the introduced initial conditions of Eq. (5.1.2).

(b) Applying the initial conditions on the main differential equation and its derivatives.

Assume the general equation of vibrational equation as:

$$f(\ddot{u},\dot{u},u,F_0\sin(\omega_0 t))=0 \tag{5.1.11}$$

Regarding the kind of vibrational system discussed earlier, a function is chosen as the answer to the differential equation from Eqs. (5.1.4) or (5.1.7), or for increasing the precision, Eq. (5.1.9) with the proper number of terms can be chosen. (Number of terms depends on the desirable precision and the calculation process's time).

Therefore, the answer to the differential equation is considered as follows:

$$u=g(t) \tag{5.1.12}$$

In other words, g(t) is the right-hand side of Eqs. (5.1.4), (5.1.7) or (5.1.9).

Afterwards, Eq. (5.1.12) will be substituted into Eq. (5.1.11) instead of u as follows:

$$f(t)=f(\ddot{g}(t),\dot{g}(t),g(t),F_0\sin(\omega_0 t))=0 \tag{5.1.13}$$

Finally, the application of initial conditions on Eq. (5.1.13) and its derivatives is expressed as:

$$f(t)=f(\ddot{g}(IC),\dot{g}(IC),g(IC),F_0\sin(\omega_0 t)), \tag{5.1.14a}$$

$$\dot{f}(IC)=\dot{f}(\ddot{g}(IC),,\dot{g}(IC),g(IC),F_0\sin(\omega_0 t)), \tag{5.1.14b}$$

$$\ddot{f}(IC)=\ddot{f}(\ddot{g}(IC),,\dot{g}(IC),g(IC),F_0\, sin(\omega_0 t)) \tag{5.1.14c}$$

Consider a cylinder that rolls without slipping on the cylindrical surface, as depicted in Fig. 5.1.1. For no slipping, we have $r\beta=R\theta$. The momentum equation is written as c, the instantaneous centre of rotation (lowest point), as follows:

$$\sum M_c=I_c\ddot{\psi}(t) \tag{5.1.15}$$

where

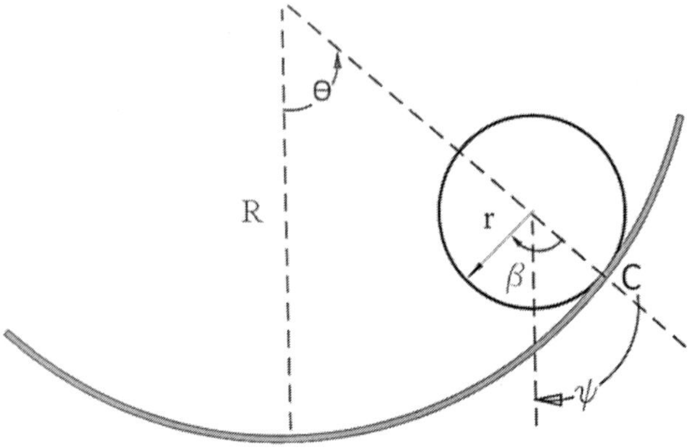

Fig. 5.1.1. Geometry of problem 1

$$\psi(t) = \beta(t) - \theta(t) = (\frac{R-r}{r})\theta(t) \tag{5.1.16}$$

$$\ddot{\psi} = \left(\frac{R-r}{r}\right)\ddot{\theta}(t) \tag{5.1.17}$$

$$I_c = \frac{3}{2}mr^2 \tag{5.1.18}$$

According to Eqs. (5.1.15) to (5.1.18), we will have:

$$-rw\sin(\theta(t)) = \frac{3}{2}mr^2\left(\frac{R-r}{r}\right)\ddot{\theta}(t) \tag{5.1.19}$$

We can rewrite the above equation in the following form:

$$\frac{3}{2}\frac{w}{g}(R-r)^2\ddot{\theta}(t) + w(R-r)\sin(\theta(t)) = 0 \tag{5.1.20}$$

The initial conditions of Eq. (5.1.20) are:

$$\theta(0) = \alpha, \quad \dot{\theta}(0) = 0 \tag{5.1.21}$$

Where α is the initial angular position of the cylinder.

In this section, we investigate Eq. (5.1.20) using an approximation for $\sin(\theta(t))$, as follows:

$$\sin(\theta(t)) = \theta(t) - \frac{\theta(t)^3}{3!}, \quad -\frac{\pi}{2} \leq \theta(t) \leq \frac{\pi}{2} \tag{5.1.22}$$

Substituting Eq. (5.1.22) with Eq. (5.1.20) and by some manipulation, Eq. (5.1.20) can be rewritten as:

$$\frac{d^2\theta(t)}{dt^2} + \frac{2g}{3(R-r)}\theta(t) - \frac{g}{9(R-r)}\theta^3(t) = 0 \tag{5.1.23}$$
$$\theta(0) = \alpha, \quad \dot{\theta}(0) = 0$$

At first, according to the above, the answer to Eq. (5.1.23) is considered as follows:

$$\theta(t) = e^{-bt}\{a\cos(\omega t + \varphi)\}, \ \theta(0) = \alpha, \dot{\theta}(0) = 0 \tag{5.1.24}$$

Now by substituting Eq. (5.1.24) with Eq. (5.1.23), $f(t)$ can easily be obtained as follows:

$$\begin{aligned} f(t) &= b^2 e^{-bt} a\cos(\omega t + \varphi) \\ &+ 2be^{-bt} a\sin(\omega t + \varphi)\omega - e^{-bt} a\cos(\omega t + \varphi)\omega^2 \\ &+ 2\frac{ge^{-bt} a\cos(\omega t + \varphi)}{3R - 3r} - \frac{g(e^{-bt})^3 a^3 (\cos(\omega t + \varphi))^3}{9R - 9r} \\ &= 3 \end{aligned} \tag{5.1.25}$$

Afterwards, by applying initial conditions, we will have:

$$\theta(0) = \alpha, \quad so \quad a\cos(\varphi) = \alpha \tag{5.1.26}$$

$$\dot{\theta}(0) = 0, \quad so \quad -ba\cos(\varphi) - a\sin(\varphi)\omega = 0 \tag{5.1.27}$$

$$\begin{aligned} f(\dot{\theta}(0)) :\ & b^2 a\cos(\varphi) + 2ba\sin(\varphi)\omega - a\cos(\varphi)\,\omega^2 \\ & + 2\frac{ga\cos(\varphi)}{3R - 3r} - \frac{ga^3(\cos(\varphi))^3}{9R - 9r} 0 \end{aligned} \tag{5.1.28}$$

$$\dot{f}\left(\dot{\theta}(0)\right):-b^3 a\cos(\varphi) - 3b^2 a\sin(\varphi)\omega + 3ba\ \cos(\varphi)\,\omega^2 + a\sin(\varphi)\omega^3 - 2\frac{gba\cos(\varphi)}{3R-er} - 2\frac{ga\sin(\varphi)\,\omega}{3R-3r} + 3\frac{ga^3(\cos(\varphi))^3 b}{9R-9r} + 3\frac{ga^3(\cos(\varphi))^2\sin(\varphi)\,\omega}{9R-9r} = 0 \tag{5.1.29}$$

Solving a set consisting of four equations with four unknowns from Eqs. (5.1.26)–(5.1.29), the constant coefficients of Eq. (5.1.24) can easily be obtained as:

$$a = -\alpha \tag{5.1.30a}$$

$$b = 0 \tag{5.1.30b}$$

$$\omega = \sqrt{-\frac{\alpha^2 g - 6g}{9R-9r}} \tag{5.1.30c}$$

$$\varphi = \pi \tag{5.1.30d}$$

By selecting the following physical values, we will have:

$$g = 9.81\frac{m}{s^2}, \alpha = \frac{\pi}{6}\ \text{rad}, \text{R} = 0.5\ \text{m}, \text{r} = 0.1\ \text{m} \tag{5.1.31}$$

$$\alpha = -\frac{\pi}{6},\quad b = 0,\quad \omega = 3.950053896,\quad \varphi = \pi \tag{5.1.32}$$

There is no damping term in this specified application, so it is clear that b=0.

Consequently, $\theta(t) = \frac{\pi}{6}\cos(3.950053896t)$ (5.1.33)

By defining two new variables, μ, and λ, the main differential equation [Eq. (5.1.23)] can be rewritten as follows:

$$\frac{d^2}{dt^2}\theta(t) + \mu\theta(t) + \lambda((\theta)^3) = 0 \tag{5.1.34}$$

Where

$$\mu = \frac{2g}{3(R-r)} \tag{5.1.35}$$

$$\lambda = \frac{-g}{9(R-r)} \tag{5.1.36}$$

In order to increase the precision, another answer for Eqs. (5.1.23) or (5.1.34) is considered as follows:

$$\begin{aligned}\theta(t) = a\cos(\omega t + \varphi_1) \\ + b\cos(2\omega t + \varphi_2) \\ + c\cos(3\omega t + \varphi_3), \theta(0) = \alpha, \dot{\theta}(0) = 0\end{aligned} \tag{5.1.37}$$

Now by substituting Eq. (5.1.37) with Eq. (5.1.34), f(t) can be obtained as follows:

$$\begin{aligned}f(t) = -a\,cos(\omega t + \varphi_1)\omega^2 \\ - 4b\,cos(2\omega t + \varphi_2)\omega^2 \\ - 9c\,cos(3\omega t + \varphi)\,\omega^3 \\ + \mu(a\cos(\omega t + \varphi_1) \\ + b\cos(2\omega t + \varphi_2) \\ + c\cos(3\omega t + \varphi_3)) \\ + \lambda(a\cos(\omega t + \varphi_1) \\ + b\cos(2\omega t + \varphi_2) + c\cos(3\omega t + \varphi_3))^3 \\ = 0\end{aligned} \tag{5.1.38}$$

Afterwards, by applying initial conditions, we will have:

$$\begin{aligned}\theta(0) = \alpha, \quad so \quad a\cos(\varphi_1) + b\cos(\varphi_2) + c\cos(\varphi_3) \\ = \alpha\end{aligned} \tag{5.1.39}$$

$$\begin{aligned}\dot{\theta}(0) = 0, \quad so \\ - a\sin(\varphi_1)\omega \\ - 2b\sin(\varphi_2)\omega - 3c\sin(\varphi_3)\omega = 0\end{aligned} \tag{5.1.40}$$

$$\begin{aligned}f\left(\dot{\theta}(0)\right) : -a\cos(\varphi_1)\,\omega^2 \\ - 4b\cos(\varphi_2)\omega^2 - 9\cos(\varphi_3)\,\omega^2 \\ + \mu(\mathrm{a}\cos(\varphi_1) + b\cos(\varphi_2) + c\cos(\varphi_3)) \\ + \lambda(\mathrm{a}\cos(\varphi_1) + b\cos(\varphi_2) \\ + c\cos(\varphi_3))^3 = 0\end{aligned} \tag{5.1.41}$$

$$\dot{f}\left(\dot{\theta}(0)\right): a\sin(\varphi_1)\omega^3 + 8b\sin(\varphi_2)\omega^3 + 27c\sin(\varphi_3)\omega^3 \\ + \mu\,(-a\sin(\varphi_1)\omega \\ - 2b\sin(\varphi_2)\omega \\ - 3c\sin(\varphi_3)\omega) \\ + \ [3\lambda\,(a\cos(\varphi_1) \\ + b\cos(\varphi_2) + c\cos(\varphi_3))^2 \\ \times(-a\sin(\varphi_1)\omega - 2b\sin(\varphi_2)\omega \\ - 3c\sin(\varphi_3)\omega)] = 0 \tag{5.1.42}$$

$$\ddot{f}\left(\dot{\theta}(0)\right): a\cos(\varphi_1)\,\omega^4 + 16b\cos(\varphi_2)\,\omega^4 \\ + 81c\cos(\varphi_3)\,\omega^4 \\ + \mu\,(-a\cos(\varphi_1)\omega^2 - 4b\cos(\varphi_2)\,\omega^2 \\ - 9c\cos(\varphi_3)\,\omega^2) \\ + [6\lambda(a\cos(\varphi_1) \\ + b\cos(\varphi_2) + c\cos(\varphi_3)) \\ \times(-a\sin(\varphi_1)\omega - 2b\sin(\varphi_2)\omega \\ - 3c\sin(\varphi_3)\omega)^2] \\ + [3\lambda(\mathrm{a}\cos(\varphi_1) + b\cos(\varphi_2) \\ + c\cos(\varphi_3))^2 \\ \times(-a\cos(\varphi_1)\,\omega^2 - 4b\cos(\varphi_2)\,\omega^2 \\ - 9c\cos(\varphi_3)\,\omega^2)\Big] = 0 \tag{5.1.43}$$

$$
\begin{aligned}
f^{(3)}\left(\dot{\theta}(0)\right):\ & -a\sin(\varphi_1)\omega^5 \\
& - 32b\sin(\varphi_2)\omega^5 - 243c\sin(\varphi_3)\omega^5 \\
& + \mu(a\sin(\varphi_1)\omega^3 + 8b\sin(\varphi_2)\omega^3 \\
& + 27c\sin(\varphi_3)\omega^3) \\
& + 6\lambda(-a\sin(\varphi_1)\omega - 2b\sin(\varphi_2)\omega \\
& - 3c\sin(\varphi_3)\omega)^3 \\
& + [18\lambda(a\cos(\varphi_1) + b\cos(\varphi_2) \\
& + c\cos(\varphi_3))\,(-a\sin(\varphi_1)\,\omega \\
& - 2b\sin(\varphi_2)\,\omega \\
& - 3c\sin(\varphi_3)\,\omega) \times (-a\cos(\varphi_1)\omega^2 \\
& - 4b\cos(\varphi_2)\omega^2 \\
& - 9c\cos(\varphi_3)\omega^2)] \\
& + [3\lambda(a\cos(\varphi_1) \\
& + b\cos(\varphi_2) + \quad c\cos(\varphi_3))^2 \\
& \times (\mathrm{a}\sin(\varphi_1)\omega^3 \\
& + 8b\sin(\varphi_2)\omega^3 \\
& + 27c\,\mathrm{a}\sin(\varphi_3)\omega^3)] = 0
\end{aligned} \tag{5.1.44}
$$

$$
\begin{aligned}
f^{(4)}\left(\dot{\theta}(0)\right):\ & -a\cos(\varphi_1)\omega^6 - 64b\cos(\varphi_2)\omega^6 \\
& - 729c\cos(\varphi_3)\omega^6 + \mu(a\cos(\varphi_1)\omega^4 \\
& + 16b\cos(\varphi_2)\omega^4 + 81c\cos(\varphi_3)\omega^4) \\
& + [36\lambda\,(-\mathrm{a}\sin(\varphi_1)\omega \\
& - 2b\sin(\varphi_2)\omega - 3c\sin(\varphi_3)\omega)^2 \\
& \times (-a\cos(\varphi_1)\,\omega^2 - 4b\cos(\varphi_2)\,\omega^2 \\
& - 9c\cos(\varphi_3)\,\omega^2)] \\
& + [18\lambda(a\cos(\varphi_1) \\
& + b\cos(\varphi_2) + c\ \cos(\varphi_3) \\
& + (-a\cos(\varphi_1)\omega^2 - 4b\cos(\varphi_2)\omega^2 \\
& - 9c\cos(\varphi_3)\omega^2)^2 \\
& + [24\lambda(a\cos(\varphi_1) \\
& + b\cos(\varphi_2) + c\cos(\varphi_3))(-a\sin(\varphi_1)\omega \\
& - 2b\sin(\varphi_2)\omega - 3c\sin(\varphi_3)\omega) \\
& \times (a\sin(\varphi_1)\omega^3 \\
& + 8b\ \sin(\varphi_2)\omega^3 + 27c\sin(\varphi_3)\omega^3)] \\
& + [3\lambda(a\cos(\varphi_1) + b\cos(\varphi_2) \\
& + c\cos(\varphi_3))^2 \\
& \times (a\sin(\varphi_1)\,\omega^4 + 8b\sin(\varphi_2)\,\omega^4 \\
& + 27c\sin(\varphi_3)\,\omega^4)] = 0
\end{aligned} \tag{5.1.45}
$$

Solving a set consisting of seven equations with seven unknowns from Eqs. (5.1.39)–(5.1.45), the constant coefficients of Eq. (5.1.37) can easily be obtained as (for the aforementioned physical values):

$$a = 0.5245656696$$

$$b = 0.0002376166154$$

$$c = 0.0007292773609$$

$$\omega = 3.974983629 \tag{5.1.46}$$

$$\varphi_1 = 0$$

$$\varphi_2 = \pi$$

$$\varphi_3 = \pi$$

As a result,

$$\begin{aligned}\theta(t) \\ &= 0.5245656696\cos(3.974983629t) \\ &- 0.0002376166154\cos(7.949967258t) \\ &- 0.0007292773609\cos(11.92495089t)\end{aligned} \tag{5.1.47}$$

According to the variation iteration method, at first, the arbitrary initial approximation is chosen as follows:

$$\theta_0(t) = \alpha\cos(\omega t) \tag{5.1.48}$$

$$\frac{d^2}{dt^2}\theta(t) = -2\frac{g\theta_0(t)}{3R - 3r} + \frac{g\left(\theta_0(t)\right)^3}{9R - 9r} \tag{5.1.49}$$

Integrating twice yields:

$$\begin{aligned}\theta(t) \\ &= -\frac{g\alpha((\cos(\omega t))^3\alpha^2 + 6\alpha^2\cos(\omega t) - 7\alpha^2 - 54\cos(\omega t) \text{ -}}{(81R - 81r)\omega^2}\end{aligned} \tag{5.1.50}$$

Equating the coefficient of $\cos(\omega t)$, in Eq. (5.1.50), we have:

$$\alpha = \frac{g\alpha(6\alpha^2 - 54)}{(81R - 81r)\omega^2} \tag{5.1.51}$$

Or

$$\omega = \frac{1}{9}\frac{\sqrt{-(6R-6r)g(\alpha^2-9)}}{R-r} \tag{5.1.52}$$

Therefore, the initial approximation will be obtained as follows:

$$\theta_0(t) = \alpha\cos(\frac{1}{9}\frac{\sqrt{-(6R-6r)g(\alpha^2-9)t}}{R-r}) \tag{5.1.53}$$

Now we construct the correction function in the following form:

$$\theta_{n+1}(t) = \theta_n(t) + \int_0^t \lambda(\frac{d^2}{d\tau^2}\theta_n(\tau) + 2\frac{g\theta_n(\tau)}{3R-3r} - \frac{g(\theta_n(\tau))^3}{9R-9r})\,d\tau \tag{5.1.54}$$

Its stationary conditions can be obtained as follows:

$$\lambda''(\tau) = 0 \tag{5.1.55a}$$

$$1 - \lambda'(\tau)(\tau = t) = 0 \tag{5.1.55b}$$

$$\lambda(\tau)(\tau = t) = 0 \tag{5.1.55c}$$

The Lagrange multiplier can therefore be simply identified as:

$$\lambda = \frac{1}{\omega}\sin(\omega(\tau - t)) \tag{5.1.56}$$

Using the above iteration formula [Eq. (5.1.54)] and the initial approximation, we can easily obtain the first component.

$$\theta_1(t) = \theta_0(t) + \int_0^t \lambda(\frac{d^2}{d\tau^2}\theta_n(\tau) + 2\frac{g\theta_n(\tau)}{3R-3r} - \frac{g(\theta_n(\tau))^3}{9R-9r})\,d\tau \tag{5.1.57}$$

Therefore, $\theta_1(t)$, can be obtained as follows:

$$\begin{aligned}
&\theta_1(t) \\
&= \alpha\cos\left(\frac{1}{9}\frac{\sqrt{-(6R-6r)g(\alpha^2-9)}t}{R-r}\right) \\
&+ \frac{\alpha^3}{(192\alpha^2-1728)\sqrt{-(6R-6r)g(\alpha^2-9)}} \\
&\times\left[8tg\alpha^2\sin\left(\frac{1}{9}\frac{\sqrt{-(6R-6r)g(\alpha^2-9)}t}{R-r}\right)\right. \\
&-72g\sin\left(\frac{1}{9}\frac{\sqrt{-(6R-6r)g(\alpha^2-9)}t}{R-r}\right)t \\
&+9\sqrt{-(6R-6r)g(\alpha^2-9)} \\
&\times\cos\left(\frac{1}{3}\frac{\sqrt{-(6R-6r)g(\alpha^2-9)}t}{R-r}\right) \\
&-9\left(\frac{1}{9}\frac{\sqrt{-(6R-6r)g(\alpha^2-9)}t}{R-r}\right) \\
&\times\sqrt{-(6R-6r)g(\alpha^2-9)}t
\end{aligned} \tag{5.1.58}$$

The next iteration will be time-consuming and complicated; therefore, we consider the above equation [Eq. (5.1.58)] as the answer of Eq. (5.1.34).

For the aforementioned physical values, the solution obtained from VIM (the first iteration) will be in the following form:

$$\begin{aligned}
&\theta_1(t) \\
&= 0.5243699095\cos(3.981450806t) \\
&+ 0.004093641228t\sin(3.981450806t) \\
&- 0.0007711337075\cos(11.94435242t)
\end{aligned} \tag{5.1.59}$$

For $\alpha = \frac{\pi}{4}$ and $\alpha = \frac{\pi}{4}$ (other different initial angular positions), the VIM solution can be obtained from Eq. (5.1.59). To obtain the AGM solution, Eqs. (5.1.39)–(5.1.45) must be solved for new initial conditions.

For $= \frac{\pi}{4}$, the AGM solution is:

$$\theta(t) = 0.7897374853\cos(3.891823467t) + 0.001993882397\cos(7.783646934t + 3.141592654) + 0.002345439545\cos(11.67547040t + 3.141592654) \quad (5.1.60)$$

And the VIM solution is:

$$\theta(t) = 0.7881071304\cos(3.902484934t) + 0.01409560347t\sin(3.902484934t) - 0.002708966921\cos(11.70745480t) \quad (5.1.61)$$

For $\alpha = \frac{\pi}{3}$.

the AGM solution is:

$$\theta(t) = 1.061717350\cos(3.781063625t) + 0.009537971050\cos(7.562127250t + 3.141592654) + 0.004981827437\cos(11.34319088t + 3.141592654) \quad (5.1.62)$$

And the VIM solution is:

$$\theta(t) = 1.054008607\cos(3.789169048t) + 0.03441098768t\sin(3.789169048t) - 0.006811055516\cos(11.36750714t) \quad (5.1.63)$$

The comparison between AGM, VIM, and the numerical methods is illustrated in Figs. 5.1.2, 5.1.3, 5.1.4, 5.1.5, 5.1.6 and 5.1.7 (for different values of initial condition a). The obtained phase plane by the AGM, VIM, and numerical solution has also been illustrated. The absolute errors compared to numerical solutions are given in Table 5.1.1.

According to the above charts and table, it is clear that in small ranges of time and initial angular positions, all utilised methods have very near precisions. But when the initial angular position is increased, the precision of AGM in comparison with VIM decreases a little. However, an approximation (the Taylor series) is applied in the equation of motion. Still, the solution obtained by AGM has acceptable high precision for the large initial angular position among VIM and the numerical solution.

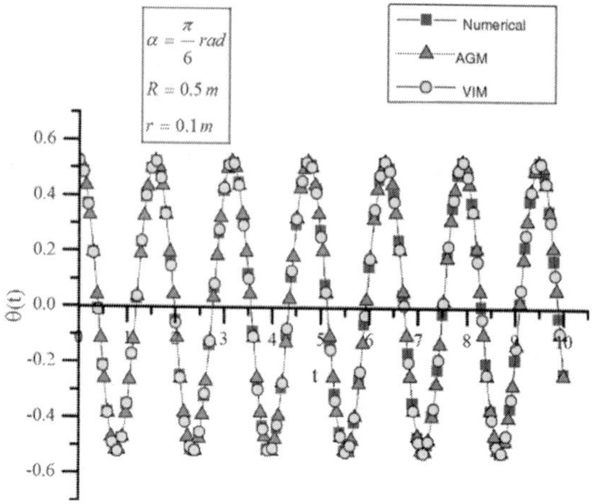

Fig. 5.1.2. Comparison between the solutions obtained by AGM, VIM, and the numerical method

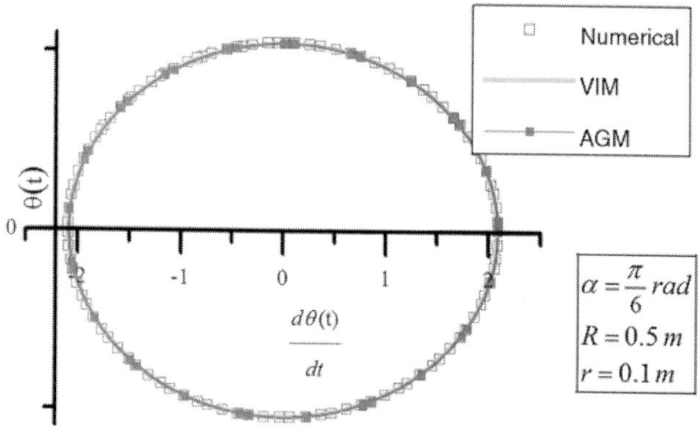

Fig. 5.1.3. Comparison between the obtained phase planes by AGM, VIM, and the numerical method

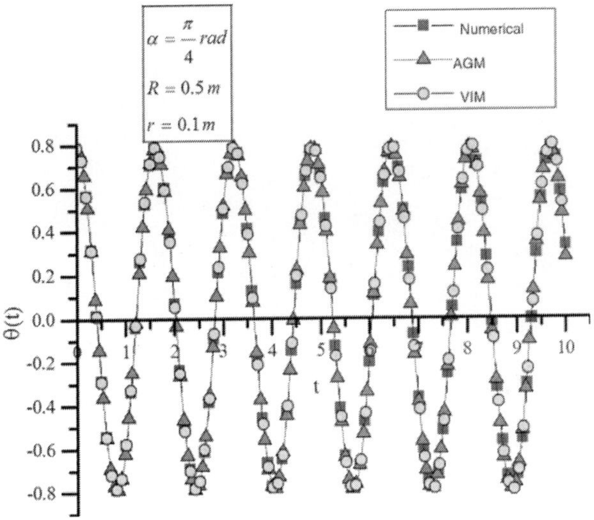

Fig. 5.1.4. Comparison between the solutions obtained by AGM, VIM, and the numerical method

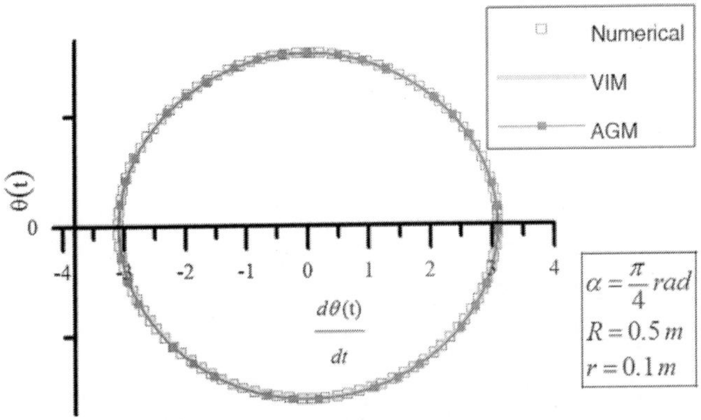

Fig. 5.1.5. Comparison between the obtained phase planes by AGM, VIM, and the numerical method

The procedure of obtaining the answer by AGM is simple and fast due to solving a desirable set of equations. Some unknowns or equations that are created depend on the needed precision. More precision needs more terms and, as a result, creates more coefficients or unknowns that must be solved.

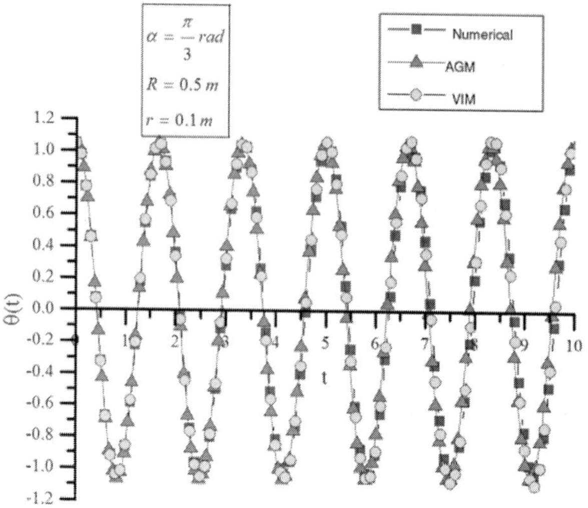

Fig. 5.1.6. Comparison between the solutions obtained by AGM, VIM, and the numerical method

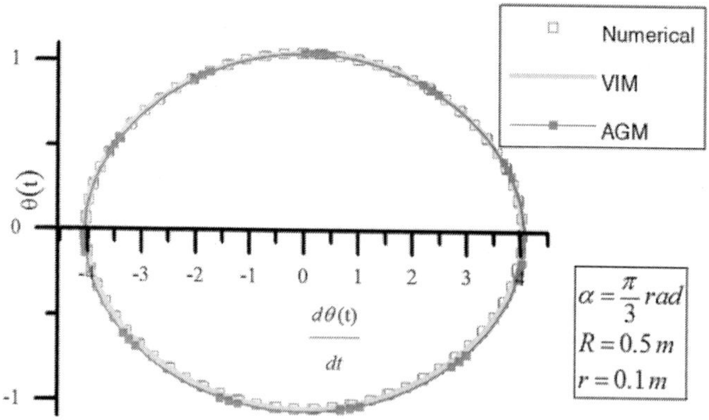

Fig. 5.1.7. Comparison between the obtained phase planes by AGM, VIM, and the numerical method

It is noteworthy that the shortage of initial conditions is terminated. We could obtain seven unknowns for an equation with only two initial conditions by using AGM. This set of equations can be solved easily and fast using computer software.

However, VIM is a powerful method for oscillatory models, but it uses the iteration technique and has its shortcomings mentioned at the beginning of the problem. The procedure for obtaining the second iteration for VIM in this specified problem is time-consuming and does not differ much from the first in its precision.

Table 5.1.1. Obtained errors in comparison with numerical solutions

$t(s)$	$\alpha = \frac{\pi}{6}$ rad		$\alpha = \frac{\pi}{4}$ rad		$\alpha = \frac{\pi}{3}$ rad	
	Error of AGM	Error of VIM	Error of AGM	Error of VIM	Error of AGM	Error of VIM
0	0.00000	0.00000	0.00000	0.00000	0.00000	0.00000
1	0.00049	0.00003	0.00309	0.00017	0.00782	0.00034
2	−0.00131	−0.00017	−0.01053	−0.00133	−0.04583	−0.00500
3	0.00145	0.00018	0.01565	0.00173	0.09018	0.00986
4	0.00012	−0.00024	−0.00762	−0.00212	−0.07511	−0.01088
5	−0.00311	−0.00008	−0.01643	0.00183	0.00155	0.01501

On the contrary, AGM calculates quickly, is straightforward, and has acceptable precision; therefore, this method can solve more complicated differential equations.

5.2. Dynamical Systems

In this problem, two nonlinear differential equations that appear in two specific dynamics problems have been analysed and solved completely.

In general, vibrational equations and their initial conditions are defined for different systems as follows:

$$f(\ddot{u}, \dot{u}, u, F_0 \sin(\omega_0 t)) = 0 \tag{5.2.1}$$

$$u(0) = A, \dot{u}(0) = 0 \tag{5.2.2}$$

Table 5.2.1. Types of vibrational systems, their equations, and answers

1) Vibrational systems without any external force		2) Vibrational systems with external force $F(t) = F_0 \sin(\omega_0 t)$	
$f(\ddot{u}, \dot{u}, u) = 0$	(2.8.3)	$f(\ddot{u}, \dot{u}, u, F_0 \sin(\omega_0 t)) = 0$	(2.8.4)
$u(t) = e^{-bt}\{a\cos(\omega t + \varphi)\}$	(2.8.5)	$u(t) = e^{-bt}\{a\cos(\omega t + \varphi)\} + d\cos(\omega_0 t + \phi)$	(2.8.6)

A total answer with constant coefficients is required to solve differential equations in AGM.

As shown in Table 5.2.1, vibrational systems generally can be divided into two parts. The constant coefficients of Eq. (5.2.5) or (5.2.6), which are {a, b, ω, φ, d, ϕ}, will compute in AGM by applying the initial conditions of Eq. (5.2.2).

(a) Applying the initial conditions on the answer of the differential equation.

$$u(t) = u(IC) \tag{5.2.7}$$

Notably, IC is the abbreviation of the introduced initial conditions of Eq. (5.2.2).

(b) Applying the initial conditions on the main differential equation and its derivatives:

Assume the general equation of the vibrational system as:

$$f(\ddot{u}, \dot{u}, u, F_0 \sin(\omega_0 t)) = 0 \tag{5.2.8}$$

According to Table 5.2.1, the suitable answer can be chosen as follows:

$$u = g(t) \tag{5.2.9}$$

Afterwards, Eq. (5.2.9) will be substituted into Eq. (5.2.8) instead of u as follows:

$$f(t) = f(g''(t), g'(t), g(t), F_0 \sin(\omega_0 t)) \tag{5.2.10}$$

Finally, the application of initial conditions on Eq. (5.2.10) and its derivatives is expressed as:

$$\begin{aligned} &f(IC) = f(g''(IC), g'(IC), g(IC), F_0 \sin(\omega_0 t)) \\ &f'(IC) = f'(g''(IC), g'(IC), g(IC), F_0 \sin(\omega_0 t)) \\ &f''(IC) = f''(g''(IC), g'(IC), g(IC), F_0 \sin(\omega_0 t)) \\ &. \\ &. \\ &. \end{aligned} \tag{5.2.11}$$

1) Consider a mass m moving without friction along a circle of radius R rotating with a constant angular velocity Ω about its vertical diameter. The geometry of the problem is depicted in Fig. 5.2.1. The forces acting on the particle are the gravitational force ω and normal reaction force N.

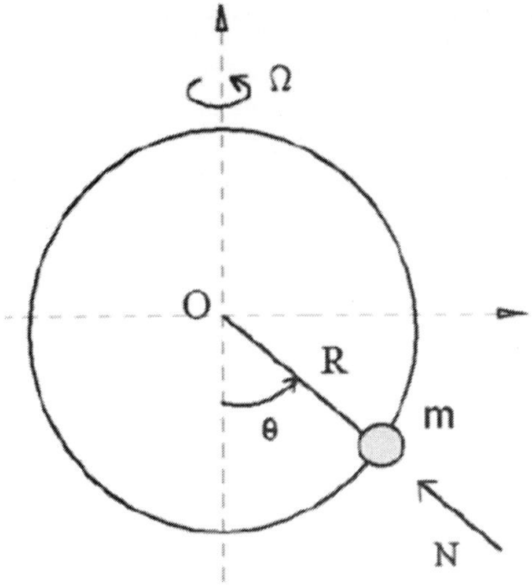

Fig. 5.2.1. Geometry of the problem

The following equation can be written:

$$\frac{d\vec{H}_O}{dt} = \sum \vec{M}_O \tag{5.2.12}$$

$$\frac{d\vec{H}_O}{dt} = \vec{r} \times m\vec{a} \tag{5.2.13}$$

$$a_r = \Omega^2 R \sin(\theta(t)) \tag{5.2.14}$$

$$a_\theta = R\ddot{\theta}(t) \tag{5.2.15}$$

According to Eq. (5.2.12) or (5.2.13), a differential equation of motion can be obtained as follows:

$$mR^2\ddot{\theta}(t) - m\Omega^2R^2\, sin(\theta(t))\cos(\theta(t)) + mgR\sin(\theta(t)) = 0 \tag{5.2.16}$$

$$\theta(0) = \alpha, \dot{\theta}(0) = 0 \tag{5.2.17}$$

Now we investigate the Eq. (5.2.16) using an approximation for sin$(\theta(t))$ and cos$(\theta(t))$ as follows:

$$sin(\theta(t)) = \theta(t) - \frac{\theta(t)^3}{3!} \tag{5.2.18}$$

$$\cos(\theta(t)) = 1 - \frac{\theta(t)^2}{2!} \tag{5.2.19}$$

Substituting Eqs. (5.2.18) and (5.2.19) into (5.2.16), and by some manipulation, Eq. (5.2.16) can be rewritten as:

$$\frac{d^2}{dt^2}\theta(t) + \varpi_0^2(t) + G(\theta(t))^3 - k(\theta(t))^5 = 0, \quad \theta(0) = \alpha, \dot{\theta}(0) = 0 \tag{5.2.20}$$

Where:

$$I = mR^2 \tag{5.2.21}$$

$$\varpi_0^2 = (\frac{mgR}{I} - \frac{m\Omega^2R^2}{I}) \tag{5.2.22}$$

$$G = (-\frac{mgR}{I} - 2\frac{m\Omega^2R^2}{3I}) \tag{5.2.23}$$

$$K = \frac{m\Omega^2R^2}{12I}) \tag{5.2.24}$$

According to the "Choosing the Answer of Governing Equation for Solving Differential Equations by AGM" Section, the answer to differential equation Eq. (5.2.20) is considered as follows:

$$\theta(t) = e^{-bt}c\cos(\omega t + \varphi), \quad \theta(0) = \alpha, \dot{\theta}(0) = 0 \tag{5.2.25}$$

Accordingly, initial conditions have been applied as follows:

$$\theta(0) = \alpha \rightarrow c\cos(\varphi) = \alpha \tag{5.2.26}$$

$$\dot{\theta}(0) = 0 \rightarrow -bc\,cos(\varphi) - c\sin(\varphi) = 0 \tag{5.2.27}$$

$$\begin{aligned} f\big(\theta(0)\big) \rightarrow b^2 c\cos(\varphi)\omega \\ &+ 2b\sin(\varphi)\omega \\ &- c\cos(\varphi)\omega^2 + (\frac{g}{R} - \Omega^2)c\cos(\varphi) \\ &+ \cdots\left(-\frac{1}{6}\frac{g}{R} + \frac{2}{3}\Omega^2\right)c^3(cos(\varphi))^3 \\ &- 1/12\Omega^2 c^5 (cos(\varphi))^5 = 0 \end{aligned} \tag{5.2.28}$$

$$\begin{aligned} &f'\left(\dot{\theta}(0)\right) \\ &\rightarrow -b^3 c\cos(\varphi) - 3b^2 c\sin(\varphi)\,\omega \\ &+ 3\,\mathrm{bc}\cos(\varphi)\omega^2 + c\sin(\varphi)\omega^3 - \left(\frac{g}{R} - \Omega^2\right)c\sin(\varphi)\,\omega \\ &- 3\left(-\frac{1}{6}\frac{g}{R} + \frac{2}{3}\Omega^2\right)c^3(\cos(\varphi))^3 b \\ &- 3\left(-\frac{1}{6}\frac{g}{R} + \frac{2}{3}\Omega^2\right)c^3(\cos(\varphi))^2\sin(\varphi)\omega \\ &+ \cdots\frac{5}{12}\Omega^2 c^5(\cos(\varphi))^5 b + \frac{5}{12}\Omega^2 c^5 (cos(\varphi))^5\sin(\varphi)\omega \\ &= 0 \end{aligned} \tag{5.2.29}$$

By solving a set which consists of four equations with four unknowns from Eq. (5.2.26) to (5.2.29), the constant coefficients of Eq. (5.2.25) can be obtained as follows:

$$\mathrm{b} = 0$$

$$\mathrm{c} = \alpha$$

$$\omega = \sqrt{-1/12\frac{-8\Omega^2 R\alpha^2 - 12g + 12\Omega^2 R + 2\alpha^2 g + \Omega^2\,\alpha^4 R}{R}} \tag{5.2.30}$$

$$\varphi = 0$$

By selecting the following physical values, we can obtain the constant coefficients of Eq. (5.2.25):

$$\alpha = 0.1\ rad, R = 0.5\ m, g = 9.81\left(\frac{m}{s^2}\right), m = 3\ kg, \Omega = \frac{\pi}{2}\ \frac{rad}{s} \quad (5.2.31)$$

$$\begin{aligned} &b = 0 \\ &c = 0.1 \\ &\omega = 4.139604773 \\ &\varphi = 0 \end{aligned} \quad (5.2.32)$$

The analytical solution can be written as follows:

$$\theta(t) = 0.1\cos(4.139604773t) \quad (5.2.33)$$

According to *VIM*, the arbitrary initial approximation that satisfies the initial conditions can be obtained as follows:

$$\theta_0(t) = \alpha\cos(\omega t) \quad (5.2.34)$$

$$\frac{d^2}{dt^2}\theta(t) = -\varpi_0^2\theta_0(t) - G\big(\theta_0(t)\big)^3 + K\big(\theta_0(t)\big)^5 \quad (5.2.35)$$

$$\frac{d^2}{dt^2}\theta(t) = -\varpi_0^2\alpha\cos(\omega t) - G\alpha^3(cos(\omega t))^3 + K\alpha^5\,(cos(\omega t))^5 \quad (5.2.36)$$

Integrating twice yields:

$$\begin{aligned}\theta(t) = \frac{1}{225\omega^5}\alpha(&-225\varpi_0^2 - 175G\alpha^2 \\ &+ 225\varpi_0^2\cos(\omega t) \\ &+ 25G\alpha^2\,(cos(\omega t))^3 \\ &+ \cdots 150G\alpha^2\,cos(\omega t) \\ &- 9K\alpha^4\,(cos(\omega t))^5 \\ &- 20K\alpha^4\,(cos(\omega t))^3 \\ &- 120K\alpha^4\cos(\omega t))\end{aligned} \quad (5.2.37)$$

Equating the coefficient of cos *(ωt)* in Eq. (5.2.37), we have:

$$\alpha = \frac{1}{225}\frac{\alpha(225\varpi_0^2 + 150G\alpha^2 - 120K)}{\omega^2} \quad or \tag{5.2.38}$$

$$\omega = \frac{1}{15}\sqrt{225\varpi_0^2 + 150G\alpha^2 - 120K} \tag{5.2.39}$$

$$\theta_0(t) = \alpha\cos(\frac{1}{15}\sqrt{225\varpi_0^2 + 150G\alpha^2 - 120K}\,) \tag{5.2.40}$$

Now we construct the correction functional, as follows:

$$\theta_{n+1}(t) = \theta_n(t) + \int_0^t \lambda(\frac{d^2}{dt^2}\theta_n(\tau) + \varpi_0^2\theta_n(\tau) + G(\theta_n(\tau))^3 - K(\theta_n(\tau))^3)d\tau \tag{5.2.41}$$

Its stationary conditions can be obtained as follows:

$$\lambda''(\tau) = 0 \tag{5.2.42a}$$

$$1 - \lambda'(\tau)\big|_{(\tau=t)} = 0 \tag{5.2.42b}$$

$$\lambda(\tau)\big|_{(\tau=t)} = 0 \tag{5.2.42c}$$

The Lagrange multiplier can therefore be simply identified as:

$$\lambda = \tau - t \tag{5.2.42}$$

Using the above iteration formula [Eq. (5.2.41)] and the initial approximation and second approximating solution are calculated as follows:

$$\theta_1(t) = \frac{\alpha}{15\left(15\overline{\omega}_0^2 - 8K\alpha^4 + 10G\alpha^2\right)}(225\overline{\omega}_0^2\cos(\omega t) - 120K\alpha^4\cos(\omega t) + 150G\,\alpha^2\cos(\omega t) - 25G\alpha^2 + \cdots 29K\alpha^4 + 25G\alpha^2(\cos(\omega t))^3 - 9K\alpha^4(\cos(\omega t))^5 - 20K\alpha^4(\cos(\omega t))^3) \tag{5.2.43}$$

For the aforementioned physical values, the result can be written as follows:

$$\theta_1(t) = 0.1\cos(4.140260156t) + 0.00001054897281 + \cdots - 0.00001054417477(\cos(4.140260156t))^3 - 0.000000004798033073(\cos(4.140260156t))^5 \quad (5.2.44)$$

According to *EBM*, Variational and Hamiltonian formulations of Eq. (5.2.20) can be easily obtained as:

$$J(\theta) = \int_0^t \left\{ -\frac{1}{2}\left(\frac{d\theta(t)}{dt}\right)^2 + \frac{1}{2}\overline{\omega}_0^2\theta(t)^2 + \frac{G}{4}\theta(t)^4 - \frac{K}{6}\theta(t)^6 \right\} dt \quad (5.2.45)$$

It is Hamiltonian and, therefore, can be written in the form:

$$H = \frac{1}{2}\left(\frac{d\theta(t)}{dt}\right)^2 + \frac{1}{2}\overline{\omega}_0^2\theta(t)^2 + \frac{G}{4}\theta(t)^4 - \frac{K}{6}\theta(t)^6 - \frac{1}{2}\overline{\omega}_0^2\theta(t)^2 - \frac{G}{4}\theta(t)^4 + \frac{K}{6}\theta(t)^6 \quad (5.2.46)$$

By choosing the trial function θ(t)=α cos(ω*t*), the following residual equation can be obtained as follows:

$$R = \frac{1}{2}\alpha^2(\sin(\omega t))^2\omega^2 + \frac{1}{2}\overline{\omega}_0^2\alpha^2(\cos(\omega t))^2 + \frac{1}{4}G\alpha^4(\cos(\omega t))^4 + \cdots - \frac{1}{6}K\alpha^6(\cos(\omega t))^6 - \frac{1}{2}\overline{\omega}_0^2\alpha^2 - \frac{1}{4}G\alpha^4 + \frac{1}{6}K\alpha^6 \quad (5.2.47)$$

If we collocate $\omega t = \frac{\pi}{4}$, the following result can be obtained as follows:

$$\omega = \frac{1}{6}\sqrt{(36\overline{\omega}_0^2 + 27G\alpha^2 - 21K\alpha^4)} \quad (5.2.48)$$

The following approximate solution can be obtained as follows:

$$\theta(t) = \alpha \cos(\frac{1}{6}\sqrt{(36\bar{\omega}_0^2 + 27G\alpha^2 - 21K\alpha^4)}) \qquad (5.2.49)$$

For the aforementioned physical values, the result can be written as follows:

$$\theta(t) = 0.1 \cos(\frac{1}{6}\sqrt{705.4371000 - 8.955043750\pi^2 t}) \qquad (5.2.50)$$

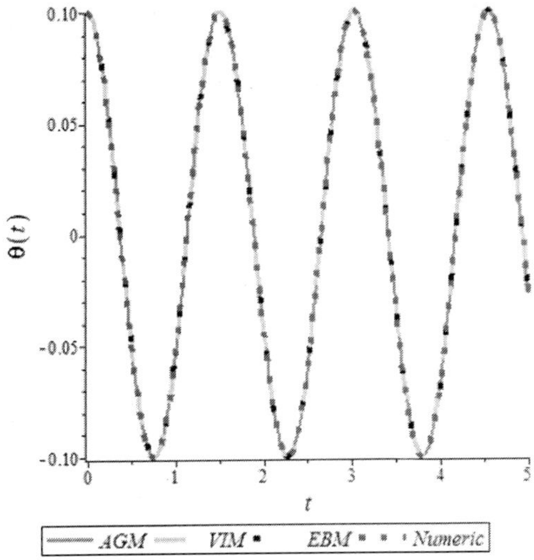

Fig. 5.2.2. Solutions obtained by AGM, VIM, EBM, and the numerical method

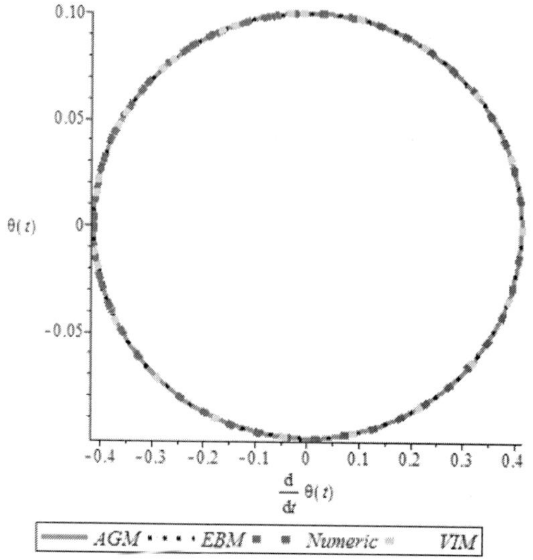

Fig. 5.2.3. Phase plane obtained by AGM, EBM, VIM, and the numerical method

The comparison between the solutions obtained by AGM, VIM, EBM, and the numeric method (Runge-Kutte 4th) is depicted in Fig. 5.2.2. The obtained phase plane by AGM, VIM, EBM, and the numeric method has been illustrated in Fig. 5.2.3.

In Figs. 5.2.4, 5.2.5, and 5.2.6, the differences between the solution obtained, θ (t), by AGM, VIM, EBM, and the numerical method have been depicted.

2) In this problem, we consider a bead of mass m that slides without friction along a wire with the shape of a parabola, $y = \lambda x2$, with an axis vertical to the earth's gravitational field, g, as shown in Fig. 5.2.7. x is the generalized coordinate of the horizontal displacement.

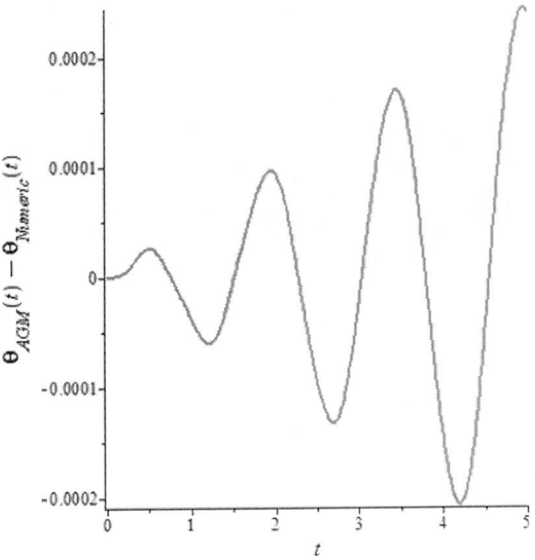

Fig. 5.2.4. Difference between the solutions obtained by AGM and the numerical method

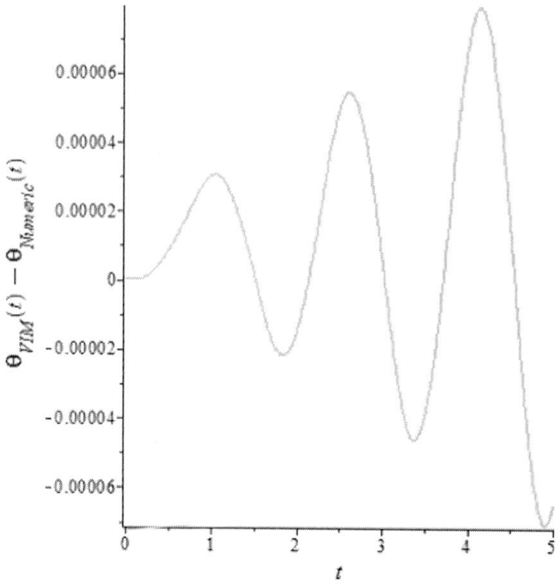

Fig. 5.2.5. Difference between the solutions obtained by VIM and the numerical method

We can write down Lagrange's equation of motion as:

$$\left(1+4\lambda^2\left(\left(x(t)\right)^2\right)\right)\frac{d^2}{dt^2}x(t)+4\lambda^2 x(t)\left(\frac{d}{dt}x(t)\right)^2 + 2g\lambda x(t) \tag{5.2.51}$$

Equation (5.2.51) can be rewritten as follows:

$$\frac{d^2}{dt^2}x(t)+\varpi_0^2 x(t)+M\left(x(t)\right)^2+Nx(t)\left(\frac{d}{dt}x(t)\right)^2=0 \tag{5.2.52}$$

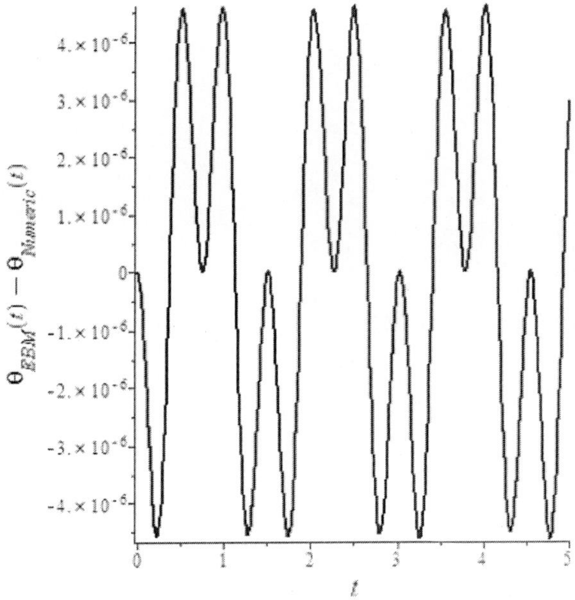

Fig. 5.2.6. Difference between the solutions obtained by EBM and the numerical method

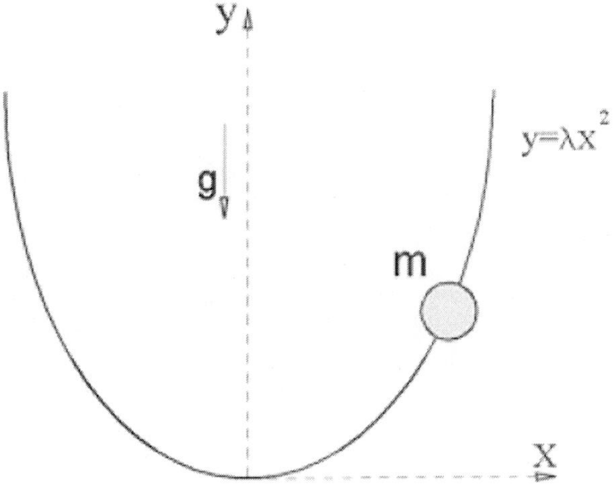

Fig. 5.2.7. Geometry of the problem

Where:

$$N = M = 4\lambda^2 \tag{5.2.53}$$

$$\overline{\omega}_0^2 = 2g\lambda \tag{5.2.54}$$

The initial conditions are defined as follows:

$$x(0) = \alpha, \qquad \dot{x}(0) = 0 \tag{5.2.55}$$

According to the "Choosing the Answer of Governing Equation for Solving Differential Equations by AGM" section, the answer to differential equation Eq. (5.2.52) is considered as follows:

$$\theta(t) = e^{-bt} c\cos(\omega t + \varphi),\ \ \theta(0) = \alpha, \dot{\theta}(0) = 0 \tag{5.2.56}$$

Initial conditions have been applied as follows:

$$x(0) = \alpha \ \rightarrow \ c\cos(\varphi) = \alpha \tag{5.2.57}$$

$$\dot{x}(0) = 0 \ \rightarrow \ -bc\cos(\varphi) - c\sin(\varphi)\omega = 0 \tag{5.2.58}$$

$$\begin{aligned} f\big(x(0)\big) \rightarrow (1 + 4\lambda^2 c^2 (\cos(\varphi))^2)(b^2 c\cos(\omega\alpha + \varphi) \\ + 2bc\sin(\omega\alpha + \varphi)\,\omega \\ - c\cos(\varphi)\omega^2 \\ + \cdots 4\lambda^2 c\cos(\varphi)(-bc\cos(\varphi) \\ - c\sin(\varphi)\omega)^2 + 2g\lambda c\cos(\varphi) = 0 \end{aligned} \tag{5.2.59}$$

$$\begin{aligned} f'\big(\dot{x}(0)\big) \rightarrow (-8\lambda^2 c^2 (\cos(\varphi))^2 b \\ - 8\lambda^2 c^2 \cos(\varphi)\sin(\varphi)\omega)(b^2 c\cos(\varphi) \\ + 2bc\sin(\varphi)\omega - c\cos(\varphi)\omega^2) + \cdots (1 \\ + 4\lambda^2 c^2 (\cos(\varphi))^2)(-b^3 c\cos(\varphi) \\ - 3b^2 c\sin(\varphi)\omega + 3bc\cos(\varphi)\omega^2 \\ + c\sin(\varphi)\omega^3) \\ + \cdots 4\lambda^2 bc\cos(\varphi)(-bc\cos(\varphi) \\ - c\sin(\varphi)\omega)^2 \\ - 4\lambda^2 c\sin(\varphi)\omega\,(-bc\cos(\varphi) \\ - c\sin(\varphi)\omega)^2 \\ + \cdots 8\lambda^2 c\cos(\varphi)(-bc\cos(\varphi) \\ - c\sin(\varphi)\omega)(b^2 c\cos(\varphi) \\ + 2bc\sin(\varphi)\,\omega - c\cos(\varphi)\,\omega^2) + \cdots \\ - 2g\lambda bc\cos(\varphi) - 2g\lambda c\sin(\varphi)\,\omega = 0 \end{aligned} \tag{5.2.60}$$

By solving a set consisting of four equations with four unknowns from Eq. (5.2.57) to (5.2.60), the constant coefficients of Eq. (5.2.56) will be obtained as follows:

$$b = 0$$

$$c = \alpha$$

$$\omega = \sqrt{2}\sqrt{\frac{g\lambda}{4\lambda^2\alpha^2 + 1}} \tag{5.2.61}$$

$$\varphi = 0$$

By selecting the following physical values, the answer can be obtained.

$$\alpha = 0.2\ m, \quad g = 9.81\left(\frac{m}{s^2}\right) \quad and \quad \lambda = \frac{\sqrt{2}}{2} \tag{5.2.62}$$

$$x(t) = 0.2\cos(3.584100054t) \tag{5.2.63}$$

According to VIM, the arbitrary initial approximation is chosen as follows:

$$x_0(t) = \alpha\cos(\omega t) \tag{5.2.64}$$

$$\begin{aligned}\frac{d^2}{dt^2}x(t) = -\varpi_0^2(t) - M\big(x_0(t)\big)^2\frac{d^2}{dt^2}x_0(t) \\ - Nx_0(t)\left(\frac{d}{dt}x_0(t)\right)^2\end{aligned} \tag{5.2.65}$$

Or

$$\begin{aligned}\frac{d^2}{dt^2}x(t) = -\varpi_0^2\alpha\cos(\omega t) - M\alpha^3(\cos(\omega t))^3\omega^2 \\ - N\,\alpha^3\cos(\omega t)\,(\sin(\,\omega t))^2\omega^2\end{aligned} \tag{5.2.66}$$

Integrating twice yields:

$$x(t) = \frac{-\alpha}{9\omega^2}(9\varpi_0^2 - 7M\alpha^2\omega^2 + 2N\alpha^2\omega^2 - 9\varpi_0^2\cos(\omega t) + M\alpha^2\omega^2(\cos(\omega t))^3 + \cdots 6M\alpha^2\omega^2\cos(\omega t) - 3N\alpha^2\omega^2\cos(\omega t) + N\alpha^2\omega^2(\cos(\omega t))^3) \tag{5.2.67}$$

Equating the coefficient of cos (ωt) in Eq. (5.2.67), we have:

$$\alpha = -\frac{1}{9}\frac{\alpha(-9\varpi_0^2 + 6M\alpha^2\omega^2 - 3N\alpha^2\omega^2)}{\omega^2} \tag{5.2.68}$$

$$\omega = 3\frac{\varpi_0}{\sqrt{9 + 6M\alpha^2 - 3N\alpha^2}} \tag{5.2.69}$$

$$x_0(t) = \alpha\cos\left(3\frac{\varpi_0 t}{\sqrt{9 + 6M\alpha^2 - 3N\alpha^2}}\right) \tag{5.2.70}$$

Now we construct the correction functional, as follows:

$$x_{n+1}(t) = x_n(t) + \int_0^t \eta\left(\frac{d^2}{d\tau^2}x_n(\tau) + \varpi_0^2 x_n(\tau) + M(x_n(\tau))^2\frac{d^2}{d\tau^2}x_n(\tau) + N(\frac{d}{d\tau}x_n(\tau))^2\right)d\tau \tag{5.2.71}$$

Its stationary conditions can be obtained as follows:

$$\eta''(\tau) = 0 \tag{5.2.72a}$$

$$1 - \eta'(\tau)|_{(\tau=t)} = 0 \tag{5.2.72b}$$

$$\eta(\tau)|_{(\tau=t)} = 0 \tag{5.2.72c}$$

The Lagrange multiplier can therefore be simply identified as:

$$\eta = \tau - t$$

Using the above iteration formula [Eq. (5.2.71)] and the initial approximation and second approximating solution are calculated as follows:

$$
\begin{aligned}
x_1(t) = \alpha\cos(\omega t) &+ \frac{1}{9}\alpha^3 M + \frac{1}{9}\alpha^3 N \\
&- \frac{64}{9}\alpha^3\left(\cos\left(\frac{1}{3}\omega t\right)\right)^9 M \\
&+ 16\alpha^3\left(\cos\left(\frac{1}{3}\omega t\right)\right)^7 M + \cdots \\
&- 12\alpha^3\left(\cos\left(\frac{1}{3}\omega t\right)\right)^5 M \\
&+ 3\alpha^3\left(\cos\left(\frac{1}{3}\omega t\right)\right)^3 M \\
&+ 16\alpha^3\left(\cos\left(\frac{1}{3}\omega t\right)\right)^7 N \\
&- \frac{64}{9}\alpha^3\left(\cos\left(\frac{1}{3}\omega t\right)\right)^9 N \\
&+ \cdots 3\alpha^3\left(\cos\left(\frac{1}{3}\omega t\right)\right)^3 N \\
&- 12\alpha^3\left(\cos\left(\frac{1}{3}\omega t\right)\right)^5 N
\end{aligned}
\tag{5.2.74}
$$

For the aforementioned physical values, the second approximating solution can be written as follows:

$$
\begin{aligned}
&x_1(t) = 0.2\cos(3.676014975t) + 0.003555555556 \\
&-0.2275555556(\cos(1.225338325t))^9 \\
&\qquad + \cdots 0.512(\cos 1.225338325t)^7 \\
&-0.348(\cos 1.225338325t)^5 \\
&\qquad + 0.96(\cos 1.225338325t)^3
\end{aligned}
\tag{5.2.75}
$$

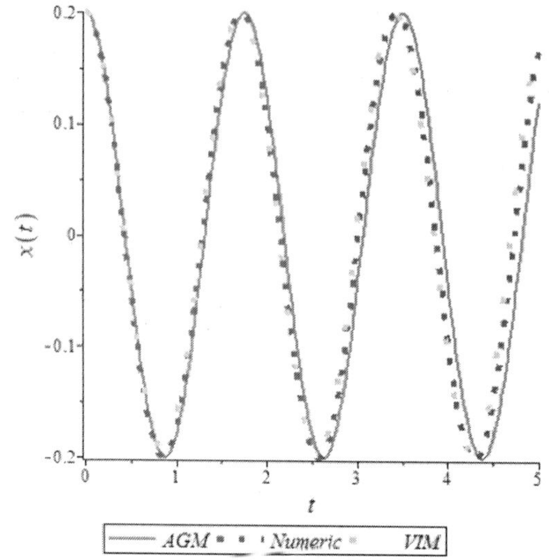

Fig. 5.2.8. Solution obtained by AGM, VIM, and the numerical method

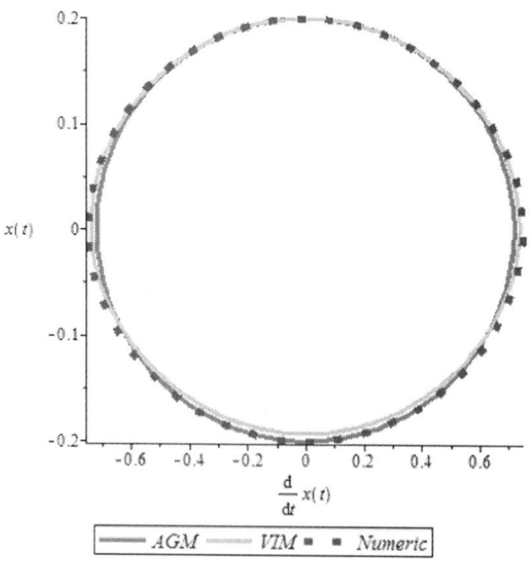

Fig. 5.2.9. Phase plane obtained by AGM, VIM, and the numerical method

The comparison between the solution obtained by AGM, VIM, and the numeric method (Runge– Kutte 4th) is depicted in Fig. 5.2.8. And in Fig. 5.2.9, the phase plane obtained by AGM, VIM, and the numerical method is illustrated.

In Figs. 5.2.10 and 5.2.11, the differences between the solution obtained ($x(t)$) by AGM, VIM, and the numerical method have been represented.

According to *EBM*, the variational and Hamiltonian formulations of Eq. (5.2.52) for $N = M = 2$ can be easily obtained as:

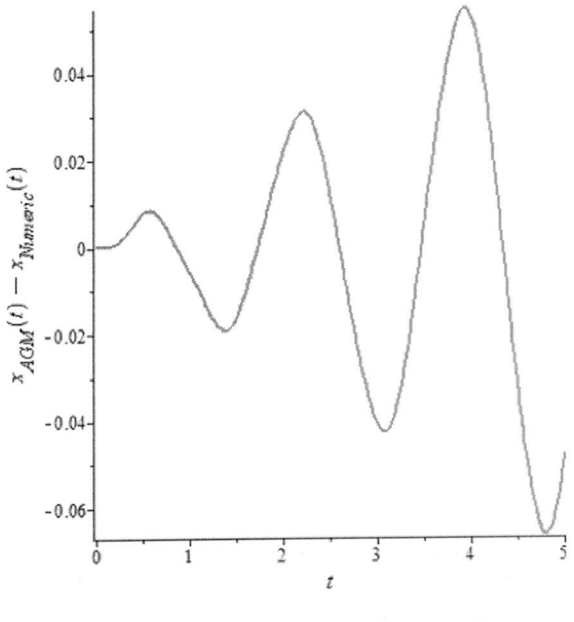

Fig. 5.2.10. Difference between the solutions obtained by AGM and the numerical method

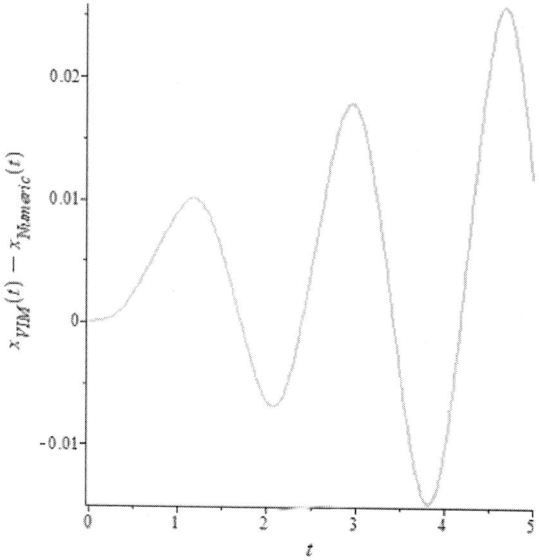

Fig. 5.2.11. The difference between the solutions obtained by VIM and the numerical method

$$J(x) = \int_0^t \left\{ -\frac{1}{2}\left(\frac{d}{dt}x(t)\right)^2 + \frac{1}{2}\varpi_0^2 (x(t))^2 + (x(t))^2 \left(\frac{d}{dt}x(t)\right)^2 \right\} dt \tag{5.2.76}$$

It is Hamiltonian and, therefore, can be written in the following form:

$$\mathrm{H} = \frac{1}{2}\left(\frac{\mathrm{d}}{\mathrm{dt}}\mathrm{x(t)}\right)^2 + \frac{1}{2}\varpi_0^2(\mathrm{x(t)})^2 + (\mathrm{x(t)})^2\left(\frac{\mathrm{d}}{\mathrm{dt}}\mathrm{x(t)}\right)^2 - \frac{1}{2}\varpi_0^2\alpha^2 \tag{5.2.77}$$

Choosing the trial function $x(t)=\alpha.\cos(\omega t)$, the following residual equation can be obtained as follows:

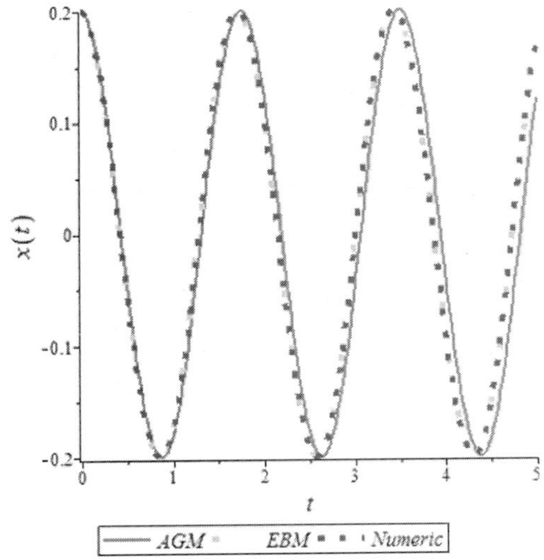

Fig. 5.2.12. The solution obtained by AGM, EBM, and the numerical method

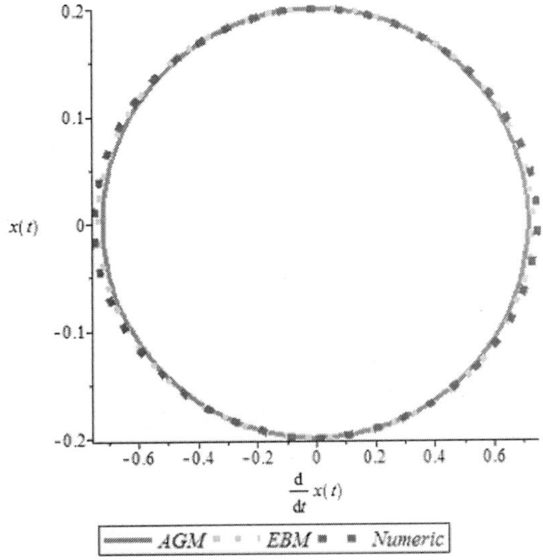

Fig. 5.2.13. Phase plane obtained by AGM, EBM, and the numerical method

$$R = \frac{1}{2}\alpha^2(\sin(\omega t))^2\omega^2 + \frac{1}{2}\varpi_0^2\alpha^2(\cos(\omega t))^2 + \alpha^4(\cos(\omega t))^2(\sin(\omega t))^2\omega^2 - \frac{1}{2}\varpi_0^2\alpha^2 \tag{5.2.78}$$

If we collocate at $\omega t = \pi/4$, the following result can be obtained as follows:

$$\omega = \frac{\varpi_0}{\sqrt{1+\alpha^2}} \tag{5.2.79}$$

The following approximate solution can be obtained as follows:

$$x(t) = \alpha\cos(\frac{\varpi_0}{\sqrt{1+\alpha^2}}) \tag{5.2.80}$$

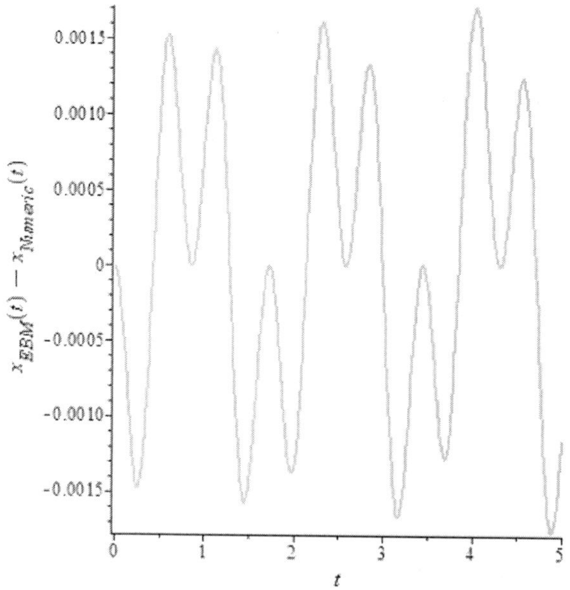

Fig. 5.2.14. The difference between the solution obtained by EBM and the numerical method

By selecting the aforementioned physical values, the answer can be written as follows:

$$x(t) = 0.2\cos(3.652374760t) \quad (5.2.81)$$

The comparison between the solutions obtained by AGM, EBM, and numeric method (Runge-Kutte 4th) is depicted in Fig. 5.2.12. Fig. 5.2.13 illustrates the phase plane obtained by AGM, EBM, and the numerical method.

Fig. 5.2.14 represents the differences between the solutions obtained x(t) by EBM and the numerical method.

According to the above figures, in the first problem, AGM has a high precision, and in the second example, it has an acceptable accuracy. The accuracy of VIM and AGM in these two specific problems is almost the same, but the precision of EBM was higher than these two methods. The variational iteration method is an effective method for nonlinear oscillators but applies an iteration technique that had been discussed previously. On the contrary, AGM is concise and simple and can obtain the answer in a short form. In these two problems solving a set of equations to obtain constant coefficients of the proposed answer was almost easy because the location point was t=0. But when faced with boundary value problems where there is another nonzero point, solving the set mentioned above of equations is cumbersome and sometimes impossible. But for more difficult nonlinear initial value problems, this method can be applied.

5.3. Unsteady Motion of Spherical Particles Falling in Non-Newtonian Fluid

An analytical investigation is applied for the unsteady motion of a rigid spherical particle in a quiescent shear-thinning power-law fluid.

Fig. 5.3.1 shows the consideration of the one-dimensional accelerated motion of a rigid spherical particle falling to an infinite extent of a power-law shear-thinning fluid. The forces acting on a falling body are gravity, buoyancy, inertia, Basset history force, virtual mass, and drag force. From the Lagrangian viewpoint, the dynamic of the particles submerged in a fluid could be obtained by integrating the forces balanced on them. The Basset force can be assumed to be negligible when the density of the spherical particle is much larger than that of the liquid. Under this condition, the equation of motion describing the falling motion of the particle can be written as,

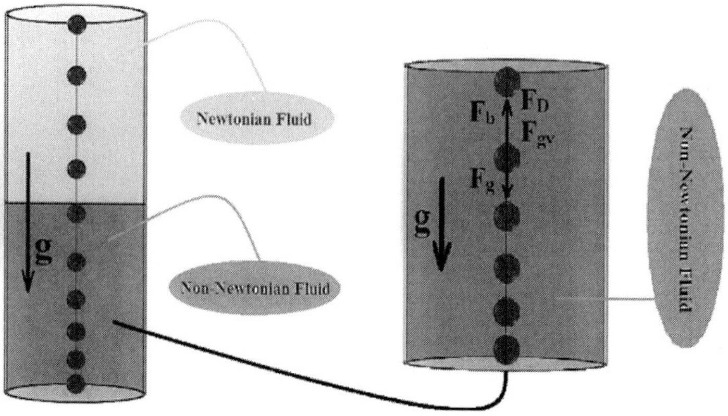

Fig. 5.3.1. Geometry of the physical model.

$$m\frac{du}{dt} = mg\left(1 - \frac{\rho}{\rho_s}\right) - \frac{\pi D^2 \rho C_D}{8}u^2 - \frac{\pi D^3 \rho}{12}\frac{du}{dt} \quad (5.3.1)$$

Where D, m, ρ_s, and C_D are the particle diameter, particle mass, particle density, and drag coefficient, respectively. From left to right, the terms represent inertia, gravity-buoyancy, drag, and virtual mass (added mass effect due to fluid acceleration around the particle). The above equation's complexity arises from the strongly nonlinear drag coefficient. Numerical or experimental results have routinely obtained the proper formulation of the drag coefficient. It is well known today that the drag coefficient for a sphere in a power-law fluid could be expressed as follows:

$$C_D = f(Re, n) \quad (5.3.2)$$

For a creeping flow region (Re << 1), the drag coefficient could be obtained from Stokes law in the following form:

$$C_D = \frac{24}{Re}X(n) \quad (5.3.3)$$

Where the Reynolds number is $Re = \rho u^2 - \frac{nDn}{K}$. n and K are the flow behaviour index and consistency coefficient, respectively. X(n), a deviation factor, was obtained by researchers via numerical or experimental results. Here, a well-correlated equation was used as follows:

$$X(n) = 6^{\frac{n-1}{n}}\left(\frac{3}{n^2+n+1}\right)^{n+1} \tag{5.3.4}$$

The correlated equation is valid for both shear-thinning (n < 1) and shear-thickening (n > 1) fluid behaviours. By substituting Eqs. (5.3.3) and (5.3.4) with Eq. (5.3.1) and by rearranging the parameters, one could give:

$$a\frac{du}{dt} + b(n)u^n - d = 0, \quad u(0) = v_0 \tag{5.3.5}$$

In which

$$a = m + \frac{1}{12}\pi D^3\rho, \qquad b(n) = 3\pi K X(n) D^{2-n}, \qquad d = mg\left(1 - \frac{\rho}{\rho_s}\right) \tag{5.3.6}$$

Eq. (5.3.5) is classified as an IVP (initial value problem) differential equation, which could be solved with suitable numerical methods such as the finite difference scheme. The numerical solution to the problem is not within the scope of this example, but the analytical solution is described in the following section.

Before presenting the results, it is necessary to provide some background knowledge about the mathematical methods employed. Therefore, this section presents some basic relationships and theories concerning the CM and the fourth-order Runge-Kutta numerical methods.

Some simple and accurate approximation techniques for solving differential equations are called weighted residual methods (WRMs). The CM, the Galerkin method (GM), and the LSM are examples of WRMs. Compared to other analytical methods, CM's many advantages make it more valuable and motivate researchers to use it for solving problems. Some of these advantages are listed below:

(a) WRMs solve the equations directly, and no simplifications are needed.

(b) They do not need any perturbation, linearization, or small parameters versus the HPM and parameter perturbation method (PPM).

(c) They are simple and powerful compared to numerical methods and achieve final results faster than numerical procedures, while their results are acceptable and have excellent agreement with numerical outcomes. Furthermore, their accuracy can be increased by increasing the statements of the trial functions.

(d) They do not need to determine the auxiliary parameter and auxiliary function versus the HAM.

(e) They are faster in solving such problems with lower accuracy than AGM.

With CM, suppose a differential operator D is acted on a function u to produce a function p:

$$D\big(u(x)\big) = p(x) \tag{5.3.7}$$

We wish to approximate u by function $\tilde{u}$, a linear combination of basic functions chosen from a linearly independent set. That is,

$$u \cong \tilde{u} = \sum_{i=1}^{n} c_i \varphi_i \tag{5.3.8}$$

Now, when substituted with the differential operator, D, the result of the operation is not p(x). Hence an error or residual will exist:

$$E(x) = R(x) = D\big(\tilde{u}(x)\big) - p(x) \neq 0 \tag{5.3.9}$$

The notion in the collocation is to force the residual to zero in some average sense over the domain. That is,

$$\int R(x)W(x)_i\, d(x) = 0 \qquad i = 0, 1, 2, \dots, n \tag{5.3.10}$$

Where the number of weight functions W_i is exactly equal to the number of unknown constants c_i in $\tilde{u}$. The result is a set of n algebraic equations for the unknown constants c_i. For CM, the weighting functions are taken from the family of Dirac δ functions in the domain. That is, $W_i(x)=\delta(x - x_i)$. The Dirac d function has the property of

$$\delta(x - x_i) = \begin{cases} 1 & if\ x = x_i \\ 0 & otherwise \end{cases} \tag{5.3.11}$$

And residual function Eq. (5.3.9) must be forced to zero at a specific point.

It is obvious that the type of the current problem is an IVP, and the appropriate method needs to be chosen. The available sub-methods in the Maple 17.0 combine the base schemes, the trapezoid, or the midpoint method. There are two major considerations when choosing a method for a problem. The trapezoid method is generally efficient for typical problems, but the midpoint method can handle harmless endpoint singularities that the trapezoid method cannot. The midpoint method, also known as the fourth-order Runge-Kutta-Fehlberg method, improves the Euler method by adding a midpoint in the step, which increases the accuracy by one order. Thus, the midpoint method is a suitable numerical technique in the present study.

The following section will illustrate the applicability of the presented methods for the nonlinear equation of motion of settling particles. In order to measure the accuracy of the results, NUM has been used here for the derived nonlinear ODE, given in Eq. (5.3.5), where the fourth-order Runge-Kutta is employed for deriving $\frac{du}{dt}$. A Maple code was used to find the numerical solution to the IVP. The adopted values of the density and consistency coefficient of the non-Newtonian fluid were ρ= 1050 kg/m3 and K = 0.5, respectively. In addition, the physical properties of particles and corresponding coefficients of Eq. (5.3.5) have been tabulated in Table 5.3.1.

In the present example, the fluid is considered non-Newtonian, and governing equations for the unsteady motion of a rigid spherical particle in a quiescent shear-thinning power-law fluid (non-Newtonian) are solved by CM and NUM. For solving Eq. (5.3.5) by WRMs, because the trial function must satisfy the initial condition in Eq. (5.3.5), so each statement in u(t) should contain t to satisfy the initial condition in t = 0. In this study, one statement is considered for the velocity profile, and as explained in the above WRMs advantages, the accuracy of results can be increased by increasing the number of statements (c_i), so

$$u(t) = c_1 t + c_2 t^2 + c_3 t^3 + \cdots + c_j t^j \tag{5.3.12}$$

Which satisfies the initial condition in Eq. (5.3.5) and by setting it into Eq. (5.3.5), residual functions, $R_i(c_1 - c_6, t)$, will be found. On the other hand, the residual functions must be close to zero.

Also, we can choose the trial function as the below statement:

$$u(t) = c_1(t - t^2) + c_2(t^2 - t^3) + \cdots + c_j(t^i - t^{j+1}) \qquad (5.3.12\text{-a})$$

By comparison between Eqs. (5.3.12) and (5.3.12-a), it is clear that Eq. (5.3.12) is simpler and more logical, so it was used as a trial function.

Table 5.3.1. Physical properties and corresponding coefficients of Eq. (5.3.5) for n = 0.5.

Particle type	D(mm)	ρ_s	m(gr)	a	b(n)	d
Plastic	0.2	1150	$4.817*10^{-6}$	$7.016*10^{-9}$	$1.911*10^{-5}$	$4.109*10^{-9}$
	0.5	1150	$7.526*10^{-5}$	$1.096*10^{-7}$	$7.552*10^{-5}$	$6.421*10^{-8}$
	1	1150	$6.021*10^{-4}$	$8.770*10^{-7}$	$2.136*10^{-4}$	$5.137*10^{-7}$
Glass	0.2	2500	$1.047*10^{-5}$	$1.265*10^{-8}$	$1.911*10^{-5}$	$5.958*10^{-8}$
	0.5	2500	$1.636*10^{-4}$	$1.980*10^{-7}$	$7.552*10^{-5}$	$9.310*10^{-7}$
	1	2500	$1.309*10^{-3}$	$1.584*10^{-6}$	$2.136*10^{-4}$	$7.448*10^{-6}$
Steel	0.2	7780	$3.259*10^{-5}$	$3.479*10^{-8}$	$1.911*10^{-5}$	$2.765*10^{-7}$
	0.5	7780	$5.092*10^{-4}$	$5.436*10^{-7}$	$7.552*10^{-5}$	$4.321*10^{-6}$
	1	7780	$4.047*10^{-3}$	$4.348*10^{-6}$	$2.136*10^{-4}$	$3.457*10^{-5}$

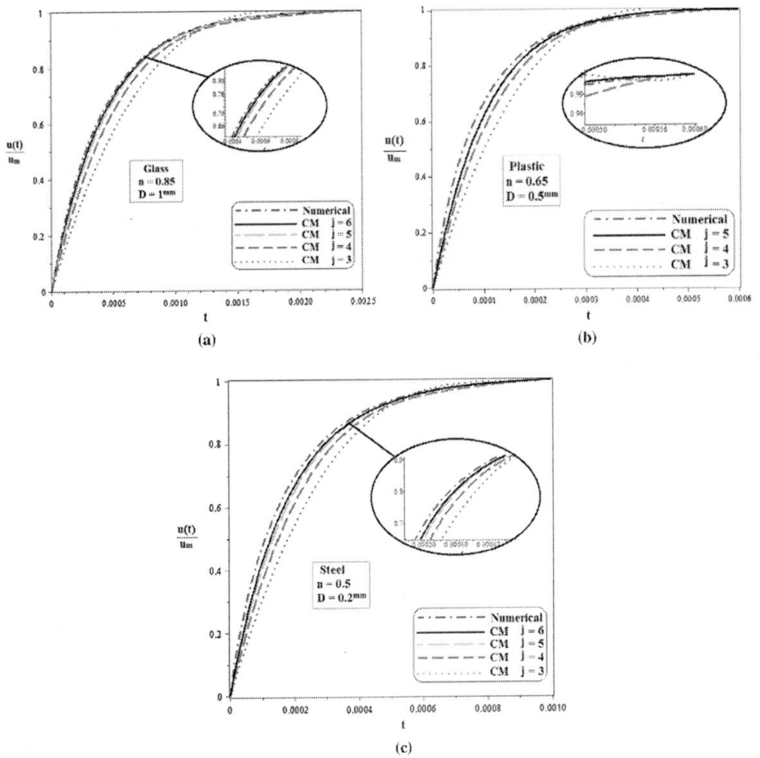

Fig. 5.3.2. CM solution for different values of j and comparing with numerical results. for (a) n = 0.85 (b) n = 0.65 (c) n = 0.5.

Six specific points in the domain t ∈ [0, tm] should be chosen to reach this aim. These points are:

$$R_i = \left(c_1 - c_j, t\right) = \frac{i, t_m}{j} \tag{5.3.13}$$

Finally, a set of jth equations with jth unknown coefficients will be obtained by substituting these points into the residual functions. After solving these unknown parameters, the velocity concentration equation will be determined. Using CM, for example, when the particle is plastic with D=1mm for n=0.85 leads to:

$j = 6:$

$$u(t) = -2.155553255 * 10^{13}t^6 + 2.373788034 * 10^{11}t^5 - 1.106742700 * 10^9t^4 + 2.854511380 * 10^{16}t^3 - 4420.787693t^2 + 4.030708825t \quad (5.3.14)$$

$j = 5$:

$$u(t) = 4.810853970 * 10^{10}t^5 - 4.581824359 * 10^8t^4 + 1.767438683 * 10^6t^3 - 3522.987025t^2 + 3.733654742t \quad (5.3.15)$$

$j = 4$:

$$u(t) = -1.007673681 * 10^8t^4 + 7.989310560 * 10^5t^3 - 2384.949638t^2 + 3.239977945t \quad (5.3.16)$$

$j = 3$:

$$u(t) = 1.927676068 * 10^5t^3 - 1210.692269t^2 + 2.506207614t \quad (5.3.17)$$

The settling velocity of a one mm-diameter glass particle with power-law fluids of n = 0.85 and a 0.5 mm-diameter plastic particle with power-law fluids of n = 0.65, and also a 0.2mm-diameter steel particle with power-law fluids of n = 0.5 versus time has been depicted in Fig. 5.3.2 (a), (b) and (c), respectively. In all the figures, the velocity is scaled with terminal velocity ($U_{terminal}$). As it is clear, the particle velocity increases until it reaches the terminal velocity, where the net force on the particle eliminates. The figures also show that increasing the value of j can enhance the region of convergence and improve the accuracy of the polynomial solution obtained by CM. Thus, the most accurate results can be obtained by increasing the value of j in the CM.

Table 5.3.2 shows the effects of different types of particles and power-law fluids on terminal velocity. The numerical results have been added to show the maximum accuracy. Because of the highest density of the steel particle, it has the highest terminal velocity, whereas the plastic particle, with the lowest density, achieves the lowest terminal velocity. In addition, terminal velocity has a dramatic upward trend when the particle diameter and power-law index parameters increase moderately (because of the increase in the particle's mass; see Eq. (5.3.1)).

Physical quantities of interest, the velocity time and acceleration time for different particles, particle diameters, and the flow behaviour indexes are shown in Figs. 5.3.3–5.3.6. Figs. 5.3.3 (a) and (b) show the effects of the flow behaviour index n on the velocity and time. Regarding Figs. 5.3.3 (a) and (b), it is obvious that in a constant time, the amount of terminal velocity in a falling procedure for both glass and steel with constant particle diameters increases by increasing the flow behaviour index n. With approaching flow behaviour to Newtonian behaviour from shear-thinning properties of the flow ($n \to 1$), the transient time to achieve terminal velocity is decreased, as indicated in Fig. 5.3.3 (a) and (b). The effects of three different particles, namely: plastic, glass, and steel, on the falling procedure are shown in Figs. 5.3.4 (a)–(c). The figures indicate the slowest time for reaching the terminal velocity of plastic particles due to their lower density.

Regarding Fig. 5.3.5, it is obvious that the time for reaching terminal velocity in a falling procedure is significantly increased with the particle size. In Fig. 5.3.6 (a), (b), and (c), it can be understood that the acceleration of particles is higher for larger particles.

Table 5.3.2. Terminal settling velocity ($\times 10^{-7}$ m/s) for different diameters of particles and several fluids.

Particle type	D (mm)	CM results			Numeri-cal results			Error %
		0.5	0.65	0.85	0.5	0.65	0.85	n =0.5
Plastic	0.2	1.772	6.781	22.625	1.733	6.753	22.544	2.25
	0.5	28.795	69.589	164.932	28.914	69.756	165.281	0.41
	1	229.736	404.505	747.348	231.047	405.336	745.285	0.57
Glass	0.2	389.187	418.997	521.890	387.877	417.369	522.887	0.34
	0.5	6091.018	4286.624	3859.046	6065.436	4270.318	3842.110	0.42
	1	48617.179	24946.056	17106.862	48346.550	24831.984	17348.164	0.56
Steel	0.2	8328.664	4444.075	3197.794	8327.324	4426.284	3183.089	0.02
	0.5	131345.961	45465.803	23441.211	130731.743	45293.424	23355.788	0.47
	1	1049239.494	26568.143	105560.9000	1045030.464	263262.304	105295.159	0.4

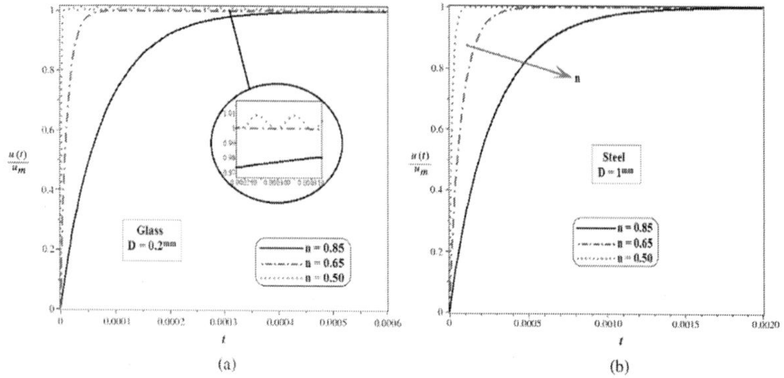

Fig. 5.3.3. Effects of the flow behaviour index (n) with a constant diameter on the velocity variation of (a) glass particle (b) steel particle.

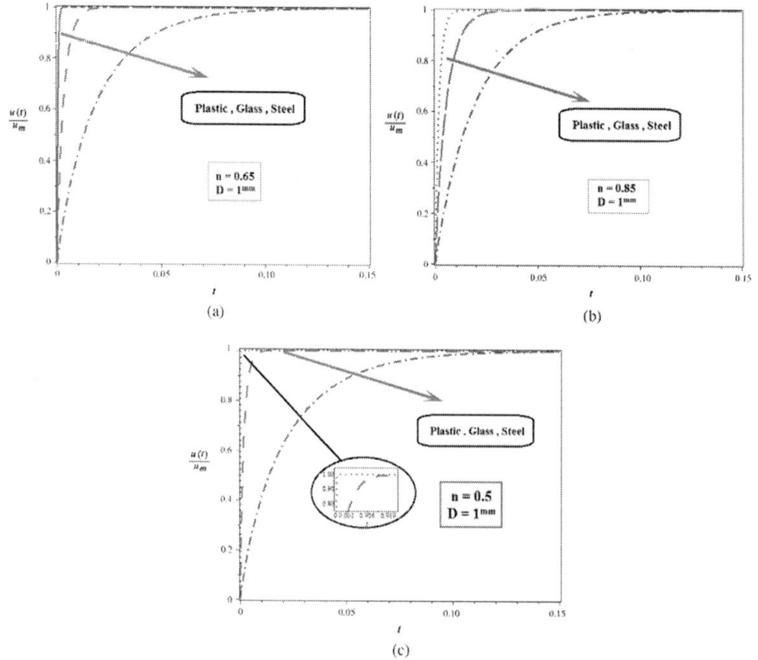

Fig. 5.3.4. Effects of the particle type on the velocity variation: (a) n= 0.65, D= 1 mm, (b) n= 0.85, D= 1 mm, and (c) n= 0.5, D= 1 mm.

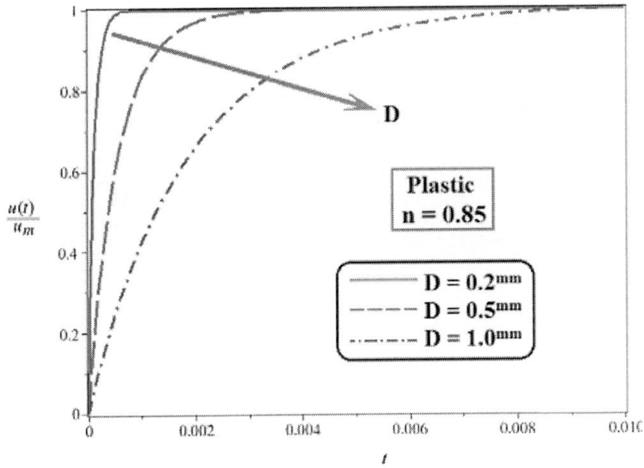

Fig. 5.3.5. Effects of the particle diameter on the velocity variation of plastic particles.

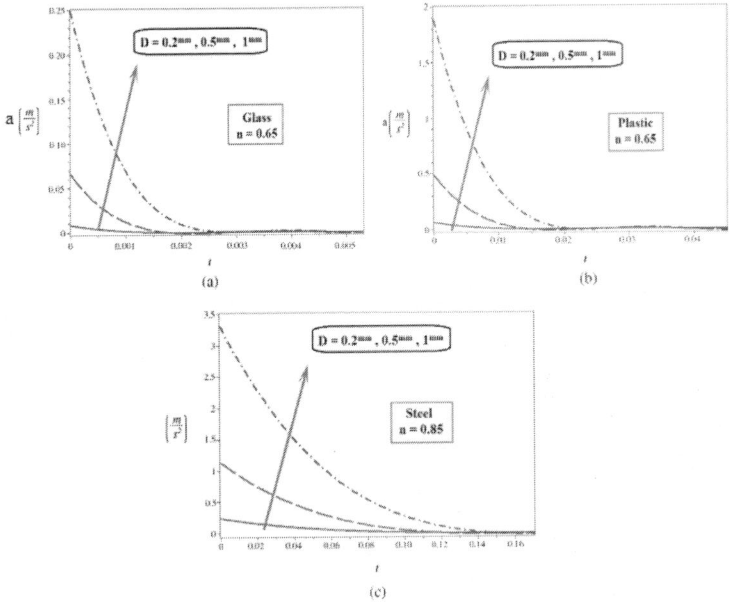

Fig. 5.3.6. Effects of the particle diameter on the velocity variation.

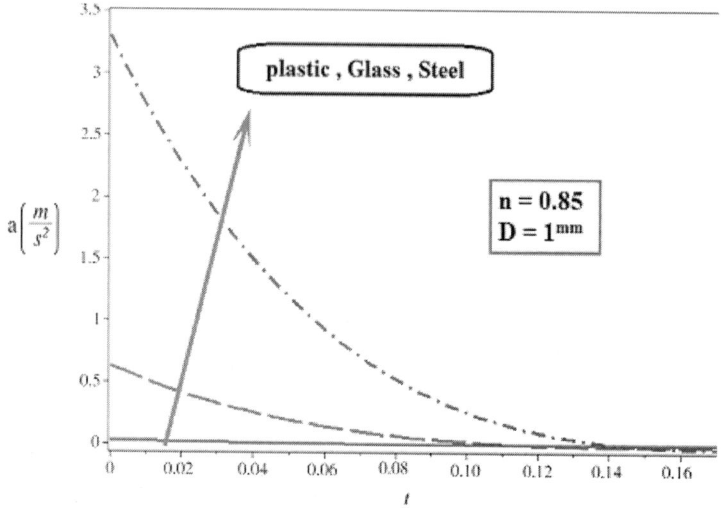

Fig. 5.3.7. Effects of the particle type on the acceleration variation (n = 0.85, D = 1 mm).

Fig. 5.3.7 indicates that the initial acceleration time required to reach a zero-acceleration state increases by increasing the particle density for a constant particle diameter (D) and flow behaviour index (n).

This work aims to apply the CM and NUM to study the strongly nonlinear differential equation with fractional power governed by the unsteady motion of falling spherical particles in a power-law non-Newtonian fluid. The current method was applied without discretisation, restrictive assumptions, or transformation. Also, this method can be used to develop valid solutions even to highly nonlinear problems and may be considered a significant refinement of the formerly developed methods. As the results are compared with the numerical method, it is clear that CM has a good agreement with NUM and provides highly accurate analytical solutions for nonlinear problems and markedly reduces the extent of the calculations required. AGM can powerfully solve such complex problems easily and more accurately. Still, CM is a faster method to solve such equations.

Moreover, the effects of various materials, sizes, and flow behaviour indexes on the transient time, terminal velocity, and acceleration have been investigated in detail. Analytical results indicated that increasing the particle size and flow behaviour index significantly increases the time of reaching the terminal velocity in a falling procedure. However, raising the particle density has the same result in the acceleration of the particle. Also, by increasing the size and density of the particle, it is clear that the time of reaching terminal acceleration in a falling procedure increases.

5.4. Heat Transfer in Rectangular Porous Fins (Si_3N_4) with Temperature-Dependent Internal Heat Generation

In this problem, heat transfer in rectangular porous fins with temperature-dependent internal heat generation is investigated, and Darcy's model for passage velocity is applied. The heat transfer governing equation is solved by two simple and accurate methods, GM and AGM.

A rectangular porous fin profile is shown in Fig. 5.4.1. The dimensions of this fin are length L, width w, and thickness t. The cross-section area of the fin is constant, and the fin has temperature-dependent internal heat generation. Also, the heat loss from the fin's tip compared with the fin's top and bottom surfaces is assumed to be negligible. Since the transverse Biot number should be small for the fin to be effective, the temperature variation in the transverse direction is neglected. Thus heat conduction is assumed to occur solely in the longitudinal direction.

Energy conservation can be written as:

$$q(x) - q(x + \Delta x) + q*.A.x = \dot{m}C_p (T(x) - T_\infty) + h(P.\Delta x)(T(x) - T_\infty) \tag{5.4.1}$$

The mass flow rate of the fluid passing through the porous material is:

$$\dot{m} = \rho . V_\omega . \Delta x . \omega \tag{5.4.2}$$

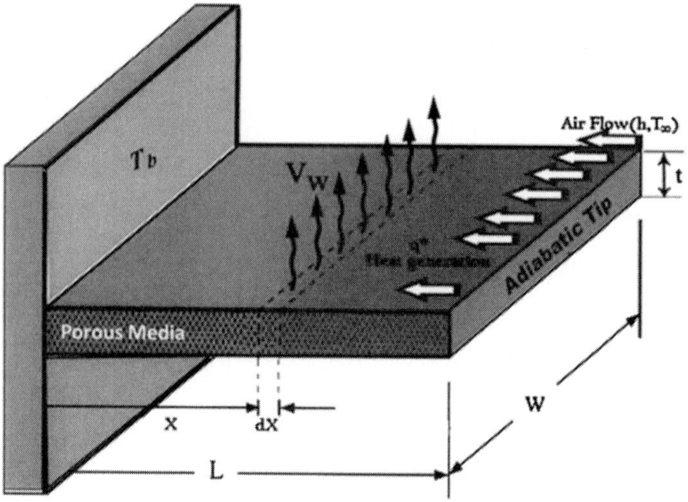

Fig. 5.4.1. Schematic of the convective porous fin with temperature-dependent heat generation

The passage velocity from Darcy's model is:

$$V_\omega = \frac{gK\beta}{\vartheta(T - T_\infty)} \tag{5.4.3}$$

Substitutions of Eqs. (5.4.2) and (5.4.3) with Eq. (5.4.1) yields:

$$\frac{q(x) - q(x - \Delta x)}{\Delta x} + q^*.A = \frac{\rho.C_p.g.K.\beta.\omega}{\vartheta}(T(x) - T_\infty)^2 + h.P(T(x) - T_\infty) \tag{5.4.4}$$

As $\Delta x \to 0$ Eq. (5.4.4) becomes:

$$\frac{dq}{dx} + q^*.A = \frac{\rho.C_p.g.K.\beta.\omega}{\vartheta}(T(x) - T_\infty)^2 + h.P(T(x) - T\infty) \tag{5.4.5}$$

Also, from Fourier's Law of Conduction:

$$q = -k_{eff} A \frac{dT}{dx} \tag{5.4.6}$$

Where A is the cross-sectional area of the fin A=(w.t), and K_{eff} is the effective thermal conductivity of the porous fin that can be obtained from the following equation:

$$k_{eff} = \phi . k_f + (1 - \phi) k_s \tag{5.4.7}$$

Where ϕ is the porosity of the porous fin. Substituting Eq. (5.4.6) with Eq. (5.4.5) leads to:

$$\frac{d^2T}{dx^2} - \frac{\rho . C_p . g . K . \beta}{t . k_{eff} . \vartheta} (T(x) - T_\infty)^2 - \frac{h . P}{k_{eff} . A} (T(x) - T_\infty) + \frac{q^*}{k_{eff}} = 0 \tag{5.4.8}$$

It is assumed that heat generation in the fin varies with temperature as Eq. (5.4.9):

$$q^* = q^*_\infty \big(1 + \varepsilon\, (T - T_\infty)\big) \tag{5.4.9}$$

where q^*_∞, is the internal heat generation at temperature T_∞. For simplicity in the above equations, some dimensionless parameters are introduced as follows:

$$\theta = \frac{(T - T_\infty)}{(T_b - T_\infty)}, X = \frac{x}{L}, M^2 = \frac{hPL^2}{k_0 A}, S_h = \frac{Da.x.Ra}{k_r}\left(\frac{L}{t}\right)^2, G = \frac{q^*_\infty}{hP(T_b - T_\infty)}, \tag{5.4.10}$$

$$\varepsilon G = \varepsilon (T_b - T_\infty)$$

Where S_h is a porous parameter that indicates the effect of the permeability of the porous medium and the buoyancy effect, so higher value of S_h indicates higher permeability of the porous medium or higher buoyancy forces. M is a convection parameter that indicates the effect of surface convecting of the fin. Finally, Eq. (5.4.8) can be rewritten as:

$$\frac{d^2\theta}{dX^2} - M^2\theta + M^2 G (1 + \varepsilon_G \theta) - S_h \theta^2 = 0 \tag{5.4.11}$$

In this research, we study finite-length fins with insulated tips. For this case, the fin tip is insulated so that there will not be any heat transfer at the insulated tip, and boundary condition will be,

$$\theta(0) = 1, \qquad \left.\frac{d\theta}{dX}\right|_{x=1} = 0 \tag{5.4.12}$$

Boundary conditions and initial conditions are required for analytical methods of each linear and nonlinear differential equation according to the physics of the problem. By the boundary conditions, the general manner of a differential equation is as follows: The nonlinear differential equation of *p* which is a function of *θ*, the parameter *θ,* which is a function of *x,* and their derivatives are considered as follows:

$$p_k : f(\theta, \theta', \theta'', \dots, \theta^m) = 0; \qquad \theta = \theta(x) \tag{5.4.13}$$

where, p_k, is the name of the nonlinear differential equation, and m is the order of the derivatives. Boundary conditions:

$$\begin{cases} \theta(0) = \theta_0, \theta'(0) = \ \theta_1, \dots, \theta^{m-1}(0) = \theta_{m-1} \\ \theta(L) = \theta_{L0}, \theta'(L) = \theta_{L1}, \dots, \theta^{m-1}(L) = \theta_{Lm-1} \end{cases} \tag{5.4.14}$$

For solving the differential equation with respect to the boundary conditions, the series of letters in the nth order with constant coefficients, which is the answer to the differential equation, is considered as follows:

$$\theta(X) = \sum_{i=0}^{n} a_i X^i = a_0 + a_1 X^1 + a_2 X^2 + \cdots + a_n X^n \tag{5.4.15}$$

The more precise answer to Eq. (5.4.13), the more choice of series sentences from Eq. (5.4.15). In applied problems, approximately five or six sentences from the series are enough to solve nonlinear differential equations. Ten sentences from the series are needed to solve the differential equation in this problem. In answer to differential Eq. (5.4.15) regarding the series from degree (*n*), there are (*n*+1) unknown coefficients that need (*n*+1) equations to be specified. The boundary conditions of Eq. (5.4.14) are used to solve a set of equations consisting of (*n*+1) ones. The boundary conditions are applied to the functions as follows:

(a) The application of the boundary conditions for the answer of differential Eq. (5.4.15) is in the form of: If $X = 0$:

$$\begin{cases} \theta\ \ (0) = a_0 = \theta_0 \\ \theta'\ (0) = a_1 = \theta_1 \\ \theta''(0) = a_2 = \theta_2 \\ \vdots \quad \vdots \quad \vdots \quad \vdots \end{cases} \tag{5.4.16}$$

And when $X = L$:

$$\begin{cases} \theta(L) = a_0 + a_1 L + a_2 L^2 + \cdots + a_n L^n = \theta_{L0} \\ \theta'(L) = a_1 + 2a_2 L + 3a_3 L^2 + n a_n L^{n-1} = \theta_{L1} \\ \theta''(L) = 2a_2 + 6a_3 L + 12a_4 L^2 + \cdots + n(n-1)a_n L^{n-2} = (\\ \vdots \quad \vdots \quad \vdots \quad \vdots \quad \vdots \quad \vdots \quad \vdots \quad \vdots \quad \vdots \end{cases} \tag{5.4.17}$$

(b) After substituting Eq. (5.4.17) into Eq. (5.4.13), the application of the boundary conditions on the differential Eq. (5.4.13) is done according to the following procedure:

$$\begin{aligned} &P0: f(\theta(0), \theta'(0), \theta''(0), \dots, \theta^m(0)) \\ &P1: f(\theta(L), \theta'(L), \theta''(L), \dots, \theta^m(L)) \\ &\vdots \quad \vdots \ \vdots \quad \vdots \quad \vdots \quad \vdots \quad \vdots \end{aligned} \tag{5.4.18}$$

Concerning the choice of n; $(n < m)$ sentences from Eq. (5.4.15) and to make a set of equations which consist of $(n+1)$ equations and $(n+1)$ unknowns, we confront several additional unknowns which are indeed the same coefficients of Eq. (5.4.15). Therefore, to solve this problem, we should derive m times from Eq. (5.4.13) according to the additional unknowns mentioned above set differential equations. Then this is the time to apply the boundary conditions of Eq. (5.4.11) on them.

$$\begin{aligned} &P_k' = f(\theta', \theta'', \theta''', \dots, \theta^{(m+1)}) \\ &P_k'' = f(\theta'', \theta''', \theta^{(IV)}, \dots, \theta^{(m+2)}) \\ &\vdots \quad \vdots \quad \vdots \quad \vdots \end{aligned} \tag{5.4.19}$$

(c) Application of the boundary conditions on the derivatives of the differential equation P_k, in Eq. (5.4.19) is done in the form of:

$$P_k' = \begin{cases} f\ (\theta'(0),\ \theta''(0), \dots, \theta^{m+1}(0)) \\ f\ (\theta'(L), \theta''(L), \dots, \theta^{m+1}(L)) \end{cases} \quad (5.4.20)$$

$$P_k'' = \begin{cases} f\ (\theta''(0),\ \theta'''(0), \dots, \theta^{m+2}(0)) \\ f\ (\theta''(L), \theta'''(L), \dots, \theta^{m+2}(L)) \end{cases} \quad (5.4.21)$$

(n+ 1) equations can be made from Eqs. (5.4.16) to (5.4.22) so that (n+ 1) unknown coefficients of Eq. (5.4.15) take for example a_0, a_1, a_2,..., a_n will be computed. The answer to the nonlinear differential Eq. (5.4.13) will be gained by determining the coefficients of Eq. (5.4.15).

For solving this given differential equation with AGM, we will have:

$$f(\theta) : \frac{d^2\theta}{dX^2} - M^2\theta + M^2G(1 + \varepsilon_G\theta) - S_h\theta^2 = 0 \quad (5.4.22)$$

In AGM, the answer to the differential equations is considered a finite series of polynomials with constant coefficients as follows:

$$\theta(X) = \sum_{i=0}^{10} a_i X^i = a_0 + a_1X^1 + a_2X^2 + \dots + a_{10}X^{10} \quad (5.4.23)$$

Notably, in the above equations, the constant coefficients of a_0 *to* a_6 are obtained by applying the introduced boundary conditions.

For Si_3N_4, the considered parameters are as follows:

$$G = 0.425, M = 1, D_a = 0.0001, \frac{L}{t} = 10$$
$$R_a = 1000, L = 1, \varepsilon_G = 0.6, S_h = 0.06, k_s = 954 \quad (5.4.24)$$

According to the above explanations, applying the boundary conditions is done in the following form:

$$\theta(0) = 1 \rightarrow a_0 = 1 \quad (5.4.25)$$

$$\theta'(1) = 0 \rightarrow a_1 + 2a_2 + 3a_3 + \dots + 9a_9 + 10a_{10} = 0 \quad (5.4.26)$$

Applying the boundary conditions on the main differential equation, which in this case study is Eq. (5.4.22), and on its derivative is done after substituting Eq. (5.4.23) into the main differential equation Eq. (5.4.22) as follows:

$$f(\theta(0)) = -0.06a_0^2 + 0.425 - 0745a_0 + 2a_2 = 0 \quad (5.4.27)$$

$$\begin{aligned} f(\theta(1)) = -0.6e - 1 \\ * (a_0 + a_1 + a_2 + a_3 + a_4 + a_5 + a_6 \\ + a_7 + a_8 + a_9 + a_{10})^2 + 89.2550 * a_{10} \\ + 71.2550 * a_9 + 55.2550 * a_8 41.2550 \\ * a_7 + 29.2550 * a_6 + 19.2550 * a_5 \\ + 11.2550 * a_4 + 5.2550 * a_3 - 0.7450 \\ * a_0 + 1.2550 * a_2 - 0.7450 * a_3 \\ + 0.425 = 0 \end{aligned} \quad (5.4.28)$$

$$f'(\theta(0)) = -012a_0a_1 - 0.745a_1 + 6a_3 = 0 \quad (5.4.29)$$

$$\begin{aligned} f'(\theta(1)) = -(0.12 \\ * (a_0 + a_1 + a_2 + a_3 + a_4 + a_5 + a_6 \\ + a_7 + a_8 + a_9 + a_{10}) \\ * (a_1 + 2 * a_2 + 3 * a_3 + 4 * a_4 + 5 * a_5 \\ + 6 * a_6 + 7 * a_7 + 8 * a_8 + 9 * a_9 + 10 \\ * a_{10}) + 712.5500 * a_{10} + 497.2950 \\ * a_9 + 330.04 * a_8 + 204.7850 * a_7 \\ + 11.5300 * a_6 + 56.2750 * a_5 \\ + 21.0200 * a_4 + 3.7650 * a_3 - 0.7450 \\ * a_1 - 1.4900 * a_2 = 0 \end{aligned} \quad (5.4.30)$$

$$f''(\theta(0)) = -0.24a_0a_2 - 0.12a_1^2 - 1.49a_2 + 24a_4 = 0 \quad (5.4.31)$$

By solving a set consisting of eleven equations with eleven unknowns from Eq. (5.4.23), the constant coefficients of Eq. (5.4.23) can easily be gained.

Constant coefficients are calculated as follows:

$$\begin{aligned} a_0 = 1, a_1 = -02989782196, a_2 = 0.19, a_3 \\ = -0.4310269333e - 1, \\ a_4 = 0.01414277321, a_5 = -0.2205026657e - 2, a_6 \\ = 0.0005315303603 \\ a_7 = -0.8077952394e - 4, a_8 = 0.00001700983062, \\ a_9 = -0.248173393e - 5, a_{10} = 2.871525433e - 7 \end{aligned} \quad (5.4.32)$$

By substituting the constant values of Eqs. (5.4.32) into (5.4.23), the differential equation is answered in the following form:

$$\theta(X) = 2.871525433 * 10^{-7} * X^{10} - 0.2481733793e - 5 * X^9 + 0.1700983062e - 4 * X^8 - 0.8077952394e - 4 * X^7 + 0.5315303603e - 3 * X^6 - 0.2205026657e - 2 * X^5 + 0.1414277321e - 1 * X^4 - 0.4310269333e - 1 * X^3 + 0.1900000000 * X^2 - 0.2989782196 * \mathrm{x} + 1.000000000 \quad (5.4.33)$$

In weighted residual methods, a trial or approximate function is chosen to approximate the independent variable, such as a displacement or a temperature, in a problem defined by a differential equation. This trial function will not generally satisfy the governing differential equation. Thus substituting the trial function into the differential equation results in a residual over the whole region of the problem.

Galerkin's method minimizes the residue obtained using the assumed mode in the governing differential equation. Let the assumed shape function of the system be written as:

$$\theta(X) = \sum_{i=1}^{n} C_i \theta_i\,(X) \quad (5.4.34)$$

Where $\theta_i\,(X)$ is the approximate solution of the differential equation, and C_i are constant coefficients. For each function, making the residual (R_i) equal to zero:

$$R_i = \int_0^1 C_i \theta_i(X) d(X) \quad (5.4.35)$$

Now we have n linear and homogeneous equations coefficients.

For solving this given differential equation with GM, we will have:

$$f(\theta){:}\ \frac{d^2\theta}{dX^2} - M^2\theta + M^2 G(1 + \varepsilon_G \theta) - S_h \theta^2 = 0 \quad (5.4.36)$$

$$\theta(X) = \sum_{i=0}^{10} C_i X^i = C_0 + C_1 X^1 + C_2 X^2 + \cdots + C_{10} X^{10} \quad (5.4.37)$$

According to the above explanations, applying the boundary conditions is done in the following form:

$$\theta(0) = 1 \rightarrow R_0 = C_0 = 1 \quad (5.4.38)$$

$$\theta'(1) = 0 \rightarrow R_1 = C_1 + 2C_2 + 3C_3 + \cdots + 9C_9 + 10C_{10} = 0 \quad (5.4.39)$$

Applying the boundary conditions on the main differential equation, which in this case study is Eq. (5.4.36), and on its derivative is done after substituting Eq. (5.4.37) with the main differential equation Eq. (5.4.36) as follows:

$$R_i = \int_0^1 \frac{df(\theta)}{dC_i} * f(\theta) d(X) \quad (5.4.40)$$

$$i = 2, 3, 4, \dots, 9, 10$$

$$R_2 = 0.1028571428e - 2 * C_1 * C_9^2 + 0.1028571428e - 2 * C_3 * C_8^2 - 2 * C_5^2 C_9 + 0.1028571428e - 2 * C_5 C_7^2 + \cdots = 0 \quad (5.4.41)$$

$$R_3 = -0.6504333333 + 0.0009818181819 * C_1 * C_9^2 + 0.0009818181819 * C_3 * C_8^2 + 0.0009818181819 * C_5^2 * C_9 + \cdots = 0 \quad (5.4.42)$$

$$\vdots$$

$$R_{10} = 0.0007200000000 * C_1 * C_9^2 + 0.0007200000000 * C_3 * C_8^2 + 0.0007200000000 * C_5^2 * C_9 + 0.0007200000000 * C_5 * C_7^2 + \cdots = 0 \quad (5.4.43)$$

By solving a set of eleven equations with eleven unknowns from Eq. (5.4.37), the constant coefficients of Eq. (5.4.37) can easily be gained.

Constant coefficients are calculated as follows:

$$C_0 = 1, C_1 = -0.2989782170, C_2 = 0.1899984502,$$
$$C_3 = -0.4308133641e - 1, C_4 = 0.01404201661,$$
$$C_5 = -0.2017857133e - 2, C_6 = 0.0005222769648,$$
$$C_7 = -0.5764668071e - 3, \tag{5.4.44}$$
$$C_8 = 0.0007951890280$$
$$C_9 = -0.5031838885e - 3, C_{10} = 0.0001215293713$$

By substituting the constant values of Eq. (5.4.44) into Eq. (5.4.37), the differential equation is answered in the following form:

$$\begin{aligned}\theta(X) = 0.1215293713e &- 3 * x^{10} - 0.5031838885e \\ &- 3 * x^9 + 0.7951890280e - 3 * x^8 \\ &- 0.5764668071e - 3 * x^7 \\ &+ 0.5222769648e - 3 * x^6 \\ &- 0.2017857133e - 2 * x^5 \\ &+ 0.1404201661e - 1 * x^4 \\ &- 0.4308133641e - 1 * x^3 \\ &+ 0.1899984502 * x^2 - 0.2989782170 \\ &* x + 1\end{aligned} \tag{5.4.45}$$

This study considers the heat transfer investigation of a rectangular porous fin (Si3N4) with air inlet flow. AGM and GM are used for solving the current problem. The effects of Darcy (Da) and Rayleigh numbers (Ra) and the temperature-dependent internal heat generation parameters are investigated on the heat transfer. The temperature distribution through the porous fin is compared with the numerical solution to validate the results. The numerical solution is performed using FlexPDE student version 6.36. FlexPDE is a script-driven system. It reads a description of the equations, domain, auxiliary definitions, and graphical output requests from a text file referred to as a “problem descriptor” or “script.”

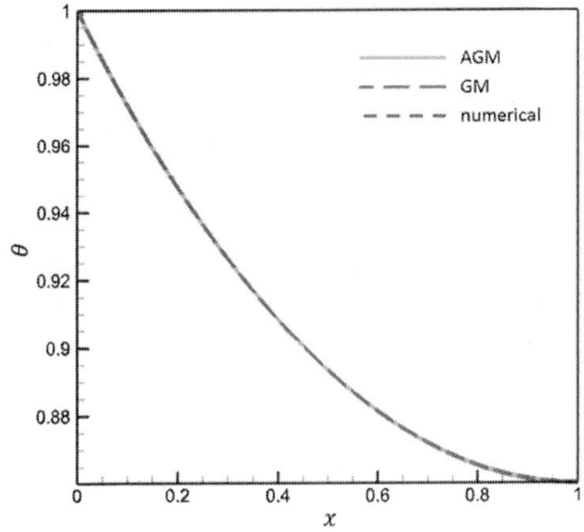

Fig. 5.4.2. Comparison of AGM, GM and Numerical results at G = 0.425, M = 1, ε_G = 0.6, S_h = 0.06, Ra = 10000 and Da = 0.0001

The FlexPDE problem descriptor language can be viewed as a shorthand language for creating Finite Element models. In Fig. 5.4.2, AGM and GM with FlexPDE numerical results have been compared.

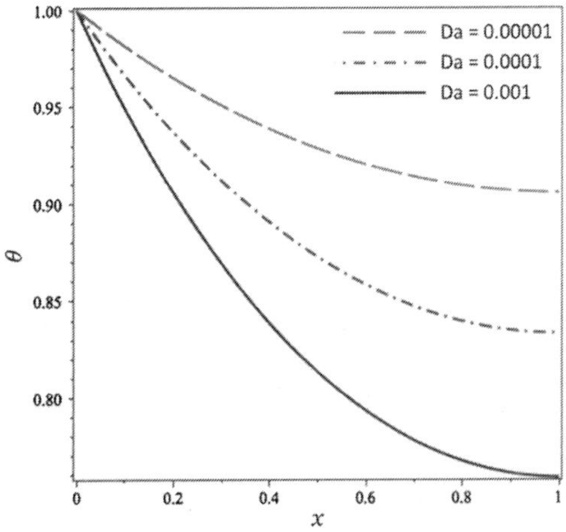

Fig. 5.4.3. Temperature distribution versus Darcy number variation at G = 0.425, M = 1, ε_G = 0.6, S_h = 0.06 and Ra = 10000

Problem descriptor language can be considered shorthand for creating finite element models. In Fig. 5.4.2, AGM and GM with FlexPDE numerical results have been compared.

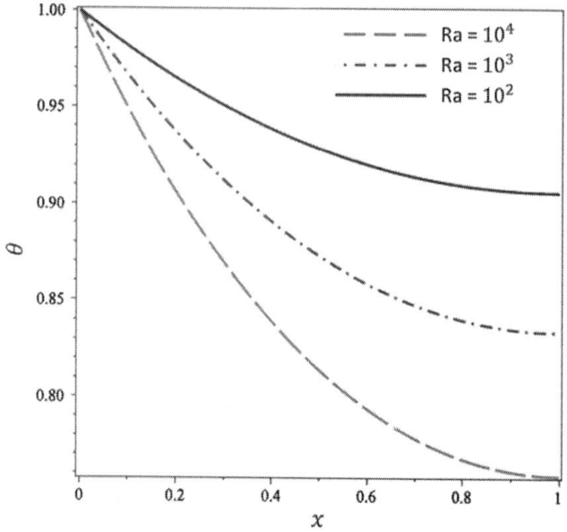

Fig. 5.4.4. Temperature distribution versus Rayleigh number variation at G=0.425, M=1, ε_G = 0.6, S_h = 0.06 and Da = 0.001

In fluid dynamics through porous media, the Darcy number (Da) represents the relative effect of the medium's permeability versus its cross-sectional area, commonly the diameter squared. The effect of the Darcy number on non-dimensional temperature is shown in Fig. 5.4.3, while the other parameters are constant. It is illustrated that the magnitude of temperature is increased by decreasing the Darcy number. Noting the definition of $Da = \frac{K}{t^2}$, the Darcy number reduces when the permeability reduces. Also, noting the permeability definition, if the bubbles of porous media are very small or poorly connected, the permeability will be low, and fluid will not flow through easily. Therefore when the Darcy number and, consequently, the permeability reduces, the collision among the fluid flow and pores of the porous screen will increase. Thus the passing fluids had more space to contact the porous media, which has internal heat generation. Consequently, the value of temperature is increased by decreasing the Darcy number.

In fluid mechanics, the Rayleigh number (Ra) for a fluid is a dimensionless number associated with buoyancy-driven flow, also known as free convection or natural convection. When the Rayleigh number is below a critical value for that fluid, heat transfer is primarily in conduction; when it exceeds the critical value, heat transfer is primarily in the form of convection. With attention to the Rayleigh number definition ($Ra = \frac{g.\beta.\Delta T.H^3}{\alpha.\vartheta}$), increasing the Ra number leads to more buoyancy force and, consequently, heat transfer rate due to the convection mechanism. So, a higher value of Ra enhances heat transfer between the solid fins and the airflow, leading to higher temperature values in the temperature profile (see Fig. 5.4.4).

The effect of temperature-dependent internal heat generation value is investigated via Fig. 5.4.5. The ε_G parameter is changed when the other parameters are constant. As mentioned before, the variation in heat generation is directly related to ε_G [see Eq. (5.4.9)]. Thus, increasing this parameter increases the heat generated through the fin. Finally, the temperature profile reaches a higher value by raising the heat generation parameter.

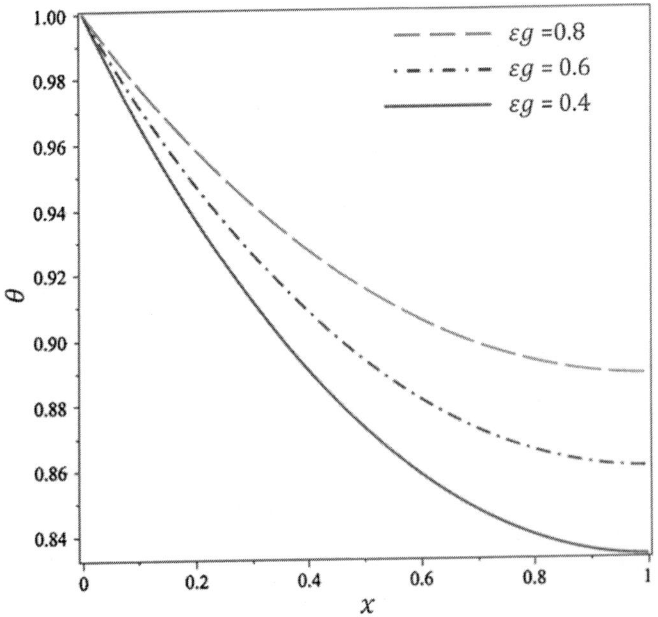

Fig. 5.4.5. Temperature distribution versus temperature-dependent internal heat generation parameter at G = 0.425, M = 1, S_h = 0.06, Ra = 10000 and Da = 0.001

Alternatively, the higher heat generation leads to higher fin temperatures because the fin must dissipate more heat to the surroundings under steady-state conditions.

In this present study, heat transfer in a rectangular porous fin (Si_3N_4) with temperature-dependent internal heat generation is studied, and Darcy's model is used for passage velocity. The governing equation was solved via AGM and GM, and the results have been compared with the FlexPDE numerical solution. The results indicate that AGM and GM provide simple, accurate, and appropriate techniques for simulating heat transfer in a porous fin with temperature-dependent internal heat generation. Also, the temperature distribution strongly depends on the Darcy and Rayleigh numbers. In this case, AGM and GM solutions are too close, but in comparison, AGM is faster in reaching the solutions. AGM illustrates that the magnitude of temperature increases with decreasing the Darcy number and, consequently, the S_h parameter and the heat generation rate are very effective in fin temperatures. When the heat generation rate increases, fin temperatures increase since more heat is dissipated to the surroundings under steady-state conditions.

List of symbols

$u(t)$	Vibrational displacement (m)
$\dot{u}(t)$	Vibrational velocity (m/s)
$\ddot{u}(t)$	Vibrational acceleration (m/s^2)
$\theta(t)$	Angular position as a function of time (rad)
A	The initial amplitude of vibration (m)
ω	Angular frequency (1/s)
$\ddot{\psi}(t)$	Angular acceleration (rad/s^2)
g	Gravity acceleration (m/s^2)
m	Mass of the cylinder (kg) or Mass (kg)
w	Weight of the cylinder (N)

I_c	Mass moment of inertia about the instantaneous centre ($kg.m^2$)
$\sum M_c$	Resultant moment about the instantaneous centre of rotation (N.m)
R	Radius of the circular path (m) or Radius of rotating circle (m)
r	Radius of the cylinder (m)
Ω	Angular velocity (rad/s)
w	Weight force (N)
N	Normal force (N)
α	Initial angular position (rad)
a	Acceleration of the mass (m/s^2)
$\sum M_O$	Resultant moment about the centre of the rotating circle (N.m)
H_O	Angular momentum about the centre of the rotating circle (N.m.s)
$x(t)$	Position of the particle as a function of time (m)
m	particle mass
u	particle velocity
g	gravity acceleration
D	particle diameter
C_D	drag coefficient
t	time
Re	Reynolds number
n	flow behaviour index

V_0	velocity at t = 0
k	consistency coefficient
$X(n)$	deviation factor
j	number of polynomial statements
u_m	terminal settling velocity
t_m	time of terminal settling velocity
CM	Collocation Method
HPM	Homotopy Perturbation Method
HAM	Homotopy Analysis Method
VIM	Variational Iteration Method
NuM	The numerical method
ρ	fluid density
ρ_s	particle density
C_p	Specific heat at constant pressure
k	Thermal conductivity
ρ	Fluid density
T	Temperature
θ	Dimensionless temperature
a	Constant
A	Section area of the fin
D_a	Darcy number
G	Generation number, dimensionless
GM	Galerkin's method

h	Convection coefficient
K_r	Thermal conductivity ratio [k_{eff}/k_f]
K	Permeability
L	Length of fin
M	Convective parameter
NUM	The numerical method
q	Conducted heat
q^*	Internal heat generation
Ra	Rayleigh number
S_h	Porous parameter
T_b	The temperature at the fin base
T_∞	Sink temperature for convection
t	The thickness of the fin
V_w	The velocity of fluid passing through the fin
w	Width of the fin
x	Axial coordinate
X	Dimensionless axial coordinate [x/L]
β	Coefficient of volumetric thermal expansion
Δ	Temperature difference
ε_G	Internal heat generation parameter
ϑ	Kinematic viscosity
φ	Porosity variable
f	Fluid properties

e_{ff}	Porous properties
s	Solid properties

References

1. Meresht, N. B., & Ganji, D. D. (2018). "Analytical scrutiny of nonlinear equation of hypocycloid motion by AGM." *Neural Computing and Applications*, 29(12), 1575-1582.

2. Meresht, N. B., & Ganji, D. D. (2017). "Solving nonlinear differential equation arising in dynamical systems by AGM." *International Journal of Applied and Computational Mathematics*, 3(2), 1507-1523.

3. Rahimi-Gorji, M., Pourmehran, O., Gorji-Bandpy, M., & Ganji, D. D. (2015). "An analytical investigation on unsteady motion of vertically falling spherical particles in non-Newtonian fluid by collocation method." *Ain Shams Engineering Journal*, 6(2), 531-540.

4. Asadian, H., Zaretabar, M., Ganji, D. D., Gorji-Bandpy, M., & Sohrabi, S. (2017). "Investigation of Heat Transfer in Rectangular Porous Fins (Si_3N_4) with Temperature-Dependent Internal Heat Generation by Galerkin's Method (GM) and Akbari-Ganji's Method (AGM)." *International Journal of Applied and Computational Mathematics*, 3(4), 2987-3000.

5. Jalili, B., Jalili, P., Shateri, A., & Ganji, D. D. (2022). Rigid plate submerged in a Newtonian fluid and fractional differential equation problems via Caputo fractional derivative. *Partial Differential Equations in Applied Mathematics*, *6*, 100452.

6. Jalili, P., Azar, A. A., Jalili, B., Asadi, Z., & Ganji, D. D. (2022). Heat transfer analysis in cylindrical polar system with magnetic field: a novel hybrid analytical and numerical technique. *Case Studies in Thermal Engineering*, *40*, 102524.

7. Jalili, P., Sharif Mousavi, S. M., Jalili, B., Pasha, P., & Ganji, D. D. (2023). Thermal evaluation of MHD Jeffrey fluid flow in the presence of a heat source and chemical reaction. *International Journal of Modern Physics B*, 2450113.

8. Jalili, P., Azar, A. A., Jalili, B., & Ganji, D. D. (2023). Study of nonlinear radiative heat transfer with magnetic field for non-Newtonian Casson fluid flow in a porous medium. *Results in Physics*, *48*, 106371.

9. Jalili, B., Roshani, H., Jalili, P., Jalili, M., Pasha, P., & Ganji, D. D. (2023). The magnetohydrodynamic flow of viscous fluid and heat transfer examination between permeable disks by AGM and FEM. *Case Studies in Thermal Engineering*, *45*, 102961.

10. Jalili, P., Azar, A. A., Jalili, B., & Ganji, D. D. (2023). The HAN method for a thermal analysis of forced non-Newtonian MHD Reiner-Rivlin viscoelastic fluid motion between two disks. *Heliyon*.